URBAN DISASTER MITIGATION

The Role of Engineering and Technology

URBAN DISASTER MITIGATION
The Role of Engineering and Technology

Edited by

F. Y. CHENG
University of Missouri-Rolla, USA

and

M.- S. SHEU
National Cheng-Kung University, Taiwan

ELSEVIER

U.K. Elsevier Science Ltd, The Boulevard, Langford Lane, Kidlington, Oxford, OX5 1GB, U.K.

U.S.A. Elsevier Science Inc., 660 White Plains Road, Tarrytown, New York 10591-5153, U.S.A.

JAPAN Elsevier Science Japan, Tsunashima Building Annex, 3-20-12 Yushima, Bunkyo-ku, Tokyo 113, Japan

---

First edition 1995

Library of Congress Cataloging in Publication Data

Urban disaster mitigation/ edited by F. Y. Cheng, M. -S. Sheu. – 1st ed.
p. cm.
Includes index.
1. Earthquake engineering–Congresses. 2. Natural disasters–Congresses. 3. Earthquake hazard analysis–Congresses. 4. Landslides–Congresses. 5. Flood control–Congresses.
I. Cheng, Franklin Y. II. Sheu, Maw Shyong.
TA654.6.U73 1995 95-12167
624.1'76–dc20 CIP

British Library Cataloguing in Publication Data
A catalogue record for this book is available from the British Library

ISBN 0 08 041920 8

Printed and bound in Great Britain by BPC Wheatons Ltd, Exeter

CONTENTS

PREFACE

Great loss of human life, structural damage, and social and economic upheaval have occurred repeatedly in recent history due to such natural hazards as earthquakes, hurricanes, landslides, floods and tsunamis. Both the US and Taiwan are threatened by such occurences and thus share a common concern to mitigate these hazards. Recent disasters of particular interest in the US are the 1989 Loma Prieta and 1994 Northridge earthquakes as well as the 1993 Midwest flood.

Several decades ago, the US embarked on natural and manmade hazards mitigation research. In the early 1980s, Taiwan launched a multiple hazards mitigation program with a sequence of five-year plans. Since then, the US and Taiwan joined together in three symposia/workshops on research and its application in multiple hazard mitigation. A fourth joint symposium workshop, *Urban Disaster Mitigation and the Role of Engineering and Technology*, was held in Chicago, July 1994. Its objectives were:

- to discuss the lessons learned from recent natural disasters
- to assess the current state of knowledge and practice in various natural hazard areas and to identify frontal joint research opportunities for advancing this knowledge
- to stimulate future cooperative research on and development of subjects of common need and importance
- to evaluate the results of Taiwan's multiple hazards research program and its practical implications for mitigation of such hazards in the US
- to continue to build the long-term bilateral scientific relationship existing between academic and practicing communities

This volume of proceedings contains state-of-the-art reports by world-renown researchers in their respective disciplines. Speakers were selected through careful screening by the Steering Committee to serve as delegates from their nation.

Sponsors of the joint symposium/workshop were National Science Foundation (NSF) in the US, under program directors Dr. Eleonora Sabadell and Dr. S.C. Liu, and National Science Council (NSC) in Taiwan. Their support is gratefully acknowledged.

Instrumental in the success of this event were a number of individuals serving on committees or with organizations.

US STEERING COMMITTEE
Franklin Y. Cheng (Chair), University of Missouri-Rolla
James E. Beavers, Martin Marietta Energy Systems Inc.
Riley M. Chung, National Institute for Standards & Technology (NIST)
Anne S. Kirmidjian, Stanford University
George C. Lee, National Center for Earthquake Engineering Research (NCEER)
State University of New York (SUNY)-Buffalo
Joanne M. Nigg, Disaster Research Center, University of Delaware
Ben Yen, University of Illinois at Urbana-Champaign

SYMPOSIUM/WORKSHOP ACTIVITIES

Symposium/Workshop Group Picture

Welcome Remarks at Opening Ceremony by Dr F. Y. Cheng,
US Delegation Chair

Welcome Remarks at Opening Ceremony by Dr M. -S. Sheu,
Taiwan Delegation Chair

Welcome Remarks at Opening Ceremony by Dr S. C. Liu, Program Director, US National Science Foundation

Banquet

RESOLUTIONS

Overall Recommendations

1. Participants unanimously recognize the importance of mitigating the potential damage resulting from natural hazards and support the continuation of cooperative research in areas of mutual concern. Research on such topics as seismology, earthquake engineering, flood control, slope stability and emergency management is considered as most urgently needed and being most likely to produce mutually beneficial results.

2. In Taiwan the ongoing program of the National Science Council on *multiple hazard mitigation* is well-planned and executed. Projects conducted under this program are beginning to produce useful information. Continuing support of this program should be provided until its successful completion. It is suggested, however, that demonstration projects be enhanced to ensure full interaction among investigators and users. Relevant results should also be made available to the American research community through an appropriate US-Taiwan joint research mechanism. Cooperation between National Center for Earthquake Engineering Research (NCEER) at SUNY-Buffalo and National Center for Research on Earthquake Engineering (NCREE) in Taiwan should be continued and enhanced.

Acknowledging the success of this symposium/workshop, participants suggest that a similar symposium/workshop be continued at intervals of two or three years.

GROUP I SEISMOLOGY

General Recommendations

The risk that earthquakes pose to society in both the US and Taiwan is growing due to increased urbanization within earthquake-affected regions of each country. Our common experience in recent damaging earthquakes underscores the need to continue basic and applied research into the causes and effects of earthquakes. Over the past several years, significant steps have been taken to modernize seismological data collections in both countries. The challenge in the coming years will be to apply these new data to the reduction of earthquake risk. Seismological research can be an effective means of achieving this goal through bilateral cooperation between researchers in our two countries to:

1. Improve our understanding of the physics of the earthquake source.

2. Make more accurate predictions of strong ground motion.

3. Develop near real-time capabilities for locating earthquakes and estimating areas of probable damage.

4. Explore the feasibility of seismic early warning systems.

Specific Cooperative Projects

1. Continued cooperation between the Central Weather Bureau and the US Geological Survey (USGS) for the development of the Taiwan Strong Motion Instrument Program (TSMIP).

2. Continued cooperation in the analysis of strong ground motion data from the SMART-1

array and SMART-2 array operated by the Institute of Earth Sciences, Academia Sinica, and expansion of activities to analyze data collected by TSMIP.

3. Creation of a working group on rapid analysis and reporting of strong earthquakes.

   a. Workshop to be held in Taiwan on rapid notification and seismic early warning systems.

   b. Creation of a seismological data exchange agreement to foster and assist joint research projects that utilize information from the new instrument capabilities of each country.

These proposed programs can be most effectively achieved through direct researcher-to-researcher contacts, including working visits to research centers, creation of task-oriented working groups, and topical workshops. We also recognize that the changing research environment, particularly the growing use of computer networks, provides new ways for researchers to collaborate, and these means should also be employed to develop and assist the creation of bilateral projects.

List of Participants

Y.T. Yeh (co-chair), National Chung Cheng University
W.L. Ellsworth (co-chair), USGS, California
H. Hwang, University of Memphis
K.W. Kuo, Central Weather Bureau
Y.B. Tsai, Pacific Gas & Electric Co.

GROUP II EARTHQUAKE ENGINEERING

General Recommendations

1. Cooperative research and scientific exchanges between NCEER in the US and NCREE in Taiwan should be continued and expanded.

2. This type of joint US-Taiwan symposium/workshop to exchange research results should be held once every two or three years.

Specific Recommendations

1. Development of site modeling using TSMIP free field data.

2. Study of structural identification/structural responses using TSMIP monitoring data from actual structures.

3. Instrumentation and analysis of real lifeline systems within TSMIP.

4. Research on structural control technology, including active, passive and hybrid systems.

5. Development of performance-based seismic design codes for buildings.

6. Study of steel building frame design practice from Northridge earthquake experience.

7. Study of residual life/strength of existing structures, due to aging, and/or after fires, earthquakes, strong winds and floods.

8. Development of non-destructive testing techniques, new materials and methodologies for repair and rehabilitation of civil infrastructures.

Note that the above recommended list of cooperative research topics is **not** inclusive or exhaustive. Additional topics may be developed as appropriate between the two countries from time to time.

List of Participants

L.R.L. Wang (co-chair), Old Dominion University

C.S. Yeh (co-chair), NCREE
A.H-S. Ang, University of California at Irvine
W.F. Chen, Purdue University
F.Y. Cheng, University of Missouri-Rolla
R.M. Chung, National Institute of Standards and Technology
W.M. Dong, Risk Management Solutions Inc.
H.M. Hwang, Memphis State University
S.C. Liu, National Science Foundation
S.T. Mau, University of Houston
L.H. Sheng, California Department of Transportation
M.S. Sheu, National Cheng-Kung University
P.S-L. Shu, Sargent & Lundy
Y.B. Tsai, Pacific Gas & Electric Co.
W.H. Tang, University of Illinois at Urbana-Champaign
Y.K. Wen, University of Illinois at Urbana-Champaign
G.C. Yao, National Cheng-Kung University

GROUP III FLOOD ENGINEERING, LANDSLIDES AND ROCKSLIDES

General Recommendations

There has been very limited joint research between the US and Taiwan on floods and landslides, mostly on an individual exchange basis, due more to lack of funds than to communication. In view of Taiwan's special geophysical characteristics of intense rain, seismicity, steep slopes and tropical weather, there is an opportunity for effective cooperation that will benefit both countries economically, socially and environmentally. This is evidenced by the substantial annual damage cost (multi-billion dollars) of floods and landslides.

Consequently, there are possible joint research areas too numerous to list. Some general research areas of common interest are as follows:

1. Forecasting and prediction of floods and landslides.

2. Effectiveness of structures and techniques for landslide and flood mitigation.

3. Data acquisition and management, including the use of GIS.

4. Multiple-hazard induced floods and landslides, such as rain-and-earthquake-induced landslides, landslide-induced floods, and flood-and-earthquake-induced dam failures.

5. Demonstration and experimental sites for the above.

6. Coastal flood and surge mitigation.

Specific Cooperative Projects

1. Research on floods:

   a. Evaluation of reliability (in both magnitude and time) of existing real-time flood forecasting methods and possible improvement in forecasting techniques.

   b. Risk of failure of flood control structures.

   c. Improved flood prediction methods for flood engineering design.

   d. Study of rainfall for improved forecasting and prediction of floods.

   e. Safety of dams.

   f. Selection and comparison of demonstration sites in Taiwan and the US.

   g. Coastal flooding due to land subsidence.

2. Research on landslides (multiple-hazards approach to landslide tasks):

   a. Prediction/hazard mapping/zone of influence of hazard, e.g. debris flow, avalanche, landslide.

   b. Study of rainfall as a cause of landslides.

   c. *Geographic information systems* to combine effects of multiple hazards.

   d. Static and dynamic stability of soil and rock slopes.

   e. Seismic displacement of soil and rock slopes.

   f. Ground motion amplification of soil and rock slopes.

List of Participants

B.C. Yen (co-chair), University of Illinois at Urbana-Champaign
S.T. Chen (co-chair), National Cheng Kung University
R.H. Chen, National Taiwan University
Frederick N-F. Chou, National Cheng Kung University
M.H. Hsu, National Taiwan University
R.L. Schuster, USGS, Colorado
T.D. Stark, University of Illinois at Urbana-Champaign
W.H. Tang, University of Illinois at Urbana-Champaign

GROUP IV SOCIO-ECONOMIC IMPACT

General Recommendations

It was recognized that the cross-national analysis of multi-hazard mitigation programs--used in Taiwan and the US--would be beneficial to both countries in their efforts to reduce the vulnerability of urban areas to natural hazard impact. Academic scholars in both countries have continuing research interests in social, economic, and policy issues that would be involved in studies on this general topic.

Research in this area would have four general purposes:

1. To understand the different strategies, policies and programs that are used by all levels of government to reduce natural hazard threats and vulnerability.

2. To identify the factors that influence the adoption and implementation of these policies, procedures and programs.

3. To identify the factors that lead to successful reduction in vulnerability of the urban social and built environments to natural hazard threats.

4. To investigate the costs and benefits--social and economic--of the various hazard reduction approaches used in both countries.

Specific Cooperative Projects

Three specific projects in the socio-economic and policy area are being recommended. While these projects are discrete and could be conducted independently, there would be great benefit in completing them sequentially, with each project building on the results of the previous effort. All three projects would provide a comprehensive analysis of the policy environment and the factors which influence it, with respect to mitigating natural hazards in major urban areas in Taiwan and the US.

1. Identification of governmental--national, state, county, municipal--policies used to reduce natural hazard threats and vulnerability in urban areas.

a. This would require the identification of a common set of natural hazards in a comparable set of urban areas to provide the basis for a cross-national study.

b. Economic and policy data would be collected on efforts undertaken or considered to make the urban environment safer.

2. Assessment of policies' effectiveness in reducing threat and vulnerability.

a. For each hazard identified in the cities selected, responsible government departments would be identified.

b. Semi-structured face-to-face interviews would be used to gather data from governmental decisionmakers with respect to specific mitigation efforts.

3. Assessment of public perception of hazards that threaten a given community and the efficacy of governmental efforts to reduce vulnerability.

a. Structured questionnaires would be developed to collect data from a random sample of the general public and from interest group (stakeholder) representatives.

b. Data would be analyzed and recommendations made.

List of Participants

C-Y.D. Chang (co-chair), National Taiwan University
J.M. Nigg (co-chair), University of Delaware
B.G. Jones, Cornell University

LOCAL ARRANGEMENTS AND HOSTS

Peter S.L. Chu, Sargent & Lundy, Chicago
Roy Wu, Director General of Coordination Council on North American Affairs (CCNAA) Office in Chicago
John Wang, Director of Science Division, CCNAA Office in Chicago

LIAISON COMMITTEE

Howard Hwang, Memphis State University
David C.C. Tung, North Carolina State University

INSTITUTIONAL SPONSORS

University of Missouri-Rolla; Rolla, MO 65401; USA
National Cheng Kung University; Tainan 70101; Taiwan

SESSION CHAIRS

Peter S.L. Chu, Sargent & Lundy, Chicago
Riley M. Chung, NIST
Weimin Dong, Risk Management Solutions Inc.
Ming-hsi Hsu, National Taiwan University
Howard Hwang, Memphis State University
C.H. Loh, National Taiwan University
Le-Wu Lu, Lehigh University
Wilson H. Tang, University of Illinois at Urbana-Champaign
John Wang, CCNAA Office in Chicago
Leon Wang, Old Dominion University
Yi-Kwei Wen, University of Illinois at Urbana-Champaign

DELEGATION CHAIRS

Franklin Y. Cheng, University of Missouri-Rolla
Maw-Shyong Sheu, National Cheng-Kung University

Editors
Franklin Y. Cheng
Maw-Shyong Sheu

Seismology

CHARACTERISTIC EARTHQUAKES AND LONG-TERM EARTHQUAKE FORECASTS: IMPLICATIONS OF CENTRAL CALIFORNIA SEISMICITY

W. L. Ellsworth

U.S. Geological Survey, MS-977, Menlo Park, CA 94025, U.S.A.

ABSTRACT

Waveform analysis of seismicity in central California reveals the routine occurrence of families of repeating earthquakes, events that rupture the same fault area in events of similar magnitude. These repeating earthquakes share many of the attributes proposed under the characteristic earthquake hypothesis for recurrence of large earthquakes including: a probability distribution for recurrence intervals with a central mean and long-tail; proportionality between length of recurrence interval and mean rate of strain accumulation; and absence of evidence for repetition before the recovery of released elastic strain energy. Temporal clustering, the principal non-Poisson component of seismicity, appears to involve successive failure of adjoining or nearby parts of the fault, without re-rupture of the already broken fault areas.

KEYWORDS

Earthquake Probabilities; characteristic earthquakes; recurrence; clustering; prediction

INTRODUCTION

The elastic rebound theory of H.F. Reid (1910), and the seismic gap hypothesis originally proposed by S.A. Fedotov (1965) and by K. Mogi (1968) form the cornerstones of long-term earthquake forecasting as it is has been applied to plate boundaries around the world. In general, the likelihood of an earthquake is related to the amount of elastic strain energy stored along the fault, with regions that have recently ruptured having low probability of failure, while those that have been locked for long periods having higher probability of failure. These ideas have been refined by numerous authors in applications on both fault-specific and global scales over the past 20 years.

In the early 1980's, the possibility that individual fault segments produced essentially but one rupture pattern, a characteristic earthquake, was proposed by several authors on the basis of both geological and seismological evidence (Schwartz and Coppersmith, 1984; Bakun and McEvilly, 1984). Under this hypothesis, the next earthquake on a particular fault segment will occur when the strain drop in the last has recovered. If correct, this model has significant practical value as it provides a vehicle for moving the estimation of seismic hazard from Poissonian probabilism into the realm of time-dependent hazard assessment (Muir-Wood, 1993). This hypothesis has played a key role in, for example, the siting of the Parkfield prediction experiment (Bakun and Lindh, 1985), and in long-term earthquake forecasts for large earthquakes on the San Andreas fault (Lindh, 1983; Sykes and Nishenko, 1994; Working Group on California Earthquake Probabilities, 1988).

More recently, the characteristic earthquake hypothesis has been drawn into question. For example, observational evidence has been found demonstrating variability in earthquake rupture pattern from cycle to cycle on the same fault segment (Thatcher, 1990; Segall and Du, 1993). Occurrence times of earthquake ruptures do not necessarily form simple, regular sequences, but may be clustered in time, as is the case for the San Andreas fault at Pallet Creek, California (Sieh *et al*; 1989). The seismic gap hypothesis has also been challenged (Kagan and Jackson, 1991), although this debate is far from resolution (Nishenko and Sykes, 1993). Indeed, theoretical fault models with uniform properties are capable of reproducing many aspects of the seismological and geological evidence for the characteristic earthquake model (Carlson and Langer, 1989).

Given these questions, it seems prudent to carefully examine the observational basis for the characteristic earthquake hypothesis, particularly in situations where the model may be tested. In this paper I summarize results from several ongoing, long-term studies of moderate-earthquakes (magnitude 4-5) and microearthquakes (magnitude 1-2) in California that address this problem. The principal findings include: 1.) Repetitive failure of the same fault area in nearly identical earthquakes is observed. This suggests that the class of earthquakes being studied may be described as characteristic earthquakes. 2.) The probability density function for the interval between earthquakes may be adequately described by a variety of long-tailed models, such as a Log Normal or a Gamma distribution. The coefficient of variation, or ratio of the standard deviation to mean return time is about 0.4. 3.) Evidence points to the state of elastic strain as the principal variable controlling recurrence time, as predicted by Reid's hypothesis. 4.) The hypothesis that temporal clusters of events typically involve re-rupturing of the same asperity is rejected. The observations show instead that temporal clusters of events usually involve failure of adjacent regions on the fault plane.

EXPANDING THE LIMITS OF INFORMATION ON THE EARTHQUAKE CYCLE

Quantitative models of the seismic cycle are severely limited by the short duration of the seismic record. Even along fast moving plate boundaries, we rarely have evidence for the recurrence of more than a handful of gap-filling earthquakes, and the level of detail we have about them degrades rapidly as the dates of earthquakes recede in time. Since we cannot easily change the length of the historical record, and paleoseismic methods cannot resolve fine details of either the timing or extent of rupture (*e.g.* Atwater *et al.*, 1991), we must turn to other data if we are to test or refine the characteristic earthquake hypothesis.

Two approaches have been suggested: numerical simulation of earthquakes, and analysis of more frequent small-magnitude earthquakes. Both methods have their merits and limitations. Computer simulations can generate millions of years of synthetic data, with detailed information available for all the microscopic variables that control the simulation. But while a particular model may capture the essence of some aspects of seismicity, such as the Gutenberg-Richter frequency-magnitude law, or Omori's law, other models based on quite different assumptions may also reproduce them, leaving predictive deductions in need of independent verification.

In California, the useable historical record is less than 150 years long, at least for events of $M \geq 6$. At present, only at Parkfield, where M 6 events rupture the same segment of the fault about every 22 years (and the next event is currently overdue), can we identify more than two ruptures of the same fault segment in historic time. Instrumental data can greatly improve our resolution of old earthquakes. Although instrumental completeness at the M 6 threshold was attained by the end of the 19th century, systematic recording of earthquakes throughout the San Andreas fault system did not commence until the early 1930's.

By studying small-magnitude earthquakes, ones for which several to many complete earthquake cycles are contained in the instrumental record, it should be possible to better define the nature of recurrence. From the

1930's until the early 1990's, Wood-Anderson seismographs systematically recorded events of M≥3.5 throughout the San Andreas fault system. These records have proven invaluable in deciphering the history of events of M≈4 (Ellsworth and Dietz, 1990). We are fortunate that this record continues in California with the new broad band instruments, but only where the old and new seismographs are located in the same vault. This is a critical factor to consider when modernizing a network which has not, unfortunately, always been appreciated. A change in station location of only a few 10's of m is sufficient to destroy the coherency between seismograms exhibited in Fig. 1.

Seismograms from the USGS telemetered network in northern and central California (Calnet) are available in digital form from the late 1970's to the present. By jointly harnessing the geometric control of Calnet for precise hypocentral location with digital analysis of the waveforms, it is now possible to routinely measure relative location differences between nearby earthquakes to a precision of ≈10 to 20 m. In other words, it is possible to achieve a location accuracy that is smaller than the dimension of the earthquake sources (M≥1.5) (Poupinet *et al.*, 1984). This means that we can quantitatively assess the geometric relationship of one earthquake to another, and can determine if they rupture the same fault area or adjacent regions (Cole and Ellsworth, 1992; Desai and Ellsworth, 1993). We can also study the evolution of the moments and rupture durations of individual earthquake sources (Vidale *et al.*, 1994), or examine the temporal stability and variability of wave propagation in the interval between repetitions (Poupinet *et al.*, 1984; Beroza *et al.*, in press).

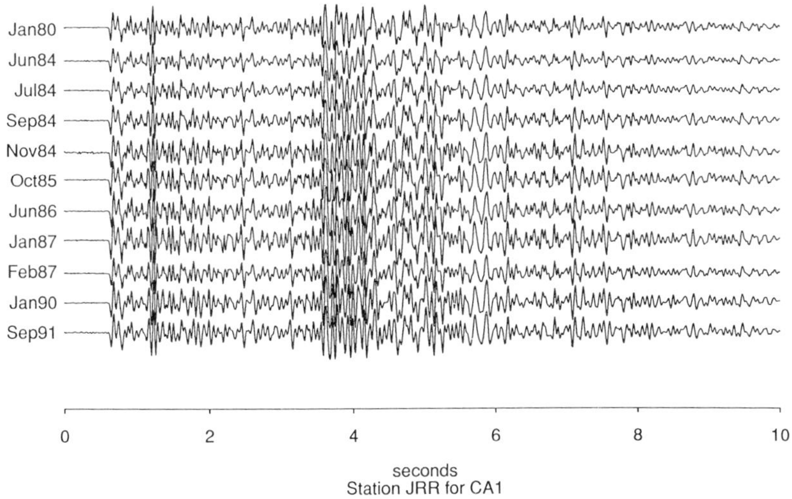

Fig. 1. Seismograms of 11 repetitions of the same M 1.4 earthquake source on the Calaveras fault, California (CA1). A total of 19 repetitions of this source have been observed between 1980-1994.

REPEATING EARTHQUAKES

The characteristic earthquake hypothesis as originally conceived by Schwartz and Coppersmith (1984) addressed the similarity of fault displacements in successive events either at a point on the fault, or on a profile along a fault segment. If adjacent fault segments should fail together one time, and separately another, they could leave the same paleoseismic record. Thus, it is possible that two rather different earthquakes could exhibit this type of characteristic behavior. The characteristic earthquake hypothesis discussed by Bakun and McEvilly (1984) considered events produced by the failure of the same region; in other words an asperity or "stuck patch" as discussed by Wesson and Ellsworth (1973). I will refer to this particular class of characteristic behavior as a "repeating earthquake."

Seismological evidence permits us to examine both of these possibilities through the comparison of waveforms. The most exacting measurement we routinely make is the distance between the centroids (center of moment release) for a pair of earthquakes with nearly identical focal mechanisms. This is done using the cross-spectral analysis methods described in Poupinet *et al.* (1984). A somewhat less exact comparison may also be made by measuring the coherency between pairs of events at a common station. Because wave propagation in the Earth has a stochastic character, coherency between two seismograms at a given frequency implies that the sources lie within a common first-Fresnel Zone, the "$\lambda/4$ Rule" (Geller and Mueller, 1980). When both the body waves and coda waves are coherent for a pair of events at frequencies of f Hz and below, the centroids of the two events will lie within $V_S/4f$ km of each other, where V_S is the shear wave velocity.

RESULTS FROM THE SAN ANDREAS FAULT SYSTEM

Application of the cross-spectral method to M 1.5-2.5 earthquakes on the San Andreas and Calaveras faults reveals the existence of numerous repeating earthquakes (Vidale *et al.*, 1994; Beroza *et al.*, in press; Fig. 1). These events are characterized by centroid separations requiring substantial, if not complete, overlap of the sources, and by nearly uniform seismic moments. The relative ease with which pairs of repeating events are identified, leads me to conclude that this is a very common type of behavior along these fast-moving (1-3 cm/yr) faults. Non-repeating behavior was not addressed by either Vidale *et al.* Or Beroza *et al.*, but has been in several other studies, described below, where we have analyzed all earthquakes in a space-time volume.

Along the San Andreas fault near Stone Canyon and Bear Valley, in the northern Gabilan Range, Ellsworth and Dietz (1990) found strong evidence for repeating earthquakes during the interval 1934- 1990. By comparing the waveforms of all M≥4 earthquakes in this region, they could document the occurrence of up to six repetitions of the same earthquake source (Table 1). For earthquakes occurring after 1970, precise locations and focal mechanisms are available from Calnet data. All earthquakes that occurred on the San Andreas proper between 1970 and 1990 also occurred one or more times between 1934 and 1969. This strongly implies that the dominant mode of generation of M≥4 earthquakes along this segment of the fault is the cyclic locking and breaking of quasi-permanent stuck patches, or asperities on the fault, as suggested by Wesson and Ellsworth (1973).

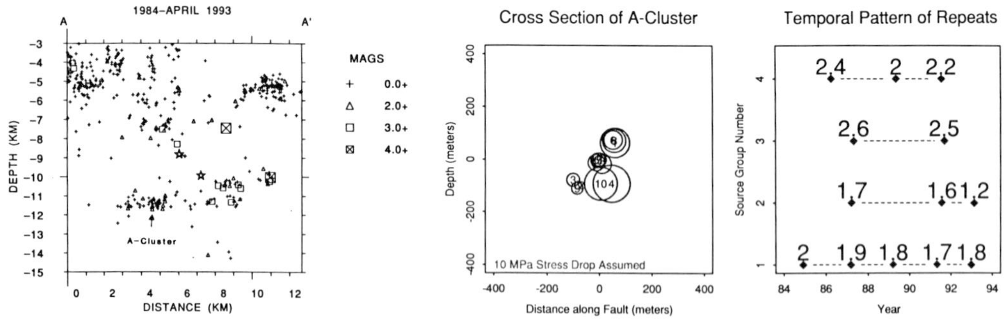

Fig. 2. Repeating earthquakes near the 1966 Parkfield hypocenter. Cross section along San Andreas fault showing seismicity 1/1984-4/1993 (Left). Location of 1966 Parkfield mainshock and foreshock (lower right and upper left stars, respectively) also shown. Cross section in the plane of the San Andreas fault surrounding one persistent cluster of activity (A-Cluster, Center) Numbers indicate sequence number. Circle size scaled to rupture dimension for a 10 MPa stress drop on a circular crack. Time-line diagram of repeats for the four identified sources (Right). Magnitudes of the events in each group are indicated.

Table 1. Stone Canyon Characteristic Earthquake Sets

SET	LATITUDE	LONGITUDE	DEPTH	DATE	M_L
	N	W	km	yrmoda hrmn	UCB
1	36-34.50	121-10.53	8.5	370217 0333	4.5
				540327 1543	4.4
				730622 0129	4.2
				841005 1616	4.0
2	36-35.04	121-10.77	5.6	510729 1053	5.0
				720224 1556	5.1
				820810 0211	4.5
				860114 0309	4.8
3	36-38.12	121-14.98	4.9	510806 0905	4.9
				720904 1804	4.7
				860531 0847	4.7
4	36-38.89	121-16.03	11.3	380927 1223	5.0
				820811 0746	4.6
5	36-40.93	121-18.42	5.5	371027 1553	4.5
				480427 2022	4.4
				560218 2358	4.2
				730115 0943	4.1
				820924 0805	4.0
				900908 1248	4.0
6	36-40.73	121-18.32	6.0	371027 2025	-
				480427 2032	-
				560219 0006	3.8
				730115 1023	3.7
7	36-41.75	121-19.19	4.7	400907 1302	4.5
				540622 1250	4.3
				891201 1237	4.4
8	36-41.49	121-19.21	4.6	400907 1038	3.5
				540622 1149	4.4
				711229 0025	3.9
				800306 1105	4.0
				900125 1404	3.9
9	36-39.78	121-17.36	7.4	381024 1339	-
				480111 0537	4.3
				650914 0909	4.0
				820831 0311	4.0
				900911 0713	4.0
10	36-35.41	121-11.82	8.4	340616 2303	4.0
				420114 0944	4.2
				561122 1643	4.2
				661010 0653	4.1
				771215 1115	4.2
				871107 1506	4.0

Along the Parkfield segment of the San Andreas fault, Desai and Ellsworth (1993) examined the systematics of the seismicity in the vicinity of the 1966 Parkfield earthquake hypocenter from 1984-1993. Repeating earthquakes were found to be the predominant process of earthquake generation within most clusters of activity. Individual clusters of seismicity (Fig. 2) generally contain several distinct earthquake sources, that were inferred to represent asperities. If their estimates of the source dimensions are approximately correct, then the asperities abut each other. Note that while the sources are in close proximity to one another, the entire group does not fail in the same temporal sequence each time. Nearest neighbor interactions between otherwise isolated groups of asperities add an element of complexity to the process.

Several extended series of repeating earthquakes have also been found along the Calaveras fault and at Parkfield on the San Andreas fault (Fig. 3) through a systematic analysis of all events in small source volumes ($\approx 1 km^3$). At Parkfield, just to the northwest of the source zone of the 1966 earthquake, multiplet PK1 was found to have remarkably little variation in either the length of the recurrence interval (coefficient of variation, *c.v.*=0.10), or in the seismic moments (*c.v.*=0.10) across 9 repeats. Along the Calaveras fault, 19 repeats of the same M 1.5 earthquake, CA1, were observed between 1980 and 1993 by Vidale *et al.* (1994). These events also have remarkably stable seismic moments, with *c.v.*= 0.15. The intervals between repeats, however, are not constant, and vary from 3 days to over 3 years. This variation in recurrence interval of nearly a factor of 500 is not random, and is likely related to the loading rate on the fault, as discussed below. The nearby multiplet CA2, with 14 repeats, also displays the same correlation between loading rate and recurrence, but has larger variability in seismic moments (*c.v.*=0.4).

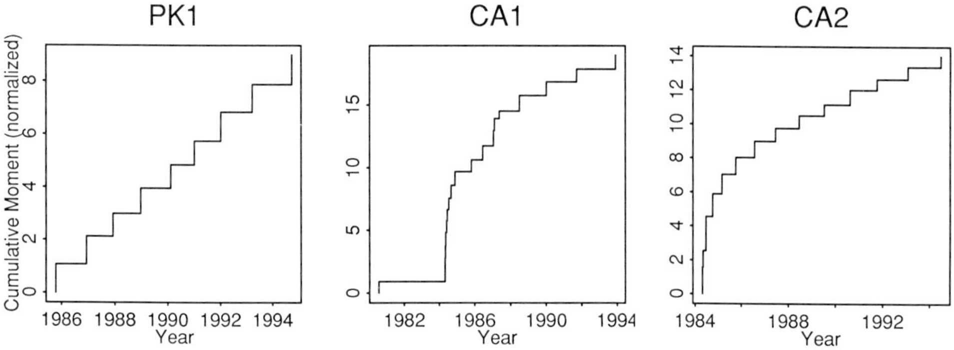

Fig. 3. Cumulative seismic moment versus time for repeating earthquake sources. Left: Multiplet PK1, located to the northwest of the 1966 Parkfield earthquake. Center: Multiplet CA1, located on the Calaveras fault at the southern end of the 1984 Morgan Hill earthquake aftershock zone. Right: Multiplet CA2, located on the Calaveras fault 4 km southeast of CA1 and further from the Morgan Hill earthquake.

IMPLICATIONS FOR PROBABILISTIC MODELS OF RECURRENCE

Because our knowledge of both the physical laws and physical state describing the Earth is so rudimentary, it is necessary to describe what limited information we possess about recurrence with simple statistical models. Types of questions we might ask about recurrence of characteristic earthquakes include the variability in repeat interval, seismic moment, slip distribution, or fault area; as well as inquiry into functional relations between the parameters. Evidence on recurrence of the type described above should be useful for examining candidate relationships between repeat interval and seismic moment, as both quantities can be measured with precision.

The simplest models assume that time to the next failure of a specific fault segment may be described by a probability density function *f(t)* (PDF), where *t* is time since the last event. Functional forms for *f(t)* used in previous studies include the Normal (Lindh, 1983; Sykes and Nishenko, 1984), Weibull (Hagiwara, 1974), and Log Normal (Nishenko and Buland, 1987) distributions. I will introduce into the discussion one additional distribution that is used to model time-to-failure data, the Gamma distribution

$$f(t) = \frac{\lambda^k t^{k-1} e^{-\lambda t}}{\Gamma(k)}$$

The Gamma distribution has properties that are intermediate between those of the Weibull/Normal and the Log Normal distributions, and thus the collected functions offer us something of a continuum of distributions that are functions of only a mean value, μ, and a coefficient of variation, *c.v.*

Fig. 4 shows *f(t)* for the five candidate distributions, for μ=1 and *c.v.*=0.5. Differences between these distributions are comparatively minor near the mean repeat time. Differences between candidate PDF's becomes important once the mean time has been exceeded. Because the parametric description of any of these PDF's is poorly constrained by data, we must acknowledge that our ignorance of *f(t)* plays an important role in characterizing the uncertainty in our estimates of conditional probability.

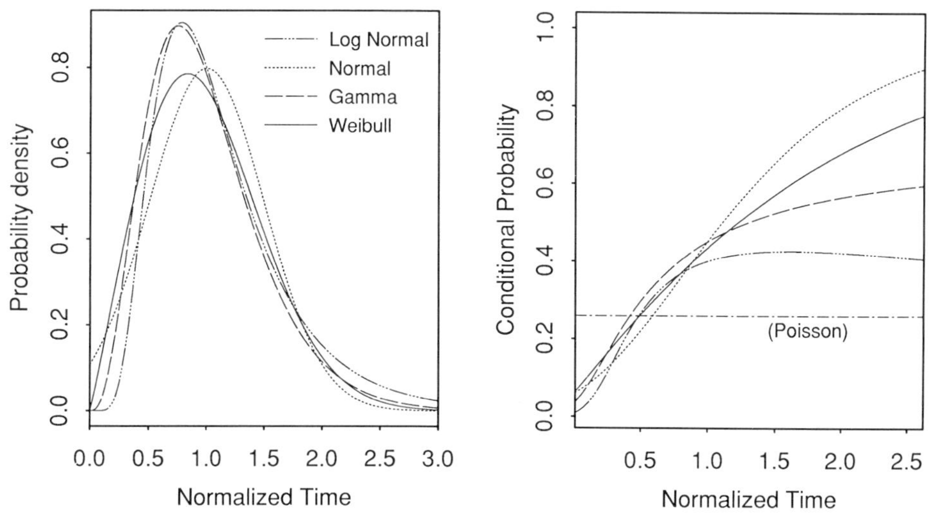

Fig. 4. Probability density functions for several models, all with μ=1 and *c.v.*=0.4 (Left). Corresponding conditional probabilities are given for a prediction interval of 0.3μ (Right). Poisson probability is also shown (μ=1).

How does the data from the San Andreas fault system constrain our choice of models? Restricting our attention to the Stone Canyon events (Table 1), we find that all candidate models fit the general trends of the observed distribution (Fig. 5). Both the Normal and Log Normal models underestimate the rate until well past the mean time, however. If we adjust these models to fit the observed cumulative distribution near the mean, the Normal model seriously overestimates the rate for $\mu \geq 1.5$, while the Log Normal model continues to underestimate it. Overall, the best fit is given by the Gamma distribution. Note also that a Poisson model may be firmly rejected. It is of some interest that the *c.v.* for these events is 0.4, considerably larger than the value determined by Nishenko and Buland (1987). Thus, for this data set at least, time elapsed since the last earthquake may contain less information than was previously assumed. We should be cautious, however, about extrapolating results from one part of a creeping fault to the general case.

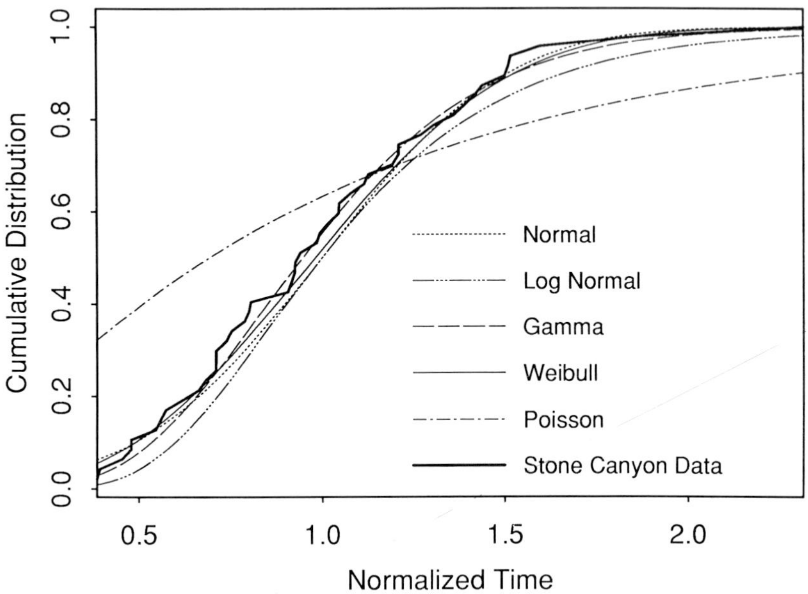

Fig. 5. Comparison between observed cumulative distribution of normalized recurrence intervals for Stone Canyon earthquakes (Table 1) and candidate probability density functions (*c.v.*=0.4). The Gamma distribution provides the best overall fit.

The correlation between mean recurrence interval and seismic moment for these data has been examined by normalizing the seismic moments and recurrence intervals within each set of Table 1. When normalized moment is compared to the normalized length of the preceding interval, we test for slip-predictable behavior. When the normalized length of the following interval is compared to normalized moment, we test for time-predictable behavior. There is a weak correlation between earthquake moment and recurrence interval for the slip-predictable model of Shimazaki and Nakata (1980), but not for the time-predictable model. In contrast, the Parkfield microearthquake multiplet PK1 (Fig. 3) exhibits classical characteristic earthquake behavior, and is both time- and slip-predictable.

STATE OF STRAIN, NOT ELAPSE TIME, CONTROLS RECURRENCE

The length of the interval between repeating events may vary in response to a variety of causes, such as the rate at which strain accumulates, abrupt changes in the strain rate, and time-dependence of the frictional strength of the fault, to name but a few possibilities. If seismic moment remains constant (and reflects constant fault slip per event on a patch that is locked between earthquakes), then over time the length of the recurrence interval should be proportional to the time required to re-accumulate the average elastic strain drop. Variations in the strain rate should produce correlated variations in earthquake rate.

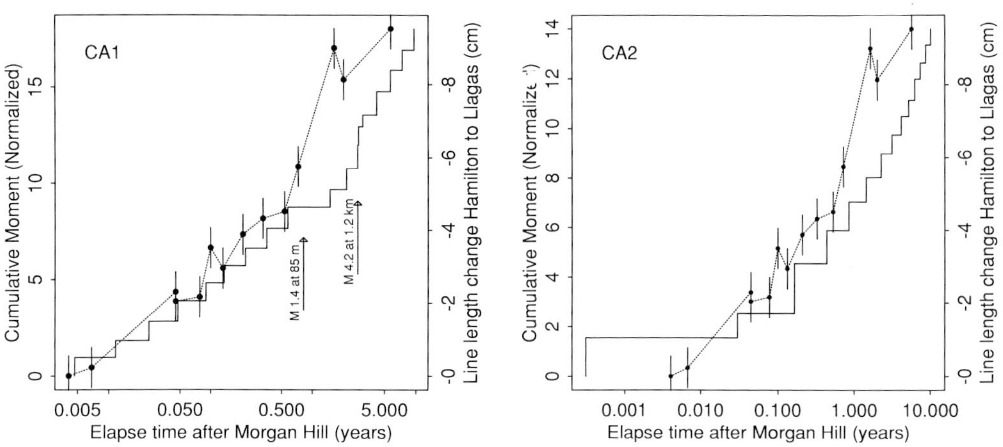

Fig. 6. Comparison between seismic moment release in repeating earthquakes and strain across the Calaveras fault as a function of the logarithm of time after the M 6.2 1984 Morgan Hill earthquake. Solid line: cumulative seismic moment release from repeating earthquake source. Dotted line: Distance between monuments crossing the Calaveras fault measured by laser ranging, relative to nominal length of 23 km, with $\pm 1\sigma$ error bars. Both multiplets show a strong response to the Morgan Hill earthquake. Multiplet CA1 (left) may also have been perturbed by two nearby earthquakes, indicated by arrows. Multiplet CA2 (right).

The two multiplets along the Calaveras fault display such a correlation with strain rate (Fig. 6). The source for CA1 is located in a small left-step of the Calaveras fault, at the south end of the M 6.2 1984 Morgan Hill earthquake rupture. CA2 is located 4 km further to the southeast, further from the earthquake. The Morgan Hill earthquake was accompanied by accelerated postseismic deformation along the Calaveras fault, as revealed by frequent geodetic observations of baselines crossing the fault (Prescott *et al.*, 1986). According to the authors, changes in the Hamilton to Llagas line (Fig. 6) imply a fault slip rate of 12 mm/day during the first two weeks of the sequence, substantially higher than the long-term slip rate of about 15 mm/year. They also infer the changing line length to reflect creep along the Calaveras fault. The two sources, CA1 and CA2, must keep-up with the displacement rate in creeping zones, since the fault is approximately slipping in block motion.

Both multiplet sources responded vigorously to the Morgan Hill earthquake (Fig. 3), and can be considered aftershocks of the event, according to standard seismological criteria. In fact, several occurrence of each source took place during the aftershock period. The time interval between successive repeats and equal increments of line length change is proportional to the logarithm of time elapsed since the main shock (Omori's law; Fig. 6). This observed correlation between the time interval required to strain the fault zone by a given amount (assuming block motion of the fault), and the time interval between successive earthquakes provides empirical support for Reid's 1910 elastic rebound theory.

For multiplet CA2, the two series track each other over four orders of magnitude change in time scale. Multiplet CA1 tracks the line length changes for the first year, then falls behind, before accelerating in 1987 (see also Fig. 3, middle). The speed-up in 1987 is clearly related to the occurrence of a nearby M 4.2 earthquake, and two CA1 events occur as part of its aftershock sequence. To test if any other earthquakes should have produced similar perturbations on the multiplet source, we searched the entire earthquake catalog for events that produced comparable or larger elastic stress changes, as resolved onto the fault plane of CA1. Only two events were found: the 1984 Morgan Hill earthquake, which increased the stress by about 1 bar; and the M 1.4 earthquake of February 19, 1985 at a hypocentral distance of 85 m, which decreased the stress by about 0.3 bars. In both cases, the estimated stress changes are rather modest. Although the 1985 event occurred near the beginning of a slow-down in earthquake production rate, we suspect that it probably was not responsible for it by itself. The M 7.0 1989 Loma Prieta earthquake was located only about 35 km from CA1 and CA2 and moved each about 0.3 bars closer to failure, but does not appear to have affected the event rate.

Changes in the rate of strain accumulation may also play a role in triggering of earthquakes, as noted above. At Parkfield, the M 4.6 earthquake of November 14, 1993 triggered aftershocks along the base of the seismic zone near the 1966 hypocenter. (It also triggered an "A-Level" alert and public warning of a possible Parkfield main shock, which did not materialize). These triggered aftershocks locate 5-10 source dimensions from the M 4.6 event. Because the aftershocks occurred in the same area as the clusters examined by Desai and Ellsworth (1993), we anticipated that the aftershocks would turn out to be additional repetitions of the already identified repeating events. To our surprise, however, all of the aftershocks in the first 3 days of the sequence located below the prior activity. Events nearest the A- cluster (Fig. 2) located 200 m below this activity, in an area where no activity had been observed. Because it is widely assumed that the fault is slipping aseismically below the seismic zone (Segall and Harris, 1987), these triggered aftershocks may be interpreted as evidence for a change in fault zone friction from velocity-strengthening (stable) behavior to velocity-weakening (unstable) behavior. Laboratory measurements of the frictional behavior of serpentinite by Reinen *et al.* (1991) display such a transition from stable to unstable sliding at a sufficiently high loading rate. While we cannot prove this to be the case at Parkfield, the occurrence of these aftershocks at new sites on the fault illustrates the point that no simple model, such as the characteristic earthquake hypothesis, can explain all modes of earthquake behavior.

TEMPORAL CLUSTERING

The most commonly observed non-Poissonian component to seismicity is the tendency of earthquake to occur in space-time clusters. In a comprehensive study of central California seismicity, Reasenberg (1985) found that a simple model based on an interaction distance and Omori's law captures the non-Poisson component of seismicity. Indeed, 50% of the events in this catalog belong to earthquake clusters.

Because clustering plays a significant role in time-dependent earthquake risk assessment (Reasenberg and Jones, 1989) it would be desirable to know how the events that form each cluster are related to one another. To address this question, Cole and Ellsworth (1992) examined all of the clusters identified by Reasenberg's algorithm at Parkfield (1984-1992), and determined the centroid separation between events. The results show that event pairs lie within the "plane" of the San Andreas fault, and that the width of the active fault zone is less than 50 m.

Another way to view the centroid separation data is on a "magnitude-separation" diagram (Fig. 7.) This diagram compares the distance between centroids for pairs of events with comparable magnitude. Because source dimension scales as the cube-root of stress drop, conclusions drawn from the diagram are rather insensitive to the stress drop used in the scaling lines. Centroid separations measured for Parkfield temporal clusters (< 1 day; open squares) indicate that sources have little overlap, and are commonly separated by several source dimensions. On the other hand, centroid separations from events inferred to be repeating earthquakes along the San Andreas and Calaveras faults near the 1989 Loma Prieta earthquake (Beroza *et al.*, in press; filled squares) must either have substantial overlap or have $\Delta\sigma > 100$ MPa.

These results imply that earthquake sequences of the foreshock/mainshock type, mainshock/aftershock type, and swarm type involve the sequential failure of distinct regions on the fault plane. There is no evidence in these data for short-time (up to 2 days) repetition of movement on the same fault patch. When we observe rapid repetition of movement on the same patch, such as seen in CA1 and CA2, it is invariably in events of much smaller magnitude than the main events in the sequence, and can be related to changes in the loading rate (Fig. 6).

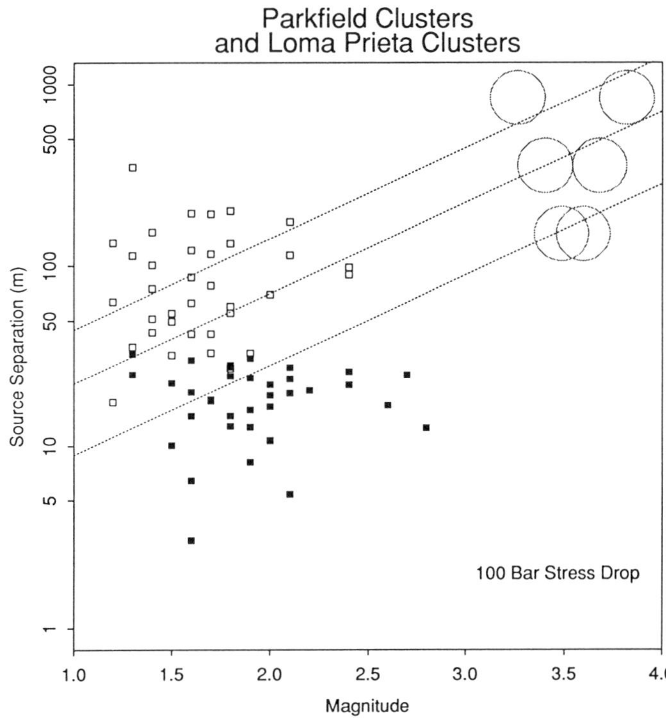

Fig. 7. Magnitude-separation diagram displaying distance between earthquake centroids determined from cross-spectral analysis versus average magnitude. Data points shown come from Parkfield temporal clusters (open squares) and repeating earthquakes near Loma Prieta (filled squares). Time separation of Parkfield events is less than 1 day. Time separation of Loma Prieta events (Fig. 1) is typically several years. Centroid separation and earthquake magnitude may be related using a simple model of two circular cracks, with specified stress drop ($\Delta\sigma$=10 MPa). Scaling lines for constant geometry of separation (or overlap) as a function of magnitude are shown for cases of 50% overlap of source areas, tangent source areas, and separation by 1 source dimension. Circles show relationship between a hypothetical pair of sources for each scaling line.

DISCUSSION

The examples of repeating earthquakes presented here illustrate the very ordinary nature of this behavior along the San Andreas fault in central California. Repeating earthquakes range in magnitude from the smallest we can reliably detect (M 0.8) to at least the size of Parkfield main shocks (M 6). One feature shared by each of the places studied here is the intermixing of seismic and aseismic slip on the fault. While it might be argued that repeating behavior arises only when this occurs, geologic evidence argues otherwise (Schwartz and Coppersmith, 1984). Because these repeating earthquakes share so many seismological characteristics with large and great earthquakes, comparison of their recurrence behaviors may be warranted.

In some cases, repeating events occur with almost clockwork precision. This appears to be the case when individual asperities are widely separated or buffered from one another, as appears to be the case at PK1 (Fig. 3). At the other extreme, temporal clustering of events often occurs when nearby source zones abut.

Examples of this behavior occur within the four sites comprising the A-Cluster at Parkfield (Fig. 2), or for Sets 2 and 3 at Stone Canyon (Table 1). The tendency for failure in one zone to trigger failure in an adjacent zone has also been commonly observed along the circum-Pacific for great earthquakes (Mogi, 1981; Thatcher, 1990).

Another property shared by recurrent ruptures in the circum-Pacific and repeating earthquakes in central California is variability in rupture pattern from cycle to cycle. Thatcher (1990) found that few, if any, recurrences in the circum-Pacific could be considered identical. In fact, many of the circum-Pacific zones that broke in a single event during one cycle broke in a series of events in another cycle. Segall and Du (1993) have also documented variations in rupture pattern for the last two M 6 Parkfield earthquakes. For most of the repeating earthquakes described here, all that we can measure at present is the variability of seismic moment within each sequence, which could be due to variations in rupture area, or to other causes (Vidale *et al.*, 1994). We can document variation in rupture pattern in the case of the last two rupture cycles for Set 2 at Stone Canyon. Here, aftershocks demonstrate that the zone failed in one event in 1972, and failed in two events the next time around (Ellsworth and Dietz, 1990).

Our ability to resolve spatial detail about repeating earthquakes in central California is primarily limited by the frequency bandwidth and dynamic range of the instruments. Use of high-frequency digital recordings, as are now available at Parkfield, should shed new light on the process, such as on the spatial distribution of seismic slip from cycle to cycle, or on the interaction geometry for neighboring asperities. By exploiting this type of information, we will be able to better determine the boundary between predictable and chaotic behavior during the earthquake cycle.

For now, we should be encouraged that the recurrence characteristics of central California repeating earthquakes have so much in common with those of much larger earthquakes. The probability distribution for their recurrence intervals is not a Poisson process, and has a well-defined mean and a relatively long tail. Although the probability gain over a Poisson process is modest, it has practical value for society. Results from small earthquakes studied here generally support the types of forecasts now being made for the San Andreas fault, such as those of the Working Group on California Earthquake Probabilities (1988), although slightly lower 30-year probabilities might be in order, if the *c.v.*=0.4 obtained from Stone Canyon (Table 1) is representative.

The absence of examples for repetition without the re-accumulation of elastic strain gives added weight to the hypothesis that once a particular fault has ruptured, it will remain safe until some significant fraction of the strain release re-accumulates. Space-time clustering of seismicity, on the other hand, represents another important process that must be taken into account in any time-dependent seismic hazard estimation analysis. Thus, just because a particular fault has failed, it does not imply that the seismic risk in the region is now low. Because interactions between neighboring asperities and variations in the rate of strain accumulation clearly have a first-order influence on recurrence behavior at all scales (M 1-9), work addressed at the mechanics and statistics of this type of interaction should extend and refine time-dependent hazard assessments well beyond the limits of the characteristic earthquake hypothesis.

CONCLUSION

Earthquake systematics in central California support the concept of characteristic earthquake behavior over a wide range of earthquake magnitude. Simple statistical models capture the essence of the recurrence interval distribution. Such models, as embedded in the characteristic earthquake hypothesis do not address all significant aspects of seismicity for hazard estimation, however. Temporal clustering represents another pervasive process, one that is amenable to study with the same waveform-based methods used here to elucidate earthquake recurrence.

ACKNOWLEDGMENTS

Discussions with and comments on the manuscript by Wayne Thatcher, John Vidale and Jim Savage are gratefully acknowledged.

REFERENCES

Atwater, B.F., Stuiver, M., and Yamaguchi, D.K. (1991). Radiocarbon test of earthquake magnitude in the Cascadia subduction zone. *Nature, 353*, 156-158.

Bakun, W.H. and McEvilly, T.V. (1984). Recurrence models and Parkfield, California, earthquakes. *Journal of Geophysical Research, 89*, 3051-3058.

Bakun, W.H. and Lindh, A. G. (1985). The Parkfield, California earthquake prediction experiment. *Science, 229*, 619-624.

Beroza, G.C., Cole, A., and Ellsworth, W.L., (in press). Stability of Coda-Wave Attenuation During the Loma Prieta, California Earthquake Sequence. *Journal of Geophysical Research*, 26 p.

Carlson, J.M., and Langer, J.S. (1989). A mechanical model of an earthquake fault. *Phys. Rev. A, 40*, 6470-6484.

Cole, A.T., and Ellsworth, W.L. (1992). Determining earthquake kinship from earthquake catalogs (abs.). *EOS, Transactions American Geophysical Union, 73*, 344.

Desai, A., and Ellsworth, W.L. (1993). Doublet analysis of recent microseismicity surrounding the 1966 Parkfield, California, Hypocenter (abs.). *EOS, Transactions American Geophysical Union, 74*, 416-417.

Ellsworth, W.L., and Dietz, L.D. (1990). Repeating earthquakes: characteristics and implications. *Proceedings of Workshop XLVI, the 7th U.S. - Japan Seminar on Earthquake Prediction, U.S. Geological Survey Open-File Report 90-98*, 226-245.

Fedotov, S.A. (1965). Regularities in the distribution of strong earthquakes in Kamchatka, the Kurile Islands, and northeastern Japan. *Akad. Nauk SSSR Inst. Fiziki Zemli Trudy, 36*, 66-93.

Geller, R.J., and Mueller, C.S. (1980). Four similar earthquakes in central California. *Geophysical Research Letters, 7*, 821-824.

Hagiwara, Y. (1974). Probability of earthquake occurrence as obtained from a Weibull distribution analysis of crustal strain. *Tectonophysics, 23*, 313-318.

Kagan, Y.Y, and Jackson, D.D. (1991). Seismic gap hypothesis: ten years after. *Journal of Geophysical Research, 96*, 21419-21431.

Lindh, A. G. (1983). Preliminary assessment of long-term probabilities for large earthquakes along selected fault segments of the San Andreas fault system in California. *U.S. Geological Survey Open-File Report 83-63*, 14 p.

Mogi, K. (1968). Some features of recent seismic activity in and near Japan. *Bulletin of the Earthquake Research Insititue, 46*, 1225-1236.

Mogi, K. (1981). Seismicity in western Japan and long-term earthquake forecasting. In: *Earthquake Prediction An International Review, Maurice Ewing Series 4* (Simpson, D.W., and Richards, P.G., eds.), pp. 43-51, American Geophysical Union, Washington.

Muir-Wood, R. (1993). From global seismotectonics to global seismic hazard. *Annali de Geofisica, 36*, 153-168.

Nishenko, S.P. and Buland, R. (1987). A generic recurrence interval distribution for earthquake forecasting. *Seismological Society of America Bulletin, 77*, 1382-1399.

Nishenko, S.P. and Sykes, L.R. (1993). Comments on "seismic gap hypothesis: ten years after" by Y.Y. Kagan and D.D. Jackson. *Journal of Geophysical Research, 98*, 9909-9916.

Prescott, W.H., King, N.E., and Gu Guohua (1986). Preseismic, coseismic and postseismic deformation associated with the 1984 Morgan Hill, California, earthquake. In: *The 1984 Morgan Hill, California Earthquake* (Bennett, J.H, and Sherburne, R.W., eds.), pp. 137-148, California Division of Mines and Geology Special Publication 68.

Reasenberg, P.A. (1985). The second-order moment of central California seismicity, 1969-1982. *Journal of Geophysical Research, 90*, 5479-5495.

Reasenberg, P.A., and Jones, L.M. (1989). Earthquake hazard after a mainshock in California: *Science, 243*, 1173-1176.

Reid, H.F. (1910). Reid, H.F., 1910, *The mechanics of the earthquake, v. 2 of The California earthquake of April 18, 1906. Report of the State Earthquake Investigation Commission*, Carnegie Institution of Washington Publication 87.

Reinen, L.A., Weeks, J. D., and Tullis, T. E. (1991). The frictional behavior of serpentinite: implications for aseismic creep on shallow crustal faults. *Geophysical Research Letters, 18*, 1921-1924.

Schwartz, D.P., and Coppersmith K.J. (1984). Fault behavior and characteristic earthquakes: examples from the Wasatch and San Andreas fault zones. *Journal of Geophysical Research, 89*, 5681-5698.

Segall, P., and Harris, R.A. (1987). Earthquake deformation cycle on the San Andreas fault near Parkfield, California. *Journal of Geophysical Research, 92*, 10511-10525.

Segall, P., and Du, Y. (1993). How similar were the 1934 and 1966 Parkfield Earthquakes?*Journal of Geophysical Research, 98*, 4527-4538.

Sieh, K.E., Stuiver, M., and Brillinger, D. (1989). Very precise dating of earthquakes at Pallet Creek and their tectonic interpretation. *Journal of Geophysical Research, 94*, 603-623.

Shimazaki, K., and Nakata, T. (1980). Time-predictable recurrence model for large earthquakes. *Geophysical Research Letters, 7*, 279-282.

Sykes, L.R. and Nishenko, S.P. (1984). Probabilities of occurrence of large plate rupturing earthquakes for the San Andreas, San Jacinto, and Imperial faults, California, 1983-2003. *Journal of Geophysical Research, 89*, 5905-5927.

Thatcher, W. (1990). Order and diversity in modes of circum-Pacific earthquake recurrence. *Journal of Geophysical Research, 95*, 2609-2623.

Vidale, J.E., Ellsworth, W.L., Cole, A., and Marone, C. (1994). Variations in rupture process with recurrence interval in a repeated small earthquake. *Nature, 368*, 624-626.

Wesson, R.L., and Ellsworth, W.L. (1973). Seismicity preceding moderate earthquakes in California. *Journal of Geophysical Research, 78*, 8527-8546.

Working Group on California Earthquake Probabilities (1988). Probabilities of large earthquakes occurring in California on the San Andreas fault. *U. S. Geological Survey Open File Report 88-398*, 62 p.

COMPARISON OF OBSERVED GROUND ACCELERATION RESPONSE SPECTRA FROM FOUR RECENT LARGE EARTHQUAKES IN CALIFORNIA WITH THE UNIFORM BUILDING CODE SEISMIC DESIGN SPECTRA

Y. -B. Tsai

Pacific Gas and Electric Company, San Francisco, CA, U.S.A.

ABSTRACT

A total of 302 horizontal- and 151 vertical-component ground acceleration response spectra observed from four recent large earthquakes in California, namely, the 1989 Loma Prieta (surface wave magnitude, M_S 7.1), 1992 Cape Mendocino (M_S 7.1), 1992 Landers (M_S 7.6), and 1994 Northridge (M_S 6.8) earthquakes, were analyzed in terms of the normalized spectrum shape and normalizing parameter. First, we used the conventional peak ground acceleration (PGA) as the normalizing parameter. The resultant normalized response spectrum shapes, Sa/PGA, were grouped separately for each earthquake to calculate the overall mean (geometric) horizontal and vertical spectra. The results showed that the observed mean spectra, both horizontal and vertical, were remarkably similar to each other among the earthquakes, even though they were located far apart in different regions of the state. When compared with the current Uniform Building Code (UBC) site-dependent Types S1, S2, and S3 design spectrum shapes, the mean horizontal spectrum shapes were found to be conservatively enveloped by the UBC Type S2 spectrum shape. After dividing the observed horizontal spectrum shapes into subsets according to their resemblance with the current UBC site-dependent design spectrum shapes, it was found that the observed mean horizontal Sa/PGA spectra for each subset were conservatively enveloped by the corresponding UBC design spectrum shape, thereby confirming the general applicability of the UBC design spectrum shapes for California. For refinement, we proposed modifications of the current UBC spectrum shapes in order to avoid excessive conservatism at high frequencies. Second, we introduced the mean spectral acceleration (MSA) as a new normalizing parameter. The resultant normalized response spectrum shapes, Sa/MSA, were used to analyze the uniform-probability dispersion of normalized spectrum shapes due to spectral peaks and troughs. The dispersion at one-standard-deviation level was found to increase slowly from a factor of about 1.2 at frequency 33 Hz to 1.4 at 1 Hz, then increases more steeply to 1.8 at 0.33 Hz for all types of spectrum shape. Third, the two normalizing parameters, PGA and MSA, were found to be positively correlated. Since the MSA is a water-level average of a response spectrum over the entire frequency band, the correlation means that the level of the entire response spectrum is also proportional to the PGA. Accordingly, the notion that large, isolated peaks on an acceleration time history may not be significant is misleading. It was further found that the MSA values of UBC Types S1, S2, and S3 records from the four earthquakes were so well mixed as to be indistinguishable at all distances. This indicates that the site coefficient factors in the current UBC provisions may not be warranted, as long as an appropriate UBC spectrum shape is picked for a site. Finally, the observed mean vertical spectrum shapes from the earthquakes were found to be less variable and to peak consistently at higher frequencies than their horizontal counterparts. Moreover, the vertical MSA was found to be about half of the horizontal MSA, even at near-source distances. These observations suggest that the current UBC provision for defining vertical design spectrum by scaling the horizontal design spectrum by a factor of two-thirds needs revision.

INTRODUCTION

This paper presents the results of an empirical study of an extraordinarily large number of observed ground motion response spectra, including 302 horizontal and 151 vertical components from 151 records from four recent large earthquakes in California. The four earthquakes were, from north to south, the 1992 Cape Mendocino (surface wave magnitude, M_S 7.1), 1989 Loma Prieta (M_S 7.1), 1994 Northridge (M_S 6.8), and 1992 Landers (M_S 7.6) earthquakes.

The purpose of this study was to compare the new data with the current Uniform Building Code (UBC) provisions on seismic ground motion for earthquake resistant design of buildings in order to gain insights for improvement. The current UBC provisions on seismic ground motion were formulated in the mid-1970's as part of an Applied Technology Council study (Applied Technology Council, 1978). Research by Seed et al. (1976) formed the basis for the provisions. It is important to update the current UBC provisions on seismic ground motion by incorporating the substantial addition of strong motion records since the 1971 San Fernando earthquake.

The current UBC horizontal seismic design spectrum is determined by combining two factors, namely, the zone and site-response factors. The former is given by the effective peak ground acceleration (ZPA) or Z factor for each of the four zones. The latter is represented in terms of a standardized response spectrum shape along with a site coefficient for each of the four soil profile types.

For the present study, we calculated from each component of recorded acceleration time history the 5%-damped acceleration response spectrum (Sa), the normalized response spectrum shapes by the conventional normalizing parameter of peak ground acceleration (PGA) and by a new normalizing parameter of mean spectral acceleration (MSA). The data were used for the following analyses:

1. We first compared the overall mean (geometric) horizontal Sa/PGA spectrum shapes for the four earthquakes with current site-dependent UBC design spectrum shapes. We further classified the normalized horizontal Sa/PGA spectra from each site according to their resemblance with the three UBC design spectrum shapes. We then compared the mean horizontal Sa/PGA spectrum shapes of each subgroup with the UBC spectrum shapes to confirm the general applicability of current UBC design spectrum shapes and to propose appropriate modifications in order to match better the new data.

2. We further calculated the mean (geometric) and dispersion of the horizontal Sa/MSA spectrum shapes of each subgroup. The results enabled us to examine the uniform-probability dispersion of response spectrum shapes due to peaks and troughs in individual spectra.

3. We next examined the attenuation of the normalizing parameter, MSA, as function of the closest distance to fault rupture surface. The results gave us new insights on the zone factor for the current UBC design spectrum. They were further combined with the results of previous item to review whether the current UBC provisions for site coefficients are reasonable.

4. The current UBC provision permits vertical ground motions to be defined by scaling horizontal ground motions by a factor of two-thirds when performing a dynamic analysis, unless otherwise justified on a site-specific basis. We compared the mean and dispersion of the vertical Sa/PGA and Sa/MSA spectrum shapes as well as the attenuation of the vertical MSA with the corresponding horizontal counterparts to see whether and what changes should be made to improve current code provisions regarding vertical ground motions.

OBSERVED RESPONSE SPECTRA FROM FOUR RECENT LARGE EARTHQUAKES IN CALIFORNIA

California has been struck by four large earthquakes since 1989. They were from north to south the M_S 7.1 Cape Mendocino earthquake of April 25, 1992, the M_S 7.1 Loma Prieta earthquake of October

17, 1989, the M_S 6.8 Northridge earthquake of January 17, 1994, and the M_S 7.6 Landers earthquake of June 28, 1992. These earthquakes had caused varying degrees of loss of lives and property. In particular, the Loma Prieta earthquake had caused 63 deaths and an estimated $6 billion loss, and the Northridge earthquakes had caused 65 deaths and $20 billion loss, due to their proximity to major urban areas.

On the positive note, an extraordinarily large number of strong motion records, both on ground and in structures, were collected from these earthquakes by the California Strong Motion Instrumentation Program (CSMIP), the U.S. Geological Survey (USGS) and major municipal (Los Angeles Department of Water and Power) and private (Southern California Edison and Pacific Gas and Electric Companies) utility companies.

A significant portion of these strong ground motion records were available in digital format for calculating the response spectra. We have assembled a total of 151 records for the present study. Each record consisted of two horizontal and one vertical orthogonal components. As summarized in Table 1, 12 records were from the Cape Mendocino, 60 from the Loma Prieta, 55 from the Northridge, and 24 from the Landers earthquakes, respectively.

We calculated the 5%-damped response accelerations (Sa) for 101 frequencies at equal increments on a logarithmic scale between 0.33 Hz and 33 Hz. This was to provide a fine and even distribution of response spectrum ordinates when plotted on a logarithmic frequency scale. The logarithmic frequency scale is preferred over the conventional linear frequency (or period) scale because it gives good resolution across the entire frequency band of the spectrum.

The 5%-damped acceleration response spectrum (Sa) obtained from each acceleration time history was further normalized by two different normalizing parameters. The first normalizing parameter was the conventional PGA, and the resultant normalized spectrum was called the Sa/PGA spectrum. The second normalizing parameter was the mean spectral acceleration (MSA), and the resultant normalized spectrum of this type was called the Sa/MSA spectrum. The MSA value was the geometric mean of 101 Sa values between frequencies 0.33 Hz and 33 Hz, representing the water-level average of Sa over the entire frequency band. In total, there were 302 horizontal components and 151 vertical components in either the Sa/PGA or Sa/MSA sets spectra.

We then classified each of the 151 records according to the resemblance of their horizontal Sa/PGA spectra with the UBC Type S1, S2, and S3 spectrum shapes. Figure 1a shows the UBC spectrum shapes and Figure 1b shows the attendant site classification criteria and site coefficients. The number of records classified in each type of UBC spectrum shape are given in Table 1 for each of the four earthquakes.

In total, there were 54, 67, 30 records whose horizontal Sa/PGA spectra resembled respectively the UBC Type S1, S2, and S3 spectrum shapes. It is noted that more than 80% of the records had UBC Types S1 and S2 spectrum shapes. It is also interesting to note in the above table that there is no clear correlation between the type of response spectrum shapes and site geology. It should be pointed out that for most of the recording sites, there is not enough information about subsurface soil profile to make site classification according to the UBC criteria.

COMPARISON OF OBSERVED Sa/PGA SPECTRUM SHAPES WITH UBC DESIGN SPECTRUM SHAPES

Figure 2a compares the mean observed Sa/PGA spectrum shapes of all horizontal components from the four earthquakes with the UBC Type S1, S2, and S3 spectrum shapes. It is reassuring that the newly observed spectrum shapes from all four earthquakes are remarkably similar to each other and are conservatively enveloped by the UBC Type S2 spectrum shape. This confirms the general applicability of the UBC design spectrum shapes for California.

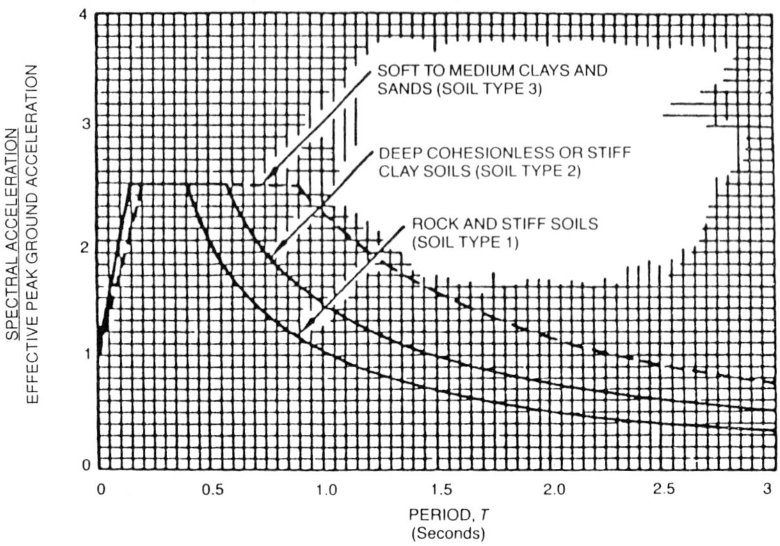

TABLE 16-J—SITE COEFFICIENTS[1]

TYPE	DESCRIPTION	*S* FACTOR
S_1	A soil profile with either: (a) A rock-like material characterized by a shear-wave velocity greater than 2,500 feet per second (762 m/s) or by other suitable means of classification, or (b) Medium-dense to dense or medium-stiff to stiff soil conditions, where soil depth is less than 200 feet (60 960 mm).	1.0
S_2	A soil profile with predominantly medium-dense to dense or medium-stiff to stiff soil conditions, where the soil depth exceeds 200 feet (60 960 mm).	1.2
S_3	A soil profile containing more than 20 feet (6096 mm) of soft to medium-stiff clay but not more than 40 feet (12 192 mm) of soft clay.	1.5
S_4	A soil profile containing more than 40 feet (12 192 mm) of soft clay characterized by a shear wave velocity less than 500 feet per second (152.4 m/s).	2.0

[1]The site factor shall be established from properly substantiated geotechnical data. In locations where the soil properties are not known in sufficient detail to determine the soil profile type, soil profile S_3 shall be used. Soil profile S_4 need not be assumed unless the building official determines that soil profile S_4 may be present at the site, or in the event that soil profile S_4 is established by geotechnical data.

Figure 1. a (Top): Normalized Seismic Design Spectrum Shapes of Uniform Building Code.
b (Bottom): UBC Site Classification Criteria and Site Coefficients.

Figure 3a compares the mean horizontal Sa/PGA spectrum shapes of the subsets of records, which resembled the UBC Type S1 spectrum shape, with the UBC spectrum shapes. It is noted from the figure that the observed Sa/PGA spectra from three of the four earthquakes are enveloped by the UBC Type S1 spectrum shape. The only exception is the mean Sa/PGA spectrum of four hard-rock records from the Landers earthquake which exceeded the UBC Type S1 spectrum shape at high frequencies. In order to accommodate all data, we propose to modify the UBC Type S1 spectrum shape by shifting its high-frequency segment toward higher frequencies, as shown in Figure 3b.

Table 1. Classification of Strong Ground Motion Records Used in This Study

Earthquake	Total No.	S1-shape	S2-shape	S3-shape
Cape Mendocino	12	3	5	4
Rock	4	2	1	1
Soil	8	1	4	3
Loma Prieta	60	24	21	15
Rock	24	9	12	3
Soil	29	15	9	5
Bay Mud	7	---	---	7
Northridge	55	23	26	6
Rock	17	10	7	---
Soil	38	13	19	6
Landers	24	4	15	5
Hard Rock	4	4	---	---
Rock	1	---	---	1
Soil	19	---	15	4
Total	151	54	67	30

Figure 4a compares the mean horizontal Sa/PGA spectrum shapes of the subsets of records, which resembled the UBC Type S2 spectrum shape, with the UBC spectrum shapes. The figure shows that the observed mean Sa/PGA spectra from all four earthquakes are enveloped by the UBC Type S2 spectrum shape, except for a spectral peak at 2.3 Hz in the mean Sa/PGA spectrum of five records from the Cape Mendocino earthquake. At further inspection, it was noted that the UBC Type S2 spectrum shape significantly exceeded the observed mean Sa/PGA spectra at high frequencies. Accordingly, we propose to modify the UBC Type S2 spectrum shape by shifting its high-frequency segment toward lower frequencies, as shown in Figure 4b.

Figure 5a compares the mean horizontal Sa/PGA spectrum shapes of the subsets of records, which resembled the UBC Type S3 spectrum shape, with the UBC spectrum shapes. The figure shows that the observed Sa/PGA spectra from all four earthquakes are enveloped by the UBC Type S3 spectrum shape, except for a spectral peak at 1.5 Hz in the mean Sa/PGA spectrum of the seven Bay-mud records from the Loma Prieta earthquake. At further inspection, it is noted that the UBC Type S3 spectrum shape significantly exceeded the observed mean Sa/PGA spectrum at high frequencies. Accordingly, we propose to modify the UBC Type S3 spectrum shape by shifting its high-frequency segment toward lower frequencies, as shown in Figure 5b.

In summary, Figure 2b compares the mean observed Sa/PGA spectrum shapes of all horizontal components from the four earthquakes with the three modified UBC Type S1, S2, and S3 spectrum shapes. It is interesting to note that the observed spectrum shapes from all four earthquakes are now more closely enveloped by the modified UBC Type S2 spectrum shape for all frequencies.

Finally, both the current UBC Types S1, S2, and S3 spectrum shapes and the proposed modifications for them can be expressed by formulas, as given in Table 2.

DISPERSION OF NORMALIZED SPECTRUM SHAPES

It was well known that the dispersion of Sa/PGA spectra is not uniform across the entire frequency band. Because the Sa/PGA spectra are all constrained to unity at the high-frequency end, the dispersion is zero at this frequency and diverges toward lower frequencies. This is evident from Figure 6a where

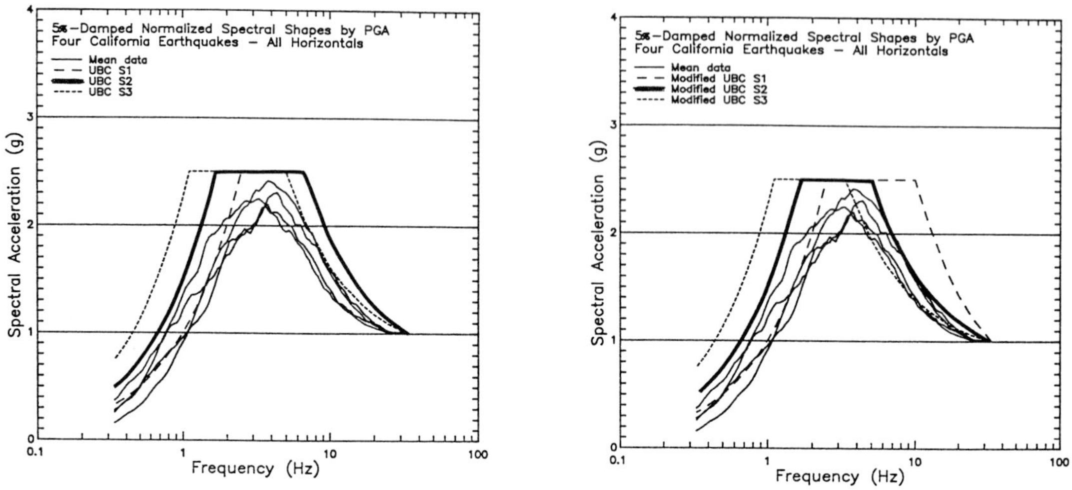

Figure 2. a (Left): Mean Horizontal Response Spectrum Shapes of All Records from Four Recent California Earthquakes with the UBC Design Spectrum Shapes.
b (Right): Mean Horizontal Response Spectrum Shapes of All Records from Four Recent California Earthquakes with Modified UBC Design Spectrum Shapes.

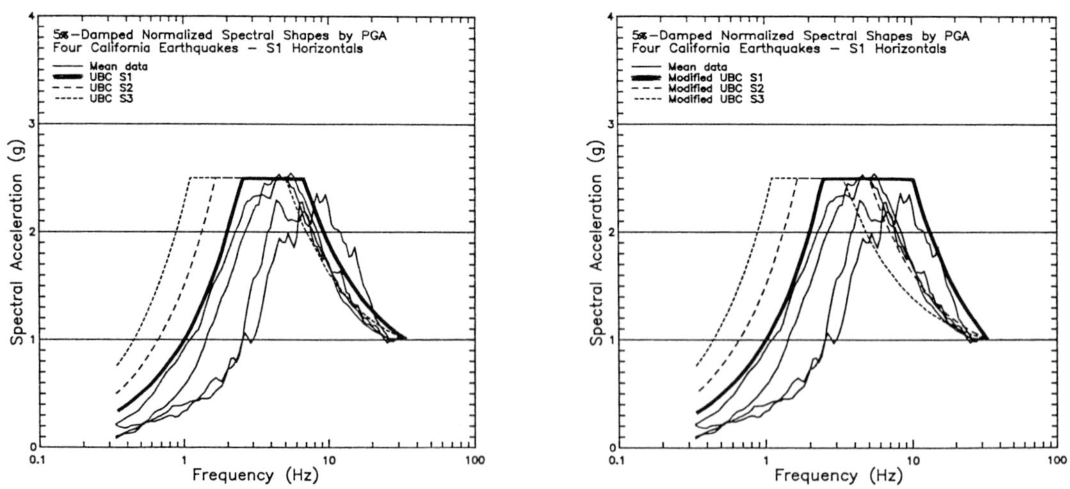

Figure 3. a (Left): Mean Horizontal Response Spectrum Shapes of UBC Type S1 Records from Four Recent California Earthquakes with the UBC Design Spectrum Shapes.
b (Right): Mean Horizontal Response Spectrum Shapes of UBC Type S1 Records from Four Recent California Earthquakes with Modified UBC Design Spectrum Shapes.

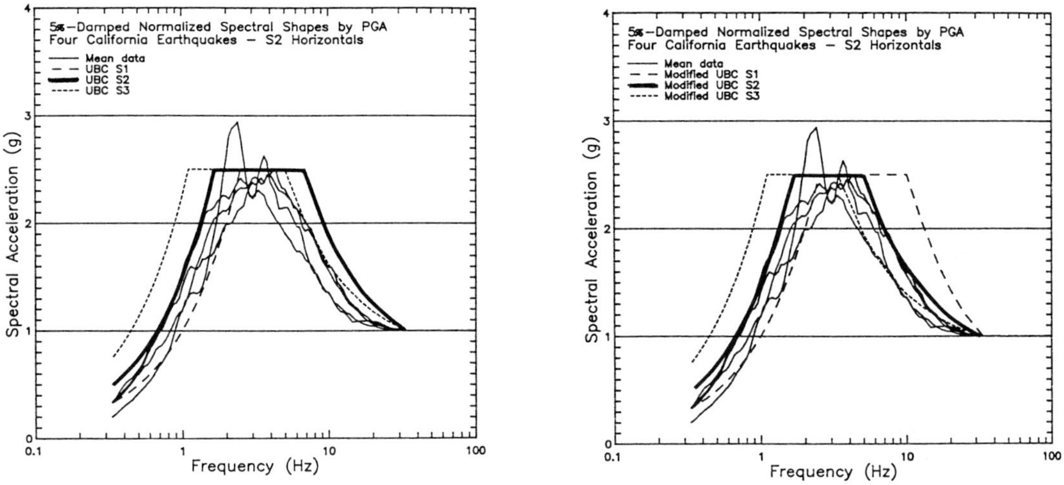

Figure 4. a (Left): Mean Horizontal Response Spectrum Shapes of UBC Type S2 Records from Four Recent California Earthquakes with the UBC Design Spectrum Shapes.
b (Right): Mean Horizontal Response Spectrum Shapes of UBC Type S2 Records from Four Recent California Earthquakes with Modified UBC Design Spectrum Shapes.

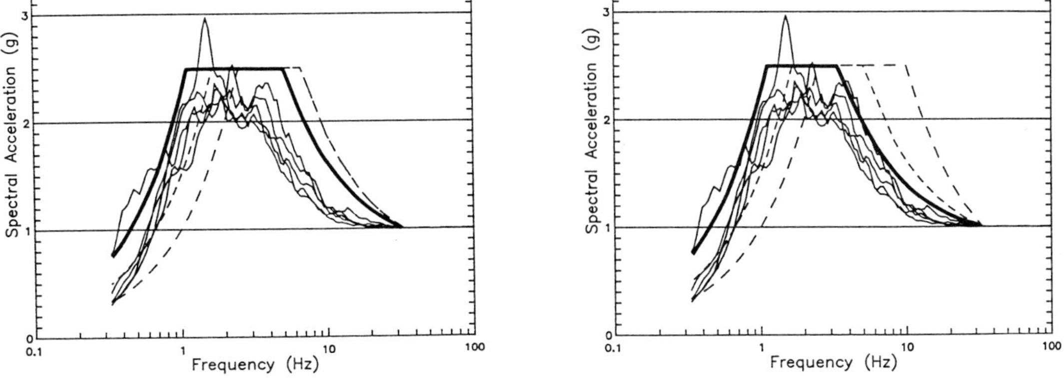

Figure 5. a (Left): Mean Horizontal Response Spectrum Shapes of UBC Type S3 Records from Four Recent California Earthquakes with UBC Design Spectrum Shapes.
b (Right): Mean Horizontal Response Spectrum Shapes of UBC Type S3 Records from Four Recent California Earthquakes with Modified UBC Design Spectrum Shapes.

the one-standard-deviation curves for the four earthquakes are plotted in terms of a multiplication factor over the mean. The figure shows that the dispersion multipliers increase from one at the high-frequency end to about 2 at the low-frequency end for three earthquakes. The outlying dispersion curve was for the Cape Mendocino earthquake based on a relatively small set of 12 records. In this case, the statistics are prone to be affected by one or two extreme records.

In order to avoid the non-uniformity problem, we propose a new normalizing parameter based on the mean spectral acceleration (MSA) calculated over the entire frequency band. Figure 6b shows the mean Sa/MSA spectra and the associated dispersion multipliers from all horizontal components of the four earthquakes. The figure shows that the mean Sa/MSA spectra are all about 0.7 times the corresponding Sa/PGA spectra. More significantly, the dispersion multipliers now have uniform probability across the entire frequency band to better reflect the dispersion of the normalized spectrum shapes. The figure shows that the dispersion multipliers increase slowly from about 1.2 at the high-frequency end of 33 Hz to about 1.4 at 2 Hz. They then increase steeply toward about 1.9 at the low-frequency end of 0.33 Hz. This means greater Sa/MSA variability at frequencies below about 2 Hz.

Figures 7a,b,c show the mean Sa/MSA spectra and the associated dispersion multipliers for the three subsets of records whose Sa/PGA resembling the UBC Types S1, S2, and S3 spectrum shapes, respectively. Regarding the mean, it is remarkable that the mean Sa/MSA spectra of larger data sets all start at about 0.7 at frequency 33 Hz, increase to peak at about 1.7 at intermediate frequencies, and then decrease to 0.2 at frequency 0.33 Hz. Regarding the dispersion, except for the dispersion curve of the small subset of only three UBC Type S1 records (six components) from the Cape Mendocino earthquake, all the other twelve dispersion curves behaved consistently by increasing slowly from a factor of about 1.1 at frequency 33 Hz to about 1.4 at frequency 1 Hz. They then increase to about 1.8 at frequency 0.33 Hz. These data indicate that the mean Sa/MSA spectrum shapes are remarkably stable and can be estimated to within a factor of 1.4 at one standard deviation level for frequencies above 1 Hz when an appropriate site-dependent UBC spectrum shape is used. The variability increases toward low frequencies below 1 Hz.

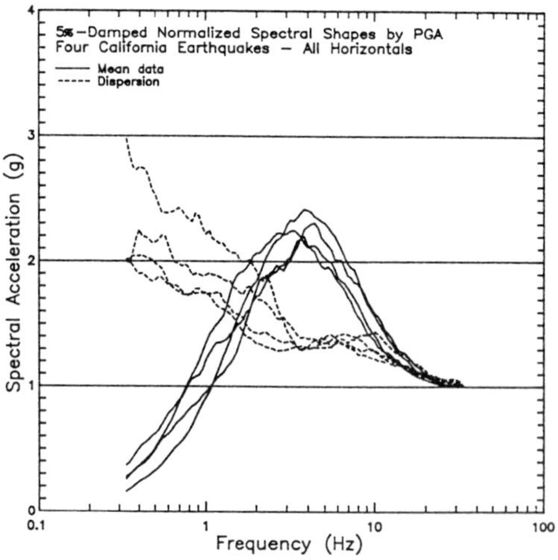

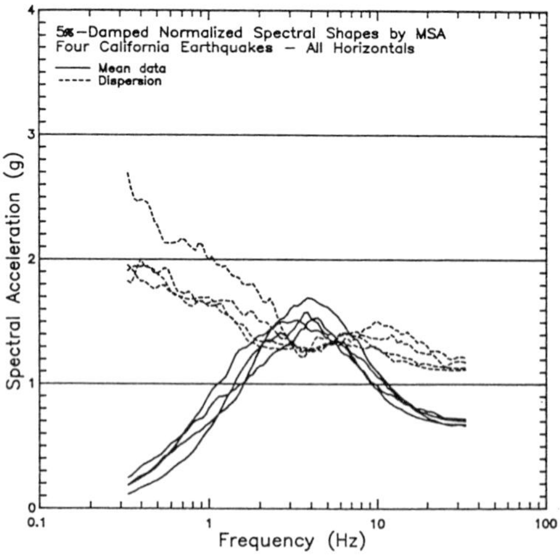

Figure 6. a (Left): Mean and Dispersion of Horizontal Response Spectrum Shapes of All Records from Four Recent California Earthquakes, Normalized by PGA.
b (Right): Mean and Dispersion of Horizontal Response Spectrum Shapes of All Records from Four Recent California Earthquakes, Normalized by MSA.

ATTENUATION OF HORIZONTAL MEAN SPECTRAL ACCELERATION (MSA)

Figure 8a plots the horizontal MSA as function of PGA. The data clearly show that the two normalizing parameters are positively correlated. This means that a large PGA will elevate the entire response spectrum, contradicting the notion that large, isolated peaks in time histories may not be significant for defining seismic ground motion.

Figure 8b plots the horizontal MSA values from all four earthquakes as function of the closest distance to fault rupture surface for the four earthquakes. It was found that the data points were so well mixed as to be indistinguishable among the earthquakes, even though the earthquake magnitudes differed significantly. From the figure, we get a mean MSA of 0.6 g at distance 10 km. When combined with one of the mean Sa/MSA spectrum shapes described in the previous section, this MSA will yield an equivalent UBC Zone 4 design spectrum. This indicates that for some sites located closer than 10 km from a fault rupture surface of a magnitude 7 earthquake, the current UBC Zone 4 design spectrum may not be adequate.

Figures 9a,b,c,d plot the horizontal MSA as function of the closest distance to fault rupture surface for each of the four earthquakes. These figures show that the MSA values of the records having the UBC Types S1, S2, and S3 spectrum shapes are well mixed for the earthquakes. The two notable exceptions are the Bay-mud records from the Loma Prieta earthquake which yields higher-than-average MSA values (Figure 9b), and the hard-rock records of the Landers earthquake which yields lower-than-average MSA values (Figure 9d). These observations indicate that the site coefficients in the current UBC provisions, as shown in Figure 1, may not be warranted for most sites, perhaps except for soft Bay-mud sites and hard rock sites.

OBSERVED VERTICAL RESPONSE SPECTRUM SHAPES AND MEAN SPECTRAL ACCELERATIONS

The current UBC provision permits the vertical ground motions to be defined by scaling the horizontal ground motions by a factor of two-thirds when performing a dynamic analysis, unless otherwise justified on a site-specific basis. The new data set presents a good opportunity to review this approach. After looking at individual vertical Sa/PGA spectra, we found that they were much less variable than their horizontal counterparts. Accordingly, we included all 151 vertical records in a single data set. Figure 10a shows the observed mean vertical Sa/PGA spectra of the subsets of records from the four earthquakes. They were remarkably similar to each other among the earthquakes, and consistently peaked at frequency about 7 Hz. In parallel with the horizontal Sa/PGA spectrum shapes, we propose a conservative enveloping spectrum shape which is also consisted of three segments. This spectrum shape is shown in Figure 10a and the corresponding formula is given in Table 2. It should be pointed out that the vertical Sa/PGA spectrum shapes peak at higher frequencies, and their low-frequency segments roll off more slowly than their horizontal counterparts.

Figure 10b shows the mean vertical Sa/MSA spectrum shapes and associated dispersion curves for the four earthquakes. Like their horizontal counterparts, the mean vertical spectra start from 0.7 at 33 Hz, increase to peak at 1.7 at 7 Hz, and then decrease to 0.2 at 0.33 Hz. The dispersion multipliers increase gently from 1.2 at 33 Hz to 1.4 at 2 Hz, then increase steeply to 1.8 at 0.33 Hz.

Figure 11a plots the vertical MSA as function of PGA. The data clearly show that the two normalizing parameters are positively correlated. This means that a large PGA will elevate the entire response spectrum, contradicting the notion that large, isolated peaks in time histories may not be significant for defining seismic ground motion. Figure 11b plots the vertical MSA as function of the closest distance to fault rupture surface. For most records, the vertical MSA is about 0.5 times the horizontal MSA. The differences in both the MSA values and Sa/MSA spectrum shapes suggest that the current UBC provision for defining the vertical design spectrum by scaling the horizontal spectrum by a factor of two-thirds needs to be revised to reflect the latest observations.

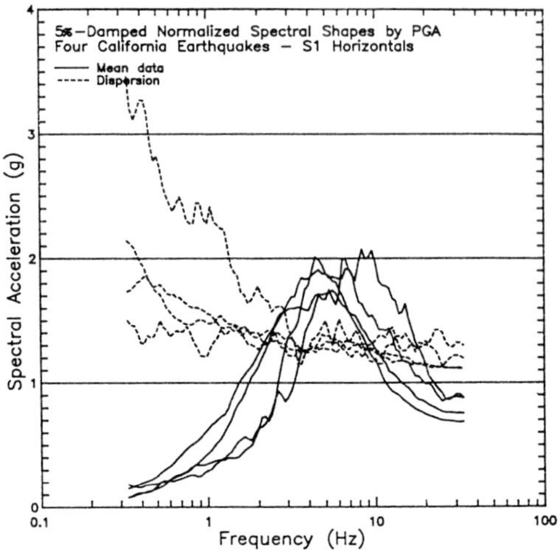

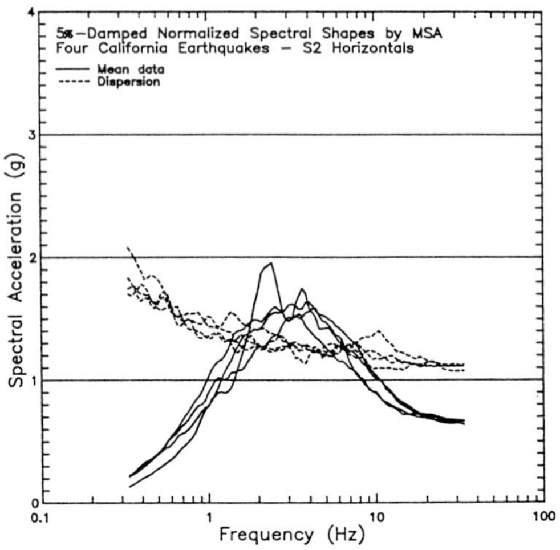

Figure 7. a (Left): Mean and Dispersion of Horizontal Response Spectrum Shapes of UBC Type S1 Records from Four Recent California Earthquakes, Normalized by MSA.
b (Right): Mean and Dispersion of Horizontal Response Spectrum Shapes of UBC Type S2 Records from Four Recent California Earthquakes, Normalized by MSA.

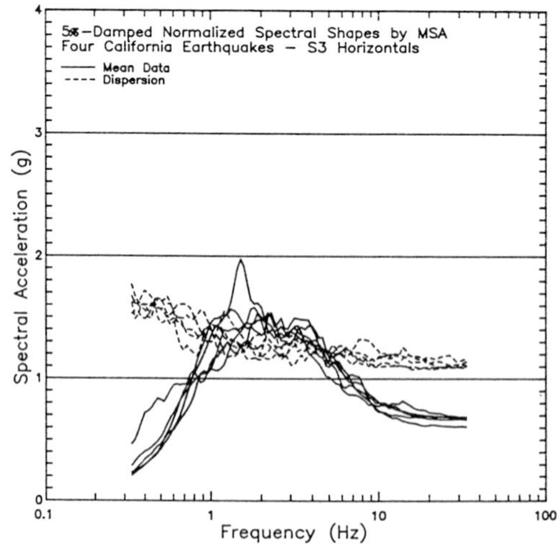

Figure 7. c: Mean and Dispersion of Horizontal Response Spectrum Shapes of UBC Type S3 Records from Four Recent California Earthquakes, Normalized by MSA.

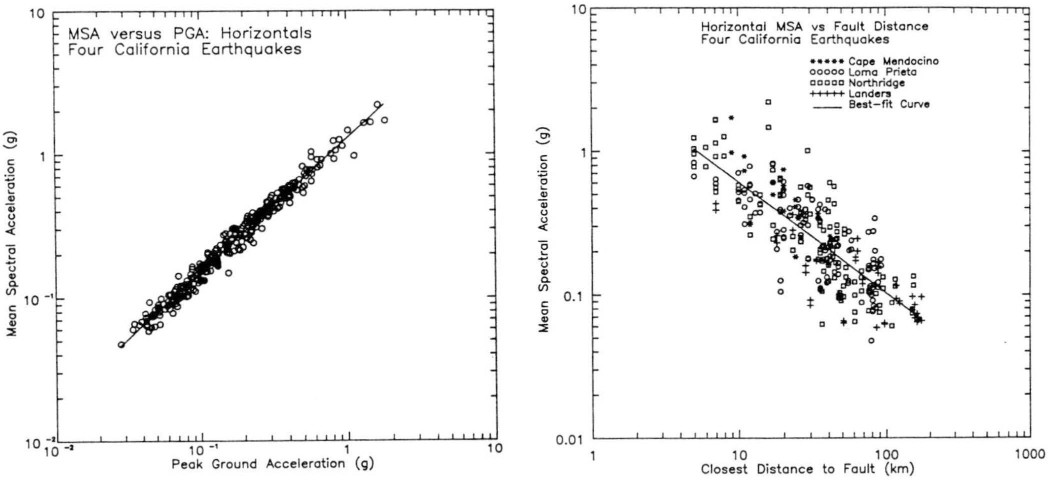

Figure 8. a (Left): Horizontal Mean Spectral Acceleration (MSA) versus Peak Ground Acceleration (PGA) for All Records from Four Recent California Earthquakes.
b (Right): Horizontal MSA versus Closest Distance to Fault for All Records from Four Recent California Earthquakes.

Table 2. Formulas for the UBC and Modified UBC Seismic Design Spectrum Shapes

Type	Period Range	UBC Spectrum Shape	Modified UBC Spectrum Shape
S1	T ≥ 0.40s	1.0/T	1.0/T
	0.40s ≥ T ≥ 0.15s	2.5	
	0.40s ≥ T ≥ 0.10s		2.5
	0.15s ≥ T ≥ 0.03s	0.625 + 12.5T	
	0.10s ≥ T ≥ 0.03s		0.357 + 21.429T
S2	T ≥ 0.57s	1.425/T	1.425/T
	0.57s ≥ T ≥ 0.15s	2.5	
	0.57s ≥ T ≥ 0.20s		2.5
	0.15s ≥ T ≥ 0.03s	0.625 + 12.5T	
	0.20s ≥ T ≥ 0.03s		0.735 + 8.824T
S3	T ≥ 0.90s	2.25/T	2.25/T
	0.90s ≥ T ≥ 0.20s	2.5	
	0.90s ≥ T ≥ 0.30s		2.5
	0.20s ≥ T ≥ 0.03s	0.735 + 8.824T	
	0.30s ≥ T ≥ 0.03s		0.833 + 5.556T
Vertical	T ≥ 0.19s		$1.09/T^{1/2}$
	0.19s ≥ T ≥ 0.11s		2.5
	0.11s ≥ T ≥ 0.03s		0.4375 + 18.75T

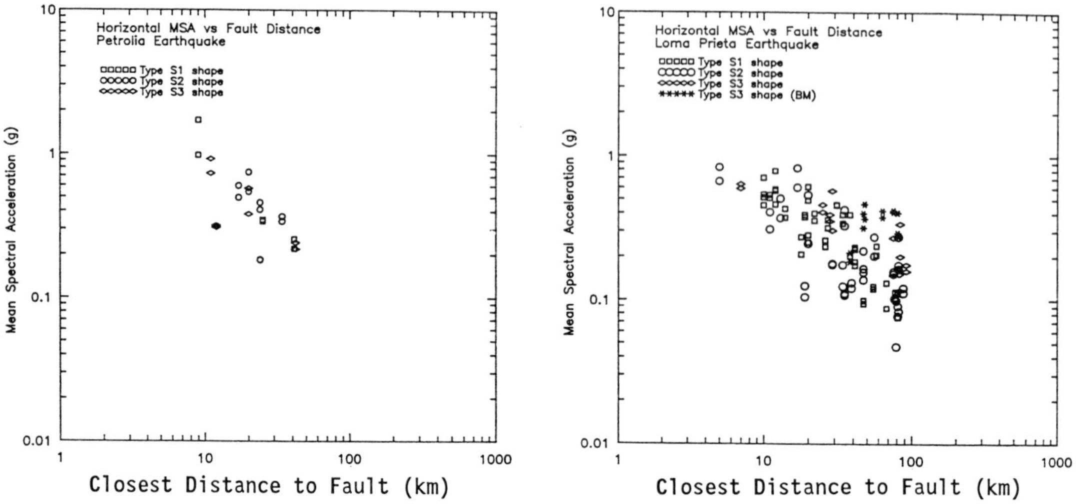

Figure 9. a (Left): Horizontal MSA versus Closest Distance to Fault for UBC Types S1, S2 and S3 Records from the 1992 Cape Mendocino, California Earthquake.

b (Right): Horizontal MSA versus Closest Distance to Fault for UBC Types S1, S2, and S3 Records from the 1989 Loma Prieta, California Earthquake.

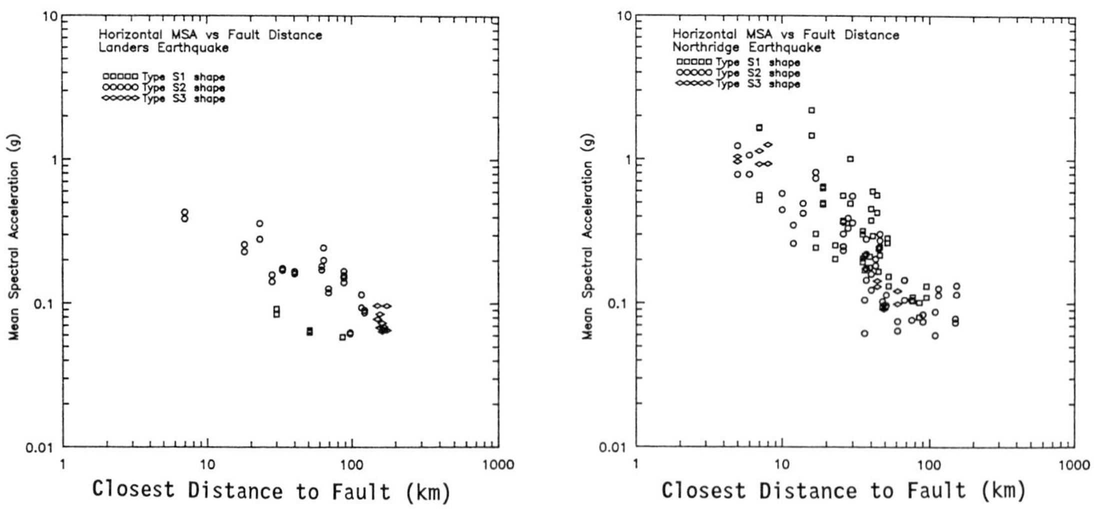

Figure 9. c (Left): Horizontal MSA versus Closest Distance to Fault for UBC Types S1, S2 and S3 Records from the 1992 Landers, California Earthquake.

d (Right): Horizontal MSA versus Closest Distance to Fault for UBC Types S1, S2, and S3 Records from the 1994 Northridge, California Earthquake.

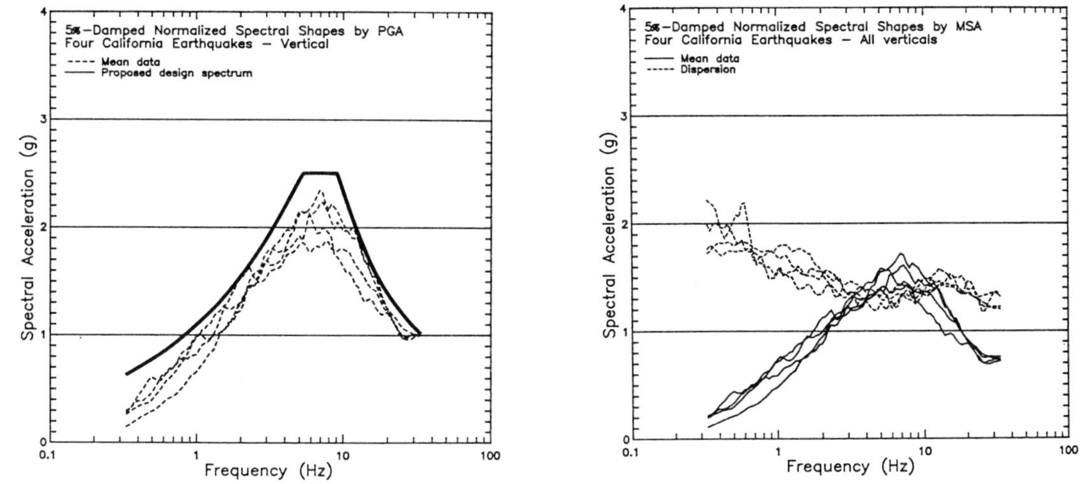

Figure 10. a (Left): Mean Vertical Response Spectrum Shapes of All Records from Four Recent California Earthquakes with the Proposed Vertical Design Spectrum Shape.
b (Right): Mean and Dispersion of Vertical Response Spectrum Shapes of All Records from Four Recent California Earthquakes, Normalized by MSA.

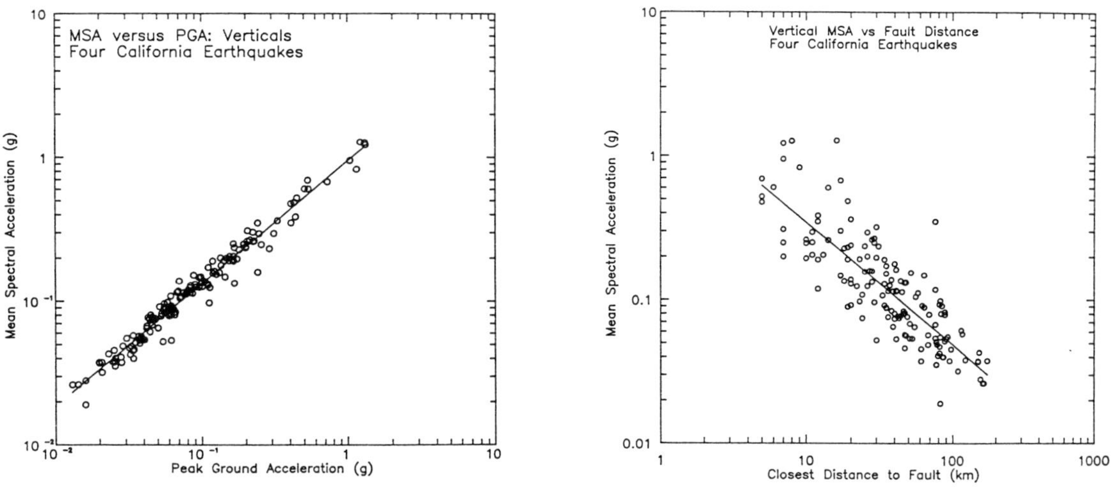

Figure 11. a (Left): Vertical Mean Spectral Acceleration (MSA) versus Peak Ground Acceleration (PGA) for All Records from Four Recent California Earthquakes.
b (Right): Vertical MSA versus Closest Distance to Fault for All Records from Four Recent California Earthquakes.

CONCLUSIONS

A large number of 302 horizontal and 151 vertical components of observed response spectra from four recent large earthquakes in California, namely, the 1989 Loma Prieta (surface wave magnitude, M_S 7.1), 1992 Cape Mendocino (M_S 7.1), 1992 Landers (M_S 7.6), and 1994 Northridge (M_S 6.8) earthquakes, were assembled and analyzed in terms of the normalized spectrum shape and normalizing parameter. The findings can be summarized as follows:

First, we used the conventional peak ground acceleration (PGA) as the normalizing parameter. The resultant normalized response spectrum shapes, Sa/PGA, from each earthquake were included in one data set to calculate the overall mean (geometric) horizontal and vertical spectra for that earthquake. The results were compared with the current Uniform Building Code (UBC) site-dependent Types S1, S2, and S3 design spectrum shapes.

It was found that the mean spectra, both horizontal and vertical, were remarkably similar to each other among the earthquakes, even though the earthquakes were located far apart in different regions of the state. The mean horizontal spectrum shapes are conservatively enveloped by the UBC Type S2 spectrum shape.

Meanwhile, the observed mean vertical spectrum shapes from the earthquakes were less variable and consistently peak at higher frequencies than their horizontal counterparts. In addition, we have found that the vertical MSA is about 0.5 times the horizontal MSA, even at near-source distances. Both observations indicate that the current UBC provision on defining vertical ground motion by scaling the horizontal ground motion by a factor of two-thirds needs revision.

We further divided the observed spectrum shapes into subsets according to the resemblance of the horizontal spectra with the current UBC site-dependent design spectrum shapes. It was found that the observed mean horizontal Sa/PGA spectra for each data subset were conservatively enveloped by the corresponding UBC design spectrum shape, thereby confirming the general applicability of the UBC design spectrum shapes for California. For further refinement, we have proposed modifications of the current UBC spectrum shapes in order to avoid excessive conservatism at high frequencies.

Second, we introduced a new mean spectral acceleration (MSA) as the normalizing parameter. The resultant normalized response spectrum shapes, Sa/MSA, were used to calculate the mean Sa/MSA spectrum and the corresponding uniform-probability dispersion due to spectral peaks and troughs over the entire frequency band. It was found that the mean Sa/MSA spectra for the UBC Types S1, S2, and S3 spectrum shapes all have the same levels. Furthermore, the dispersion multiplier at one-standard-deviation level increases slowly from about 1.2 at frequency 33 Hz to 1.4 at 1 Hz, then increases more steeply to 1.8 at 0.33 Hz. This is true for all types of mean horizontal spectrum shape as well as for the mean vertical spectrum shape.

Third, we examined the relation between PGA and MSA. We found that the two parameters are closely correlated. Since the MSA represent a water-level average of a response spectrum over the entire frequency band, this close correlation means higher response spectral level for a higher PGA. Accordingly, the notion that an isolated, large peak on an acceleration time history may not be significant is not supported by the present data set. It was further found that the MSA values of UBC Types S1, S2, and S3 records were well mixed at all distances. This indicates that the site coefficient factors in the current UBC provisions may not be warranted, as long as an appropriate UBC spectrum shape is picked for a site. However, it should be pointed out that no systematic correlations was apparent between the observed horizontal response spectrum shape and site surface geology of the recording sites. Therefore, correct identification of appropriate site-dependent spectrum shape for a site poses the biggest challenge.

ACKNOWLEDGMENTS

This study was sponsored by the Research and Development Department of the Pacific Gas and Electric Company, as part of a project to study the effects of strong ground motion on substations. It was made possible by the enthusiastic support of the program manager, Donna S. Lindquist. The author wish to thank his colleagues, Lloyd Cluff, Woody Savage, and Edward Matsuda for their support. Thanks are also due to Dennis Ostrom of Southern California Edison and Ron Tognazzini of the Los Angeles Department of Water and Power for providing several valuable strong ground motion records. Most of the strong ground motion records used in this study were provided by the California Strong Motion Instrumentation Program and the U.S. Geological Survey.

REFERENCES

Applied Technology Council (1978). Tentative provisions for the development of seismic regulations for buildings, ATC 3-06 Report, Palo Alto, California, 505p.

California Strong Motion Instrumentation Program (1989). Plots of the processed data for the interim set of 14 records from the Santa Cruz Mountains (Loma Prieta), California earthquake of 17 October 1989, California Division of Mines and Geology, Office of Strong Motion Studies, Report No. OSMS 89-08, 142p (First interim report, December 13, 1989).

California Strong Motion Instrumentation Program (1992). Preliminary processed strong-motion data for the Landers earthquake of 28 June 1992 (Release No.1), Report No. OSMS 92-11, California Strong Motion Instrumentation Program, 63p.

Darragh, R., T. Cao, C. Cramer, M. Huang and A. Shakal (1992). Processed CSMIP strong-motion records from the Cape Mendocino/Petrolia earthquake of April 25, 1992: Release No.1, Report No. OSMS 92-12, California Strong-Motion Instrumentation Program, 130p.

Darragh, R., T. Cao, C. Cramer, F. Su, M. Huang and A. Shakal (1992). Processed CSMIP strong-motion records from the Landers, California earthquake of 28 June 1992: Release No.2, Report No. OSMS 92-13, California Strong-Motion Instrumentation Program, 155p.

Darragh, R., T. Cao, M. Huang and A. Shakal (1993). Processed CSMIP strong-motion records from the Landers, California earthquake of 28 June 1992: Release No.3, Report No. OSMS 93-01, California Strong-Motion Instrumentation Program, 104p.

Darragh, R., T. Cao, C. Cramer, M. Huang and A. Shakal (1994). Processed CSMIP strong-motion records from the Northridge, California earthquake of January 17, 1994: Release No.1, Calif. Dept. of Conservation, Div. Mines and Geology, Office of Strong Motion Studies, Report No. OSMS 94-06B, February 4, 1994, 48p.

Darragh, R., T. Cao, C. Cramer, V. Graizer, M. Huang and A. Shakal (1994a). Processed CSMIP strong-motion records from the Northridge, California earthquake of January 17, 1994: Release No.3, Calif. Dept. of Conservation, Div. Mines and Geology, Office of Strong Motion Studies, Report No. OSMS 94-09, March 15, 1994, 78p.

Darragh, R., T. Cao, C. Cramer, V. Graizer, M. Huang and A. Shakal (1994b). Processed CSMIP strong-motion records from the Northridge, California earthquake of January 17, 1994: Release No.4, Calif. Dept. of Conservation, Div. Mines and Geology, Office of Strong Motion Studies, Report No. OSMS 94-10, April 25, 1994, 110p.

Darragh, R., T. Cao, M. Huang and A. Shakal (1994a). Processed CSMIP strong-motion records from the Northridge, California earthquake of January 17, 1994: Release No.2, Calif. Dept. of Conservation, Div. Mines and Geology, Office of Strong Motion Studies, Report No. OSMS 94-08, February 24, 1994, 48p.

Darragh, R., T. Cao, M. Huang and A. Shakal (1994b). Processed data for Pacoima - Upper left abutment from the Northridge, California earthquake of January 17, 1994. Calif. Dept. of Conservation, Div. Mines and Geology, Office of Strong Motion Studies, Report No. OSMS 94-12A, June 8, 1994, 10p.

Darragh, R., T. Cao, M. Huang and A. Shakal (1994c). Preliminary processed data for Tarzana - Cedar Hill Nursery A from the Northridge, California earthquake of January 17, 1994. Calif. Dept. of Conservation, Div. Mines and Geology, Office of Strong Motion Studies, Report No. OSMS 94-12B, September 9, 1994, 10p.

Hawkins, H.G., D.K. Ostrom and T.A. Kelly (1993). Comparison of strong motion records, Landers earthquake 28 June 1992, near Lucerne Valley, California (abstract), Seismological Research Letters, v. 64, no. 1, p. 19.

Huang, M.J., T.Q. Cao, U.R. Vetter, and A.F. Shakal (1990a). Second interim set of CSMIP processed strong-motion records from the Santa Cruz Mountains (Loma Prieta) earthquake of 17 October 1989, California Strong Motion Instrumentation Program Report No. OSMS 90-01.

Huang, M.J., T.Q. Cao, U.R. Vetter, and A.F. Shakal (1990b). Third interim set of CSMIP processed strong-motion records from the Santa Cruz Mountains (Loma Prieta) earthquake of 17 October 1989, California Strong Motion Instrumentation Program Report No. OSMS 90-05.

International Conference of Building Officials (1994). Uniform Building Code, International Conference of Building Officials, Whittier, California.

Lindvall Richter Benuska Associates (1994). Processed LADWP Power System strong-motion records from the Northridge, California earthquake of January 17, 1994, Los Angeles, California, 172p.

Maley, R., A. Acosta, F. Ellis, E. Etheridge, L. Foote, D. Johnson, R. Porcella, M. Salsman, and J. Switzer (1989). U.S. Geological Survey strong-motion records from the northern California (Loma Prieta) earthquake of October 17, 1989, U.S. Geological Survey Open-File Report 89-568.

National Strong-Motion Program Staff (1992). Selected accelerograms from U.S. Geological Survey stations that recorded the Landers (Yucca Valley) and Big Bear earthquakes of June 28, 1992, U.S. Geological Survey, 22p.

Seed, H.B., C. Ugas, and J. Lysmer (1976). Site-dependent spectra for earthquake-resistant design, Bull. Seism. Soc. Am., Vol 66, No.1, pp. 221-244.

Shakal, A., M. Huang, M. Reichle, C. Ventura, T. Cao, R. Sherburne, M. Savage, R. Darragh, and Petersen, C. (1989). CSMIP strong-motion records from the Santa Cruz Mountains (Loma Prieta), California, earthquake of 17 October 1989, California Strong Motion Instrumentation Program Report No. OSMS 89-06, Division of Mines and Geology, Sacramento, California.

Shakal, A., R. Darragh, M. Huang, T. Cao, R. Sherburne, R. Sydnor, P. Malhotra, C. Cramer, J. Wampole P. Fung and C. Petersen (1992). CSMIP strong-motion records from the Petrolia, California earthquakes of April 25-26, 1992, Report No. OSMS 92-05, California Strong Motion Instrumentation Program, 74p.

Shakal, A., M. Huang, T. Cao, R. Sherburne, R. Sydnor, P. Fung, P. Malhotra, C. Cramer, F. Su, R. Darragh and J. Wampole (1992). CSMIP strong-motion records from the Landers, California earthquake of 28 June 1992, Report No. OSMS 92-09, California Strong Motion Instrumentation Program, 330p.

Tsai, Y.B. and F.W. Brady (1994). Observed changes in soil-structure interaction effects of a fully embedded reactor containment structure during moderate and strong earthquakes, Proceedings of the Fifth U.S. National Conference on Earthquake Engineering, v. IV, 25-34, Earthquake Engineering Research Institute.

UNCERTAINTY ANALYSIS ON SEISMIC HAZARD EVALUATION

C. -H. Loh* and Y. T. Yeh**†

** Department of Civil Engineering, National Taiwan University, Taipei, Taiwan*
*** Institute of Earth Sciences, Academia Sinica, Taipei, Taiwan*
† Institute of Seismology, National Chung Cheng University, Chia-I, Taiwan

ABSTRACT

The probabilistic assessment of seismic ground motion has gained wide acceptance as a method of expressing earthquake hazard. In this study, a variety of mathematical representations have been used to model the seismic source, the seismicity associated with those sources, and the seismic ground motion. The importance of the uncertainties inherent in these models and in the parameters used to define them have been recognized in the determination of realistic seismic hazard. The purpose of this study is to determine the uncertainty in seismic hazard estimation resulting from uncertainties in models and parameters used as input to the analysis. An application of this analysis to the Taiwan region is presented. With the aim of determining uncertainties, the following subjects are discussed:

(a) Effect of dispersion on the estimation of Peak Ground Acceleration (PGA) attenuation form — It is assumed that the normalized weighted residual on PGA dispersion follows log normal distribution. In hazard calculation the required probability should include the effect of this uncertainty. Discussion on the modification of distribution function of the required probability is presented.

(b) Effect of spectral amplitude probability distribution on uniform hazard response spectrum — An uniform hazard response spectrum was generated and discussed by using different methods for parameter estimation on the distribution of spectral accumulation.

(c) Effect of local site condition on the seismic hazard estimation — Seismic hazard is calculated based on site-dependent PGA attenuation. For some specific site condition, modification on hazard probability should be done. In this study, equivalent single layer soil model was first developed. A combination of soil amplification function of the model and the random vibration analysis on extreme value was adopted to correct seismic hazard estimation while the uncertainty of local site condition was considered.

KEYWORDS

Seismic hazard analysis; Peak Ground Acceleration (PGA); response spectrum.

INTRODUCTION

Taiwan is a member of the Ryukyu-Philippine island arc chain which rims the western border of the Pacific Ocean. The area has frequently suffered severe damage from destructive earthquake throughout her historic line. To establish the seismic design criteria for major projects, it is important that full use be made of all pertinent historical evidence of seismic activity in the Taiwan area and corrected these information with regional tectonics, geology factors, and soil conditions. For the above reason, a systematic and comprehensive seismic hazard analysis had been examined. The principal objective of a seismic hazard analysis is to estimate the likelihood that different levels of ground shaking intensity will be experienced at a specific site. Generally, a seismic hazard analysis integrates the contributions of all possible earthquakes and calculates the probability that the selected ground motion parameters will be exceeded within the specified exposure time. Based on the proposed peak ground acceleration (PGA) attenuation model, the result of a seismic hazard analysis for each individual site is displayed as a seismic risk curve — a plot of annual probability of exceedance versus a specified ground motion parameters. The analysis must account for the spatial and temporal randomness of earthquake occurrences, and the uncertainty in ground motion predictions for events of different magnitudes at random distances. The methodologies for performing a probabilistic seismic hazard analysis as described by Cornell (1968), McGuire (1976), Der Kireghian and Ang (1977) and Loh *et al.*, (1991a, b, 1992) are well established in engineering practice. For such analysis, the most important issue is to determine the relationship of PGA and earthquake characteristics. In addition to the above-mentioned factor, the probability density function relating the ordinates of a response spectra had been studied to present an alternative method for generating uniform-hazard response spectra (Loh *et al.*, 1994).

The purpose of this study is to describe the current methodology of seismic hazard analysis in Taiwan area and to discuss the uncertainties in relating to seismicity as well as local site conditions on the estimation of seismic hazard.

SEISMIC HAZARD ANALYSIS FOR TAIWAN AREA

From the engineering point of view, it is nature to select the ground motion parameters that are closely related to the safety of engineering structure. The simplest to select is the PGA. Then based on that, to construct the acceleration risk map with different exceedance probability serves as a basis for design purpose. The procedure for a seismic hazard analysis of a specific site involves the following three steps:

(1) Determination of model parameters and physical relationships appropriate to the analysis: The analysis includes the determination of seismic zones, source type models, *b* values, occurrence rates, upper bound magnitude and focal depth. Based on the geological structure, subduction model and seismicity, 16 source zones and 5 sources zones (Fig. 1) are defined as area sources for shallow (depth $\leq$ 35 km) and deep (depth > 35 km) earthquakes (Loh *et al.*, 1992). In addition to the area sources defined previously, some of the active faults are also introduced as line-sources for the seismic hazard analysis.

(2) Developing attenuation relation for PGA: The mathematical relationship used for modeling PGA attenuation had been discussed by many researchers (Campbell, 1981; Joyner and Boore, 1981; Sadigh, 1993). The functional form used for modeling the attenuation of PGA of Taiwan data for hard site condition is expressed (as shown in Fig. 2) as follows:

$$Y = 0.0267\, e^{1.345M}\,(R + 0.2138\, e^{0.7499M})^{-1.7329} \qquad (1)$$

where Y is PGA (in "g"), M is the earthquake magnitude and R is the hypocentral distance. The root-mean square error of natural logarithm of PGA, $\sigma_{\ln \varepsilon}$, is about 0.69. The attenuation relationship is significantly affected by the scatter of the recorded PGA values. In calculating the hazard curves, the estimated

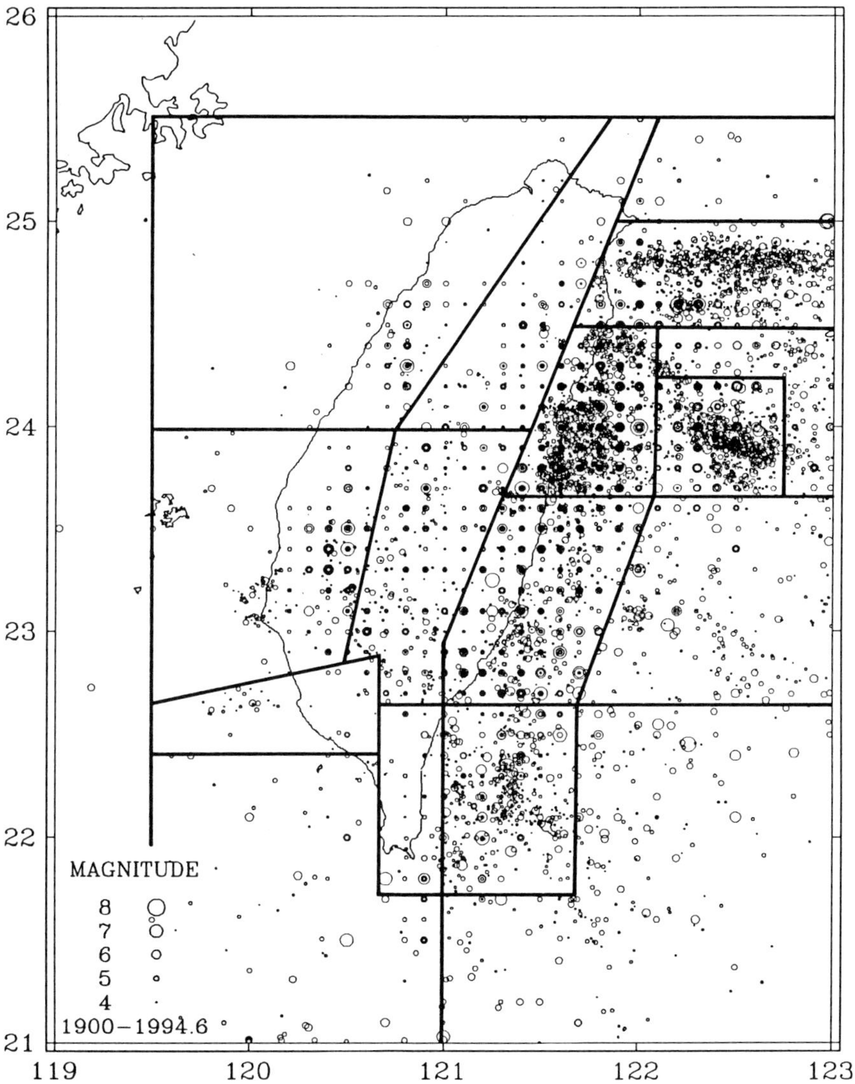

Fig. 1a Epicenters of earthquakes (1900-1994.6) with $M \geq 4.5$ and the zoning scheme for shallow earthquakes (depth ≤ 35 km)

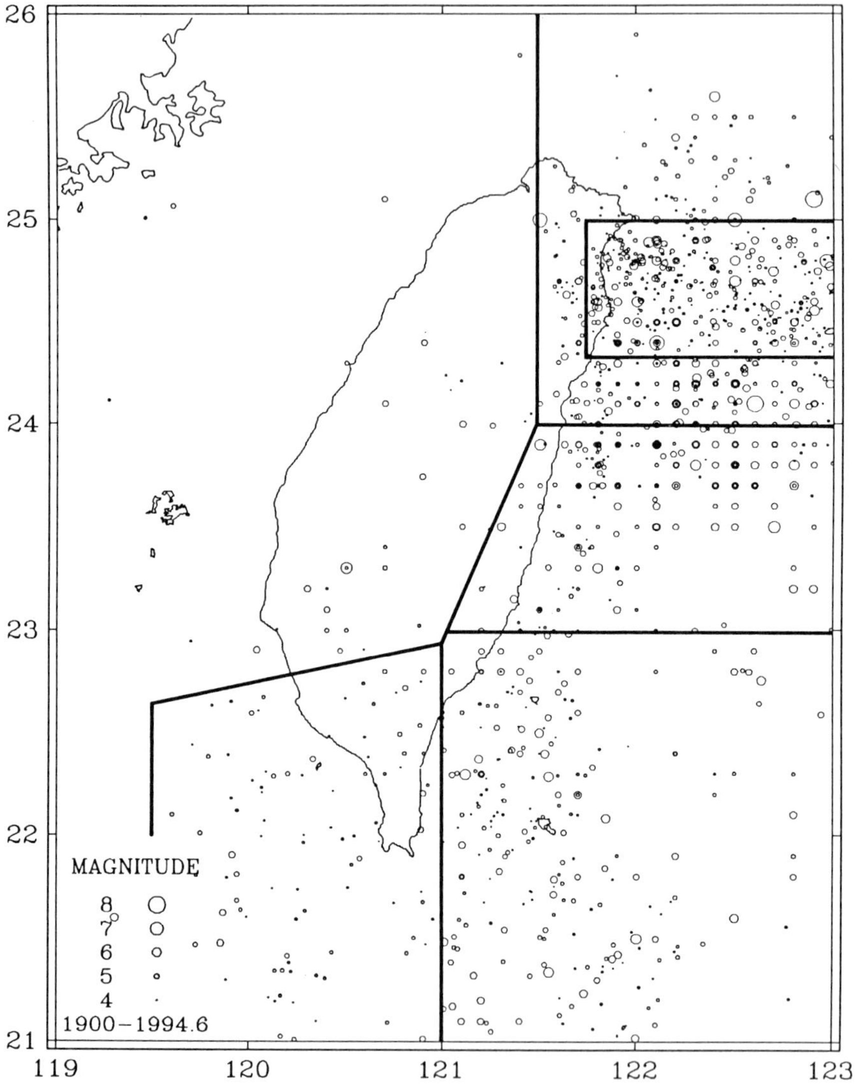

Fig. 1b Epicenters of earthquakes (1900-1994.6) with M ≥ 4.5 and the zoning scheme for deep earthquakes (depth > 35 km).

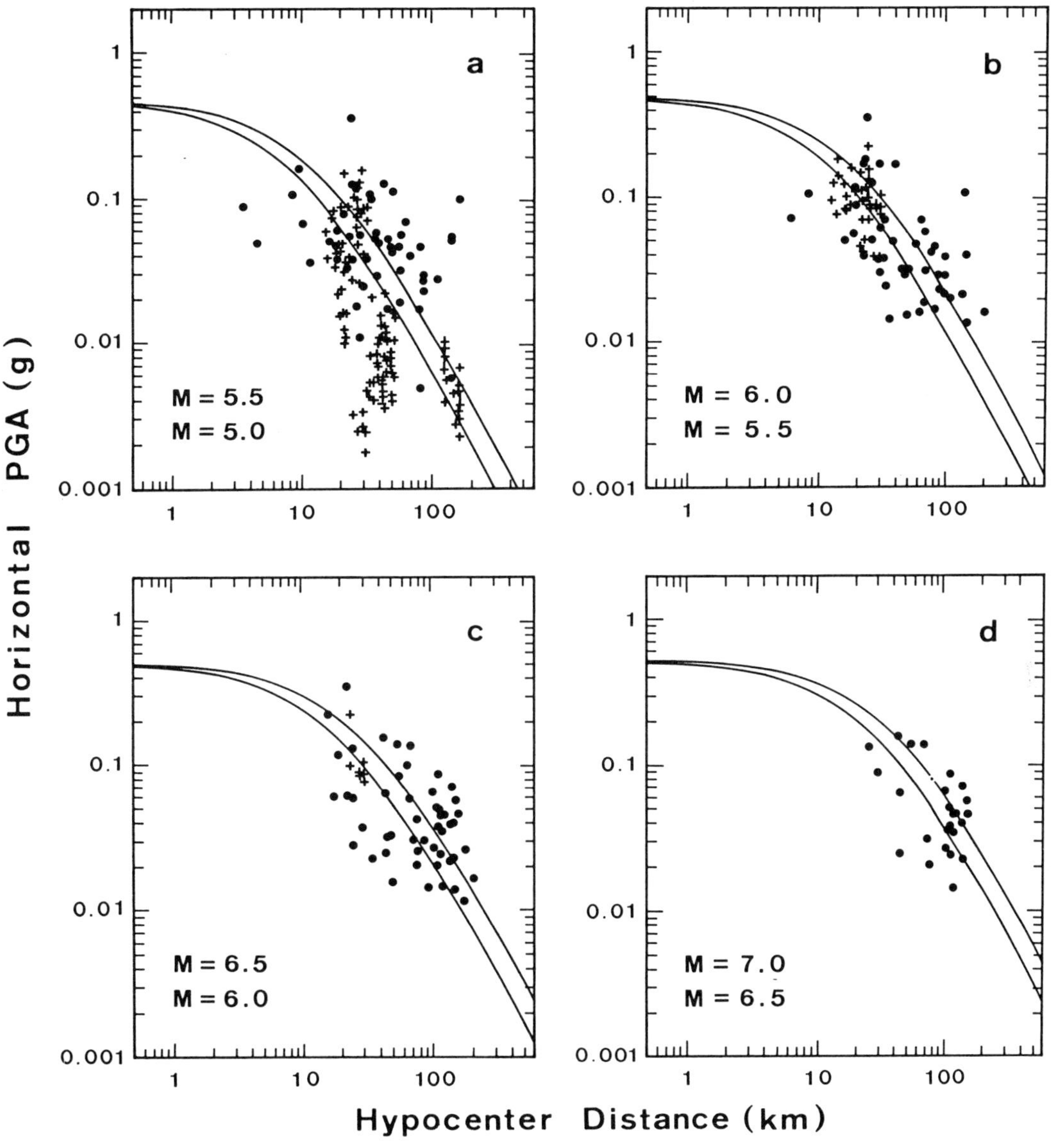

Fig. 2 Plots of PGA value with respect to hypocenter distance. The regressional PGA attenuation form is also shown: (a) magnitude between 5.0 and 5.5, (b) magnitude between 5.5 and 6.0, (c) magnitude between 6.0 and 6.5, (d) magnitude between 6.5 and 7.0.

dispersion of PGA must be considered in the analysis. We define the actual intensity Y_a as $Y_a = NY$ and N is the correction necessary on the predicted intensity Y. The required probability, including the effects of the dispersion, becomes (Der Kiureghian and Ang, 1977):

$$P\,[Y_a > y] = \int_0^{\infty} P(Y > y \mid n)\, f_N(n)\, dn \tag{2}$$

where $f_N(n)$ is the probability density function of N. It is assumed that N is a random variable and follows log-normal distribution. Because of limited date recorded within short distance, different functional forms could be selected to predict the PGA value at near field.

(3) Constructing iso-acceleration map of Taiwan area: The area under consideration in developing seismic hazard map was divided into 74 quadrilaterals with side length of one-fourth degree of latitude and longitude, respectively. The seismic hazard was calculated for every center point of this mesh. Using the fault-rupture model (for shallow zone) and point source model (for deep zone), the PGA contours related to a certain risk level can be plotted. Figure 3 shows the iso-acceleration map of the Taiwan area for return period of 475 years.

Since the fault-rupture model is used in hazard calculation for shallow source, the shortest distance between the site and fault must be consistent with the distance R defined in PGA attenuation equation. Therefore, the hypocentral distance should be modified as the shortest distance in the final regressional analysis of PGA attenuation equation. Discussion on this will be shown in the following section.

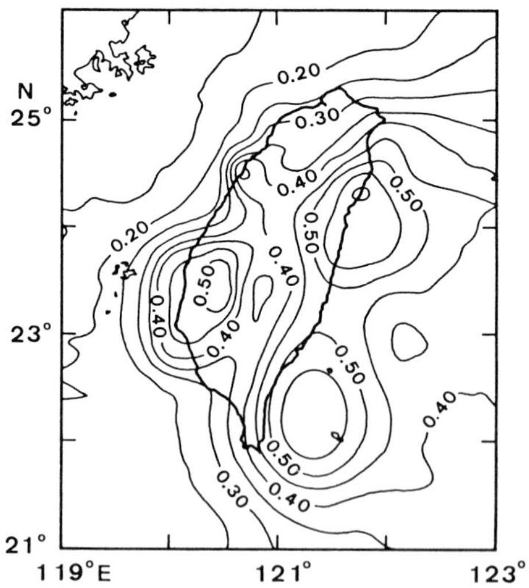

Fig. 3 Iso-acceleration map of Taiwan for return period of 475 years (dispersion of PGA attenuation equation is not considered).

UNIFORM-HAZARD RESPONSE SPECTRA — AN ALTERNATIVE APPROACH

In the development of response spectra for use in designing seismic resistant structures, it has been a common practice to establish first sets of response spectra normalized to 1 *g* and then scale them down to a specified PGA level in its design applications. It should be recognized that the resulting spectra do not represent the same probability of exceedance over the full frequency range of interest (Or say, not an uniform hazard). To generate site-specific uniform-hazard response spectra, an alternative method is presented.

To develop this alternative method, let A denote the random PGA. Its annual probability of exceedance function, $Q(A)$, is given directly by the seismic hazard curve for PGA; thus, the corresponding probability distribution and density functions can be easily obtained numerically using the relations:

$$P(A) = 1 - Q(A) \quad and \quad p(A) = \frac{dP(A)}{dA} \tag{3}$$

Further, let S denote the random normalized acceleration response spectral value $S_{na}(T, \xi)$ for a specified period T and damping ratio ξ. Having obtained a set of mean normalized acceleration response spectral curves, denoted by $\mu = \mu\ (T, \xi)$, and the corresponding set of variance curves, denoted by $\sigma^2 = \sigma^2(T, \xi)$, the probability density function for random variable S can be expressed using the log-normal form:

$$p(S) = \frac{1}{\sqrt{2\pi}\,\eta S} exp\ [-\frac{1}{2}(\frac{ln\ S - \lambda}{\eta})^2\ ], \qquad 0 < S < \infty \tag{4}$$

in which η and λ are shape and scale parameters, respectively. These two modal parameters can be estimated from data by using three different estimation methods: method of moment, maximum-likelihood method and non-linear regression analysis.

It is reasonable and practical to assume that random variables A and S are statistically independent. In such case, their joint probability density function is of the simple form:

$$p(A,S) = p(A)\, p(S) \tag{5}$$

This joint probability density function can now be used to obtain the desired probability distribution function for random variable $S_a(T, \xi) \equiv A \times S$ representing the non-normalized acceleration response spectral value for period T and damping ratio ξ. This action is taken using the relation

$$P(C) \equiv \text{Prob}[S_a > C] = \int_{S=0}^{\infty} \int_{A=0}^{C/S} p(A,S)\ dA\ dS \tag{6}$$

Substituting Eq. (5) into Eq. (6), gives (Loh, 1994)

$$P(C) \equiv \int_{S=0}^{\infty} p(S)\ [\int_{A=0}^{C/S} p(A)\ dA]dS \tag{7}$$

since C is used as a dummy variable for S_a, the probability distribution function for $S_a(T, \xi)$ is identically equal to $P(C)$, *i.e.*,

$$P(S_a) = P(C) \tag{8}$$

Having obtained the probability distribution functions $P(S_a)$ for the full range of discrete values of T and for a specified discrete value of ξ, uniform-hazard acceleration response curves are obtained directly therefrom.

Based on the discussion of seismic hazard curve, design spectrum, and the uniform hazard response spectra, three seismic hazard curves can be developed (as shown in Fig. 4) at a specific site:

Curve 1: The traditional seismic-hazard curve was generated based on PGA attenuation equation. It plots the annual probability of exceedance with respect to different PGA.

Curve 2: Based on the numerical calculation of Eq. (7), Curve 2 was generated with a specific damping ratio and a structural period. The annual probability of exceedance for a given acceleration response spectra can be plotted as $1 - P_r\,[S_a < C\,]$.

Curve 3: For design purpose with a given annual probability of exceedance, the PGA value was selected from hazard curve (Curve 1). This PGA value was used to scale down the normalized design spectrum at a specific period. It means that the PGA value in Curve 1 will be used to scale the normalized acceleration spectra $S_{na}(T, \xi)$ so as to obtain the traditional design response spectra value:

$$C_d\,(T, \xi) = A \times S_{na}(T, \xi) \tag{9}$$

Curve 3 is generated from Eq. (9) for different probability of exceedance with a specified $S_{na}\ (T, \xi)$ at a given structural period.

If one can obtain the hazard curve as indicated in Curve 2 of Fig. 4 for different structural period, then by giving a specified annual probability of exceedance, the uniform hazard response spectra at a site can be obtained.

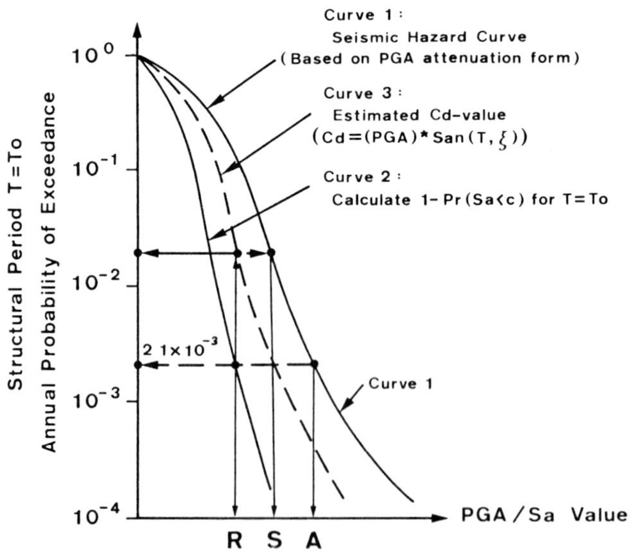

Fig. 4 Schematic diagram showing the calculation of uniform hazard PGA value.

In traditional seismic design, Curve 3 was used to determine the base shear coefficient (specified the annual probability of exceedance). The general iso-acceleration map for return period of 475 years was generated from Curve 1. Both Curve 1 and Curve 3 do not represent the uniform hazard for designing a structure. Taking an example, as shown in Fig 4, if the annual probability of exceedance is chosen as 2.1×10^{-3} (= 475 years return period), from Curve 1 the PGA value corresponding to this return period is given at point A. If the uniform hazard response spectrum was used, Curve 2 must be adopted to determine the S_a value, which corresponds to the same return period, and the S_a value is given at point R. Generally, the value at point R is the base shear coefficient for design. If the design method is used (*i.e.*, Curve 3), this S_a value given at point R corresponds to the annual probability of exceedance at point e. The PGA-value that will be used to generate iso-intensity map can be read from Curve 1 with the annual probability of exceedance shown at point e. In this figure, point S is the PGA value for structure design with return period of 475 years. It means that if 475 years return period was selected to design a structure, the selected PGA value from Curve 3 will be less than that was selected from Curve 1. Generally speaking, if Curve 1 is used for selecting the PGA value to design a structure, it will give a larger return period. Figures 5a and 5b show the iso-acceleration map of Taiwan corresponding to uniform hazard response spectra with return period of 475 years at $T = 1.0$ sec. The dispersion consideration for PGA attenuation equation is 2σ and for S_a is 1.0σ (Fig. 5a) and 2.0σ (Fig. 5b), respectively.

DISCUSSION ON SEISMIC HAZARD PARAMETERS

As expected, much of the effort in seismic hazard analysis is devoted to the determination of the modal parameters and physical relationships of seismicity, including the determination and assessments of all the associated uncertainties. In this paper, two parameters related to seismic hazard analysis are discussed: PGA attenuation equation and the seismic zoning scheme.

Ground motion attenuation:

The attenuation shown in Eq. (1) is based on the regressional analysis of Taiwan data (on hard site condition). The distance parameter R was defined as the hypocentral distance. In hazard calculation, the fault-rupture model was used for shallow earthquakes to calculate the exceedance probability in which the shortest distance between the site and fault-rupture was adopted. Difference between the hazard model and the PGA attenuation on the definition of distance is obvious. In order to be consistent with the fault-rupture model, the shortest distance must be used in the PGA attenuation equation instead of hypocentral distance. The following procedures are adopted to determine the shortest distance between the site and the fault-plane so that the distance R can be redefined in developing PGA attenuation equation for shallow events:

(1) From the recorded data of an earthquake, the fault plane solution can provide information on the direction of the strike and the dip angle of the fault.
(2) Using the distribution of hypocenters of aftershocks, one can determine the boundary of the source area.
(3) The shortest distance between the recorded station to the nearest boundary of the source rupture area is determined. It is defined as the shortest distance between source and site.

Figure 6 shows the comparison on the plot of PGA attenuation with respect to distance for the data of Dec. 15, 1993 earthquake in southwestern Taiwan. The open circles are plotted with respect to the hypocentral distance and dots are plotted with respect to the shortest distance. The attenuation equation shown in Eq. (1) is also plotted for comparison. It is obvious that the predicted PGA from Eq. (1) will need modification.

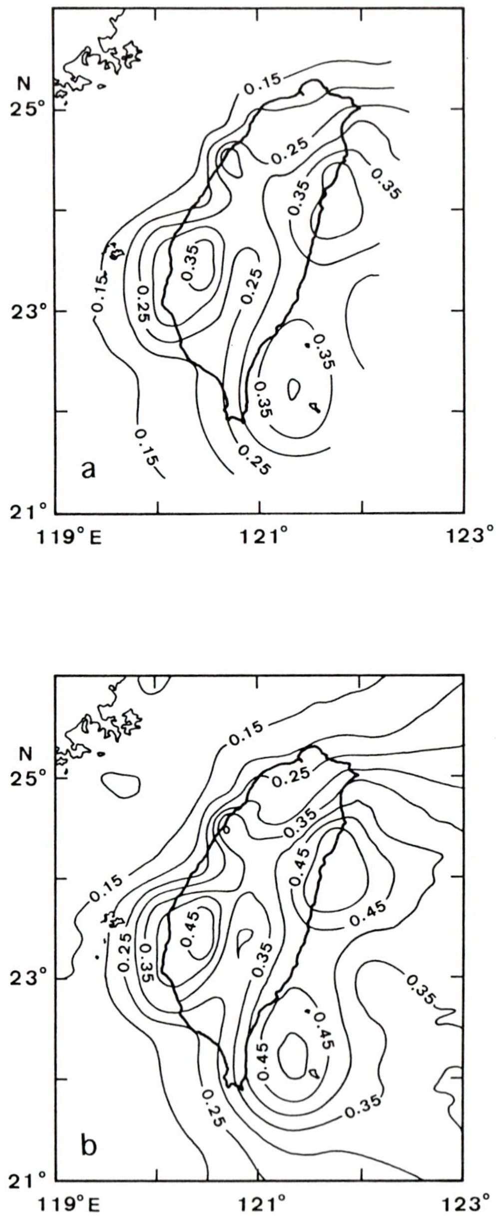

Fig. 5 Iso-acceleration maps of Taiwan for return period of 475 years: (a) dispersion consideration for PGA is 2.0s and for S_a is 1.0s, (b) dispersion consideration for PGA is 2.0s and for S_a is 2.0s.

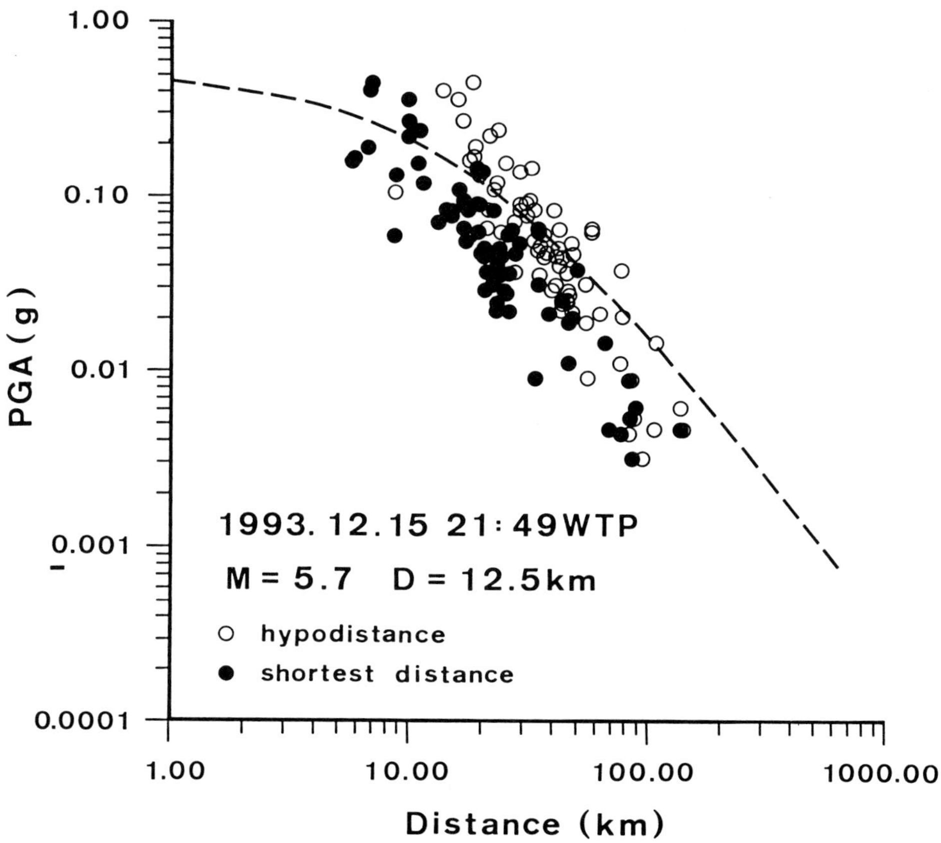

Fig. 6a Plots of PGA with respect to distance for Dec. 15, 1993 earthquake (for all data). The open circles are plotted for hypocentral distance and the dots are for shortest distance.

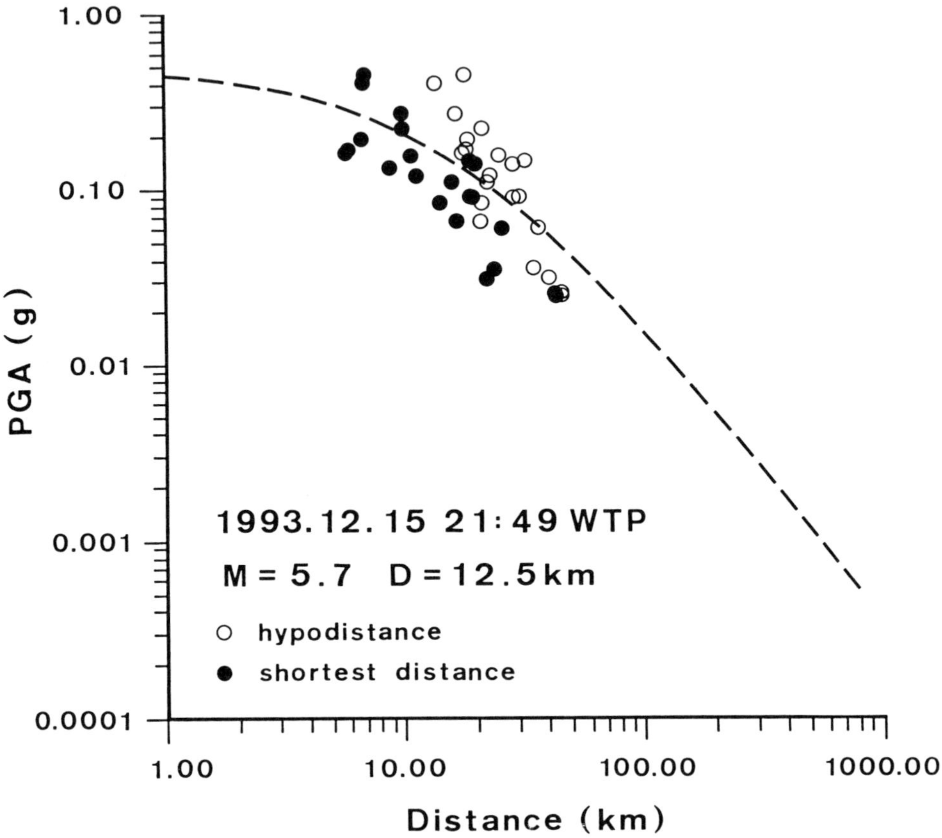

Fig. 6b Plots of PGA with respect to distance for Dec. 15, 1993 earthquake (for hard site data). The open circles are plotted for hypocentral distance and the dots are for shortest distance.

Zoning Scheme in SHA:

Except the discussion on the PGA attenuation equation, zoning scheme is also an important issue in seismic hazard calculation. Different zoning scheme may (or may not) have significant influence on the results of seismic hazard analysis. Besides the zoning scheme shown in Fig. 1 two other types of zoning schemes for shallow earthquakes are used to perform the seismic hazard analysis. The zoning scheme B (Fig. 7) includes more zones and the zoning scheme E (Fig. 8) is generated by considering larger seismic zones, and the western seismic zone is divided into two subzones. In our study, the effect of zoning scheme to the final hazard calculation is not significant.

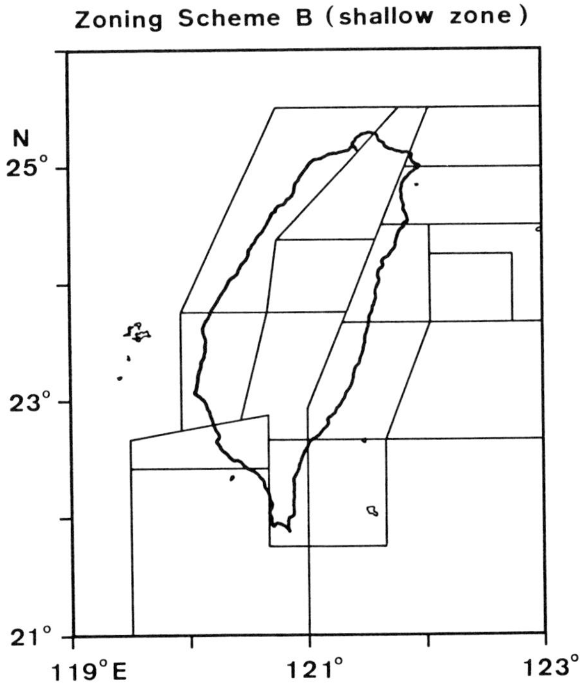

Fig. 7 Zoning scheme B for shallow zone.

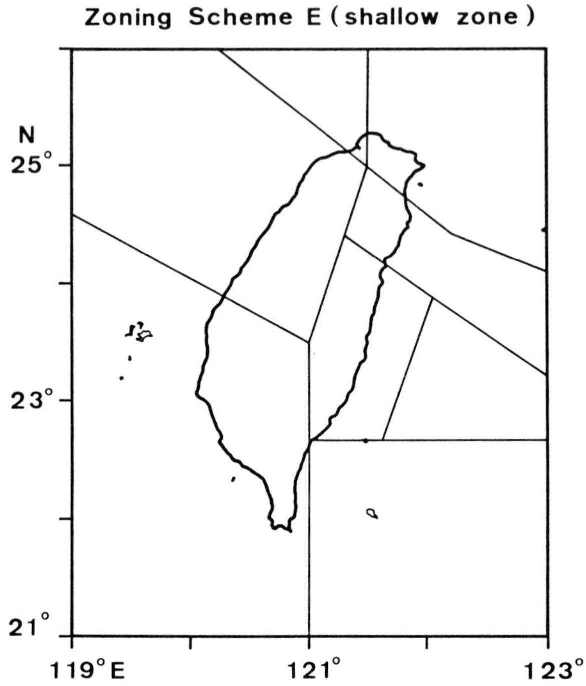

Fig. 8 Zoning scheme E for shallow zone.

EFFECT OF LOCAL SOIL NONLINEARITY ON SEISMIC HAZARD EVALUATION

It is generally agreed that a particular surface acceleration reflects to some degree of the characteristics of the near-surface soil layers at the recording site. To evaluate the seismic hazard at that particular site, it is necessary to include the effect of local soil amplification on the analysis. The most important one is to include the effect of nonlinear soil amplification and to evaluate the annual probability of exceedance at that particular soil site especially for strong ground shaking. Figure 9 shows the schematic representation of seismic hazard analysis in relating to the response of soil site on which the effect of soil nonlinearity is considered. The procedures of analysis are listed as follows:

Step 1: Estimate the annual probability of exceedance for different PGA value at a specific site (outcrop bedrock). The hard site design spectrum is also used so as to generate the spectral compatible time history. The simulated accelerogram was normalized to a specified PGA with return period T.

Step 2: Deconvolve the surface ground motion to a specified depth beneath the ground surface. Use this deconvolved seismic wave as input motion to a specific site where the soil properties are known in each soil layer.

Step 3: The nonlinear model of each soil layer system is developed (accuracy of the soil model must be evaluated). The surface ground motion is calculated by using the deconvolved seismic wave as input from the base. The relationship between outcrop PGA value or downhole PGA value to that of soil surface PGA is developed.

Step 4: From the hazard curve of hard site, given an annual probability of exceedance, the PGA value on the specific soil site can be estimated by using the relation between hard site PGA and soil site PGA from Step 3. Figure 10 plots the annual probability of exceedance at a site with respect to outcrop PGA and soil site PGA.

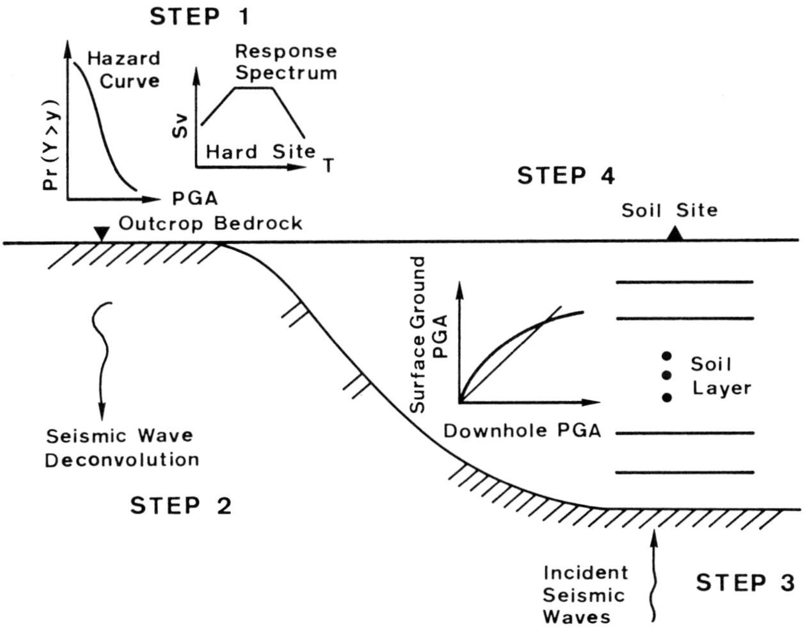

Fig. 9 Schematic representation of seismic hazard analysis in relating to the response of soil site on which the effect of soil nonlinearity is taken into consideration.

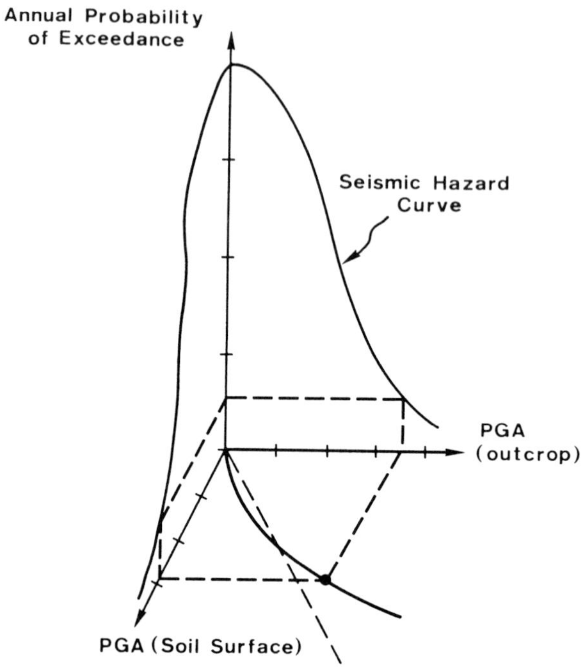

Fig. 10 Three dimensional plot of annual probability of exceedance at a site with respect to outcrop PGA, soil site PGA, and liquefaction damage index.

REFERENCES

Campbell, K. (1981). Near-source of peak horizontal acceleration. *BSSA*, 71, No.6, 2039-2070.

Cornell, C.A. (1968). Engineering seismic risk analysis. *BSSA*, 58, No.5, 1583-1606.

Der Kiureghian, A. and A. H-S. Ang. (1977). A fault-rupture model for seismic risk analysis. *BSSA*, 67, No.4, 1173-1194.

EERI Committee on Seismic Risk. (1989). The basics of seismic risk analysis. *Earthquake Spectra*, 5. No.4, November, 675-702.

Joyner, W.B. and D.M. Boore. (1981). Peak horizontal acceleration and velocity from strong-motion records including records from the 1979 Imperial Valley, California Earthquake. *BSSA*, 71, 2011-2038.

Loh, C.H., Y.T. Yeh, W.Y. Jean and Y.H. Yeh. (1991a). Seismic hazard analysis in the Taiwan area using a bounded fault-rupture model. *BSSA*, 81, No.1, 265-272.

Loh, C.H., *et al.* (1991b). Probabilistic seismic risk analysis in the Taiwan area based on PGA and spectral amplitude attenuation formula. *Engineering Geology*. 30, 277-304.

Loh, C.H., F.S. Bau, and Y.T. Yeh. (1992). Seismic hazard and uncertainty analysis of the Taiwan area. *Soil Dynamics and Earthquake Engineering*. 11, 497-508.

Loh, C.H., W.Y. Jean, and J. Penzien. (1994). Uniform-hazard response spectra – An alternative approach. *Earthquake Engineering and Structural Dynamics*. 23, 433-445.

Sadigh, K., *et al* (1993). Specification of long-period ground motions: updated attenuation relationships for rock site conditions and adjustment factors for new-fault effects. *ATC* 17-1.

TAIWAN STRONG MOTION INSTRUMENTATION PROGRAM (TSMIP) AND PRELIMINARY ANALYSIS OF SITE EFFECTS IN TAIPEI BASIN FROM STRONG MOTION DATA

K. -W. Kuo,* T. -C. Shin* and
K. -L. Wen**

** Seismology Center, Central Weather Bureau, Taipei, Taiwan*
*** Institute of Earth Sciences, Academia Sinica, Taipei, Taiwan*

ABSTRACT

Incorporating with "The ROC 6-Year National Development Plan", Central Weather Bureau proposes a Taiwan Strong Motion Instrumentation Program to enhance Strong-motion Observation in Taiwan area. The program will install up to 600 digital free field strong motion instruments, and 400 x 3 digital channels of strong motion monitoring systems in nine metropolitan areas, including: Taipei, Hsinchu, Taichung, Chiayi, Tainan, Kaohsiung, Ilan, Hualien, and Taitung. The main purposes is to study the characteristic of ground motion at different geological conditions, and the response of different type man-made structures. All results can be applied to improve the design spectrum and building codes of current use.

Till 1994, There are 43 strong motion stations has been installed in Taipei basin. Several earthquakes has been recorded by the strong motion instruments since 1992. The first event with magnitude 6.2 triggered by more than thirty stations was used to study the characteristics of the basin effects on ground motions. In this analysis, the characteristics of the ground motions are studied both in time and frequency domains for realizing the site effects of the Taipei basin.

KEYWORDS

earthquakes; ground motion; free field strong motion instruments; Strong-motion Observation Network; Taiwan Strong Motion Instrumentation Program; frequency domains; time domains; site effects; basin effects; Taipei basin.

INTRODUCTION

Taiwan is located on the Circum-Pacific seismic belt. Safeguarding life and property from destructive effects of earthquakes is a major concern for the people of the region. The damaging effects of earthquakes are primarily caused by strong ground shaking. Reduction of life and property losses from strong ground shaking requires conscientious application of construction codes and earthquake resistant design, enforcement of adequate land-use policies, and implementation of appropriate retrofit measures. Implementation of these mitigation measures must be based large part recordings of large earthquakes at close-in distances. It is the strong motion data in distance ranges from 0 to 100 km for large earthquakes. Those are crucial for earthquake resistant design, understanding the source mechanism of earthquakes, and realizing the propagation of seismic waves from source to site, including local site effects.

During the Ming and Ching Dynasties (1624-1896), Taipei basin had been struck by at least four damaged earthquakes (Hsu, 1983). Among them, a magnitude of 7.3 large earthquake occurred closed to Taipei city in 1909 and caused severe damage. In the recent year, the rapid growth of economic environment including high-rise building, public communication systems, and rapid transit systems, increase the damage potential. Taking the safety and investment of structures into account, the need of local seismic ground motion records is significantly important, especially at the stage of design. It also provide information for the improvement of seismic resistance of old buildings. Recently from the records of the surface strong motion observation, the studies show the preliminary analysis. Hope it can be a good reference for other studies.

GEOLOGY OF THE TAIPEI BASIN

Taipei basin is a triangle shape, with Shanchia, Kuantu, and Nankang at the southwestern, northwestern and eastern corners. The Tatun volcano group is located along the northern fringe of the basin, and the Chishinshan with an elevation of 1,120 meters is the highest place. The Linkou tableland on the west of the basin has an altitude ranging between 200 and 250 meters. To the southeast, there are foothills of several hundred meters high. The land surface of Taipei basin is almost flat and tilting gently to the northwest. The total area of the Taipei basin with an altitude below 20 meters is 243 square kilometers. The Keelung River is east-west direction and the Dahang Creek and the Hsintien Creek down from south goes through the basin center then turn north-west to the ocean (Fig. 1).

Because the basin is filled with the unconsolidated sediments, the subsurface geology of the Taipei basin can only be established by the information obtained from boring, electrical and seismic prospecting. Fig. 2 is the contour map of the basement of Taipei basin (Wang Lee *et al.*, 1978).

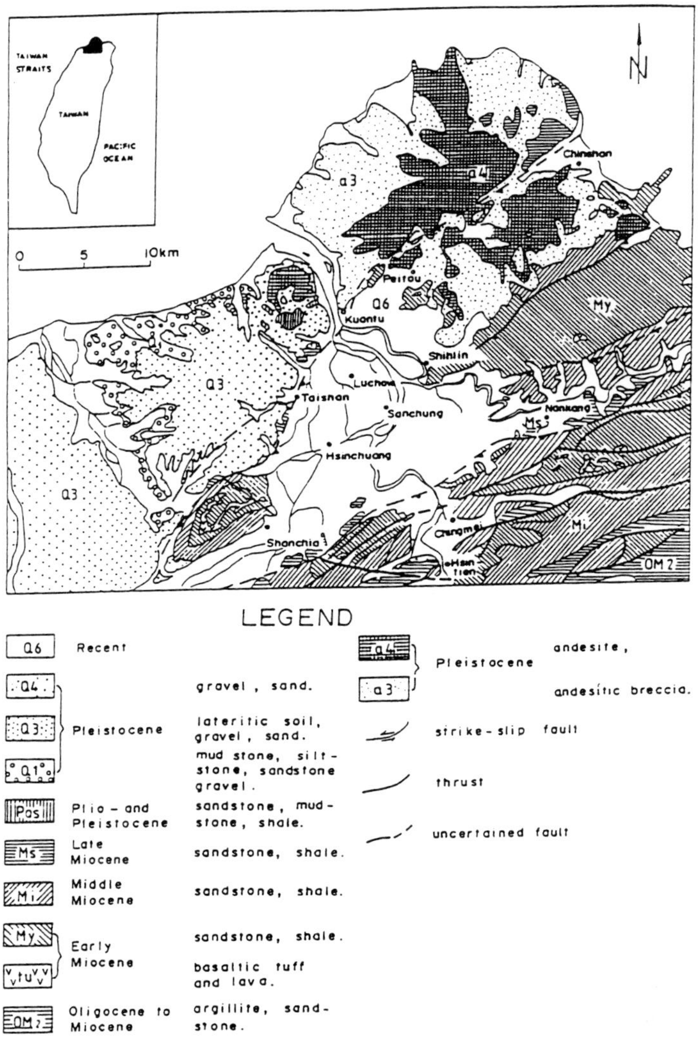

Fig. 1. Geological map of northern Taiwan (Wang Lee et al., 1978).

For the Taipei basin, only the P wave velocity structure is fully investigated. Table 1 lists the results from reflection seismic survey (Wang *et al.*, 1993). In the table, it list the results from refraction survey by Fong *et al.* (1976). The shear wave velocity strata is available only at some shallow strata. Lin *et al.* (1980), used down-hole method, measured 10 sites and the basin center area. The results show that the shear wave velocity of Sungshan Formation is lower than 370 m/s. Lee (1988) studied the relationship between N value and Vs. The relationship can be used in Sungshan Formation. Yeh *et al.* (1994) analyzed the seismic wave arrival time from the downhole records provide the V_p and V_s in the Wugu area.

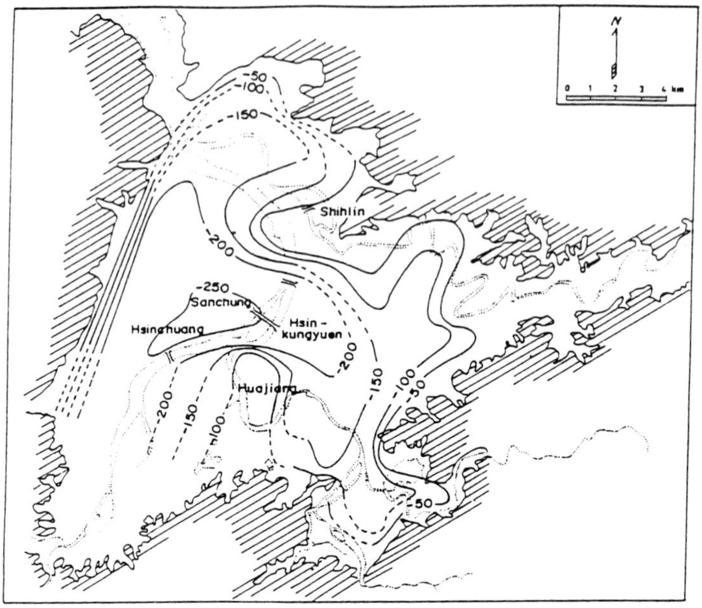

Fig. 2. Contour map of the basement top (Wang Lee et al., 1978).

Table 1 P wave velocity structure of Taipei basin (Wang *et al.*, 1993)

Depth (m)	Velocity (m/sec)	Velocity* (m/sec)	Layer
0- 10	400	350	Soil
10- 60	1550	1500	Sungshan F.
60-120	2050	1260	Chingmei F.
120-180	2400	2410	U. Hsinchuan F.
180-240	3100		L. Hsinchuan F.
>240	3800	3640	Nankang F.

*: Fong et al. (1976).

STRONG MOTION OBSERVATION

Taiwan Strong Motion Instrumentation Program (TSMIP) is a six-year plan from 7/1/1991 to 6/30/1997, executed by the Seismology Center, Central Weather Bureau (Shin, 1993). The TSMIP has 4 major program elements:

(1) Installation and maintenance of up to 600 digital free field strong motion instruments, and up to 400 x 3 digital channels of strong motion monitoring systems in representative buildings and other large structures. Calibration of all the installed instruments and execution of quality-control of all data output.

(2) Operation of a Central Weather Bureau Data Management Center for data archiving and dissemination.

(3) Funding research making use of the TSMIP data and other related data with research result readily applicable to social needs.

(4) Development of an earthquake early warning system.

The main purpose of this program is to study the characteristics of ground motion in different geological conditions, and to study the responses of different types of man-made structures. All results can be used to improve the design spectrum and building codes of current use. Fig. 3 shows the distribution of 43 strong motion stations in the Taipei Basin.

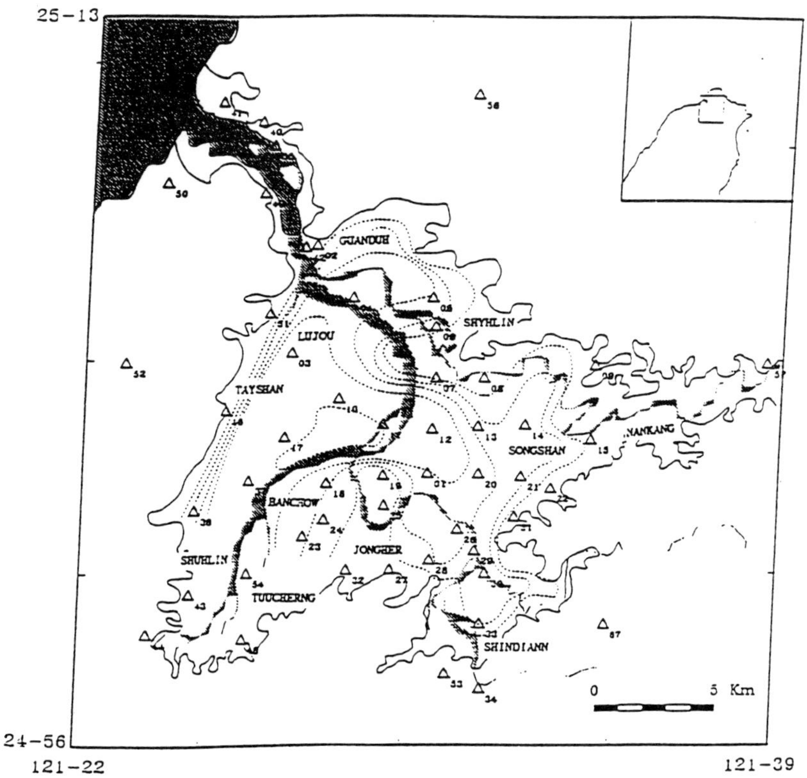

Fig. 3. Taipei strong motion observation network.

Each strong ground motion station has the same design. A small fiber-glass house covered on a concrete plate. The strong motion instrument is a force-balance accelerometer (IDS-3602S) made by TERRA Technology Corp. or A900 by Geotech. It has 16 bits resolution, can record the ground motion up to 2g. All stations have AC power. When the power system shutdown by earthquake or other problem, the DC power of the recording system still can operate about 4 days.

RESULTS AND DISCUSSIONS

Several earthquakes had been recorded by this network since the installation of accelerographs. The first event that triggered more than thirty stations was listed in Table 2. In this study, the characteristics of the ground motions are studied both in time and frequency domains for realizing the site effects in the Taipei Basin area.

Table 2 Earthquake parameters recorded by surface network

Origin Time	Epicenter	Depth (km)	M_L	Records*
1994.06.05 01:09:30.1	121.84 E 24.46 N	5.3	6.2	33

*: Recorded station numbers.

Figure 4 is peak ground acceleration (PGA) contour map in the Taipei basin. In this Figure, the dash lines show the basement depth contour as that in Fig. 2. Bold lines are the contour of peak ground motions. From the PGA contour in Fig.4 shows that it has no correlation with the basin structure. The records are filtered at the frequency band from 0.1 to 0.8 Hz. The low frequency band PGA contour shows in Fig. 5. From this low frequency band contour, we can find the contours high are not the same as that in Fig. 4. The high PGA areas are along the Dahang Creek and the Keelung river and consist with geological structures of the Taipei basin. One of the contour high is located in the western part of the basin center (shown in dash line). Our low frequency response seems very consist with the geological structure.

To understand the amplification effects in the frequency domain. We transferred the time domain records to frequency domain and calculated the spectra's ratios with respect to that of rock site TAP016. Figs. 6a and 6b show the results of 0.48 and 2.91 Hz, respectively. From these figures, we can know that different frequency will have different amplification in the Taipei Basin. Frequency responses in both frequencies show the variation can be about 5 times different. The contours of Fig. 6a are more correlated with geological structure and low frequency PGA contour in Fig. 5.

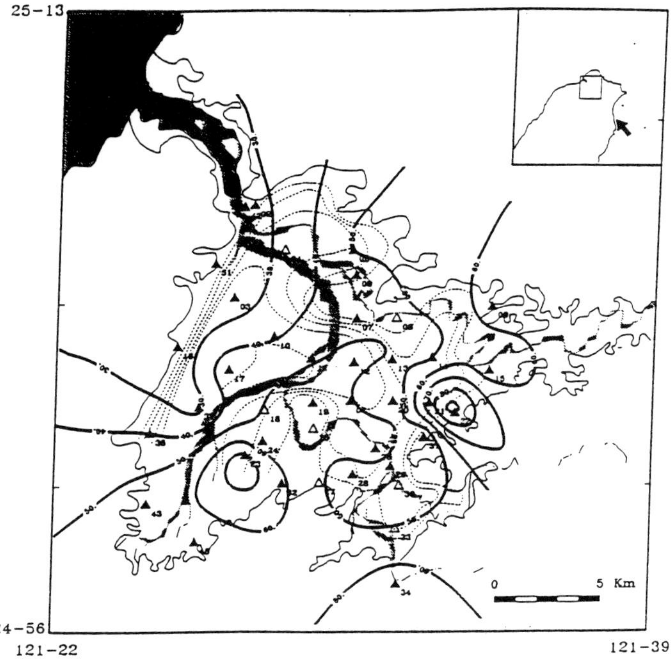

Fig. 4. PGA contour map in the Taipei Basin.

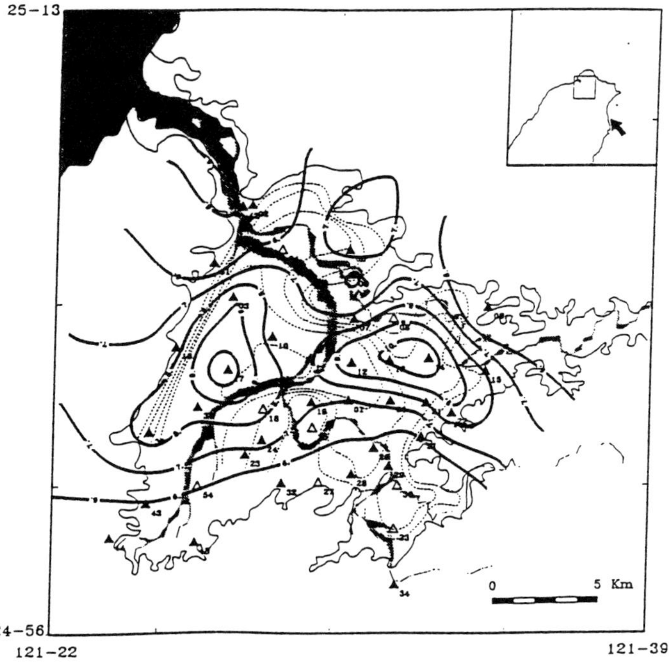

Fig. 5. PGA contour map after filtered from 0.1 to 0.8 Hz.

(a)

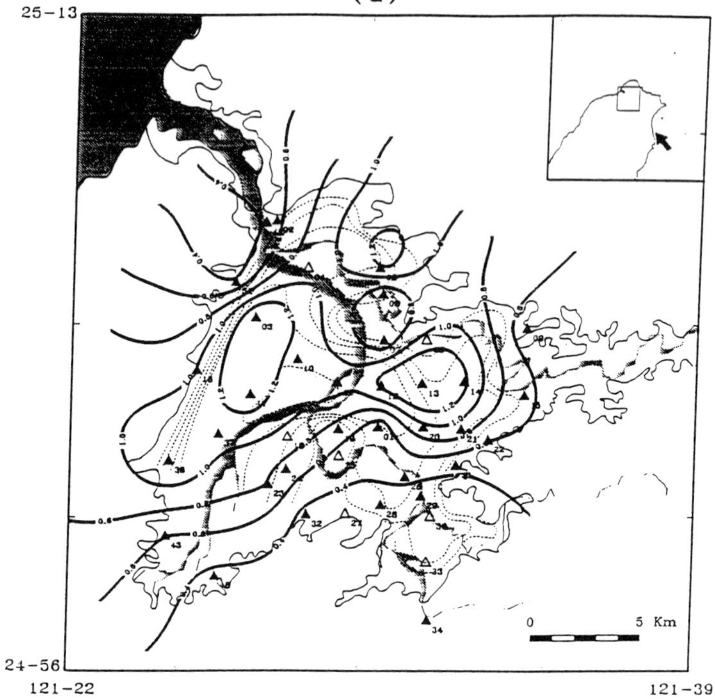

(b)

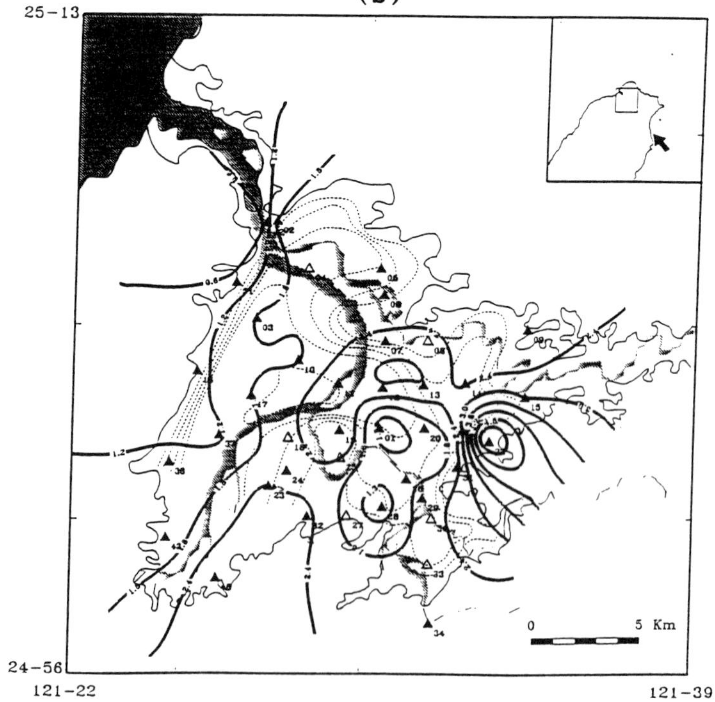

Fig. 6. Spectra ratio contours at (a) 0.48 Hz, (b) 2.91 Hz.

(a)

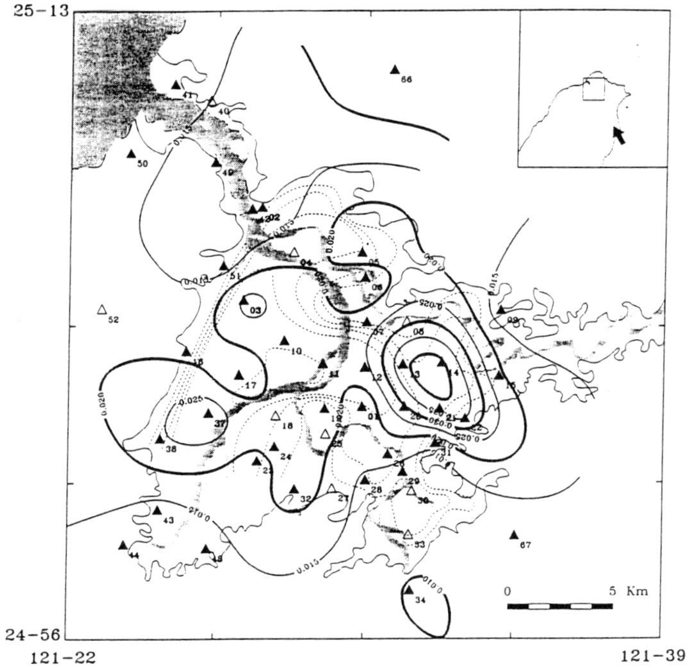

(b)

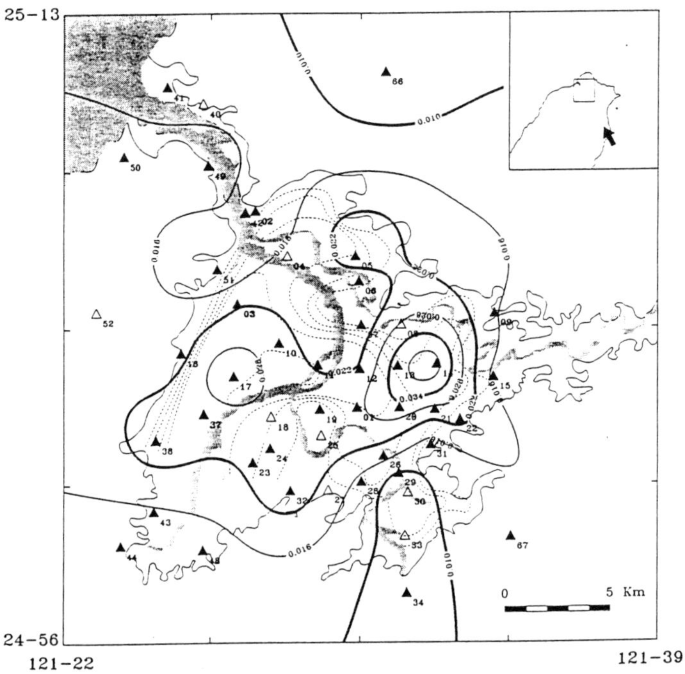

Fig. 7. Acceleration response spectra contours at 1.69 seconds with 5% damping ratio (a) EW component, (b) NS component.

(a)

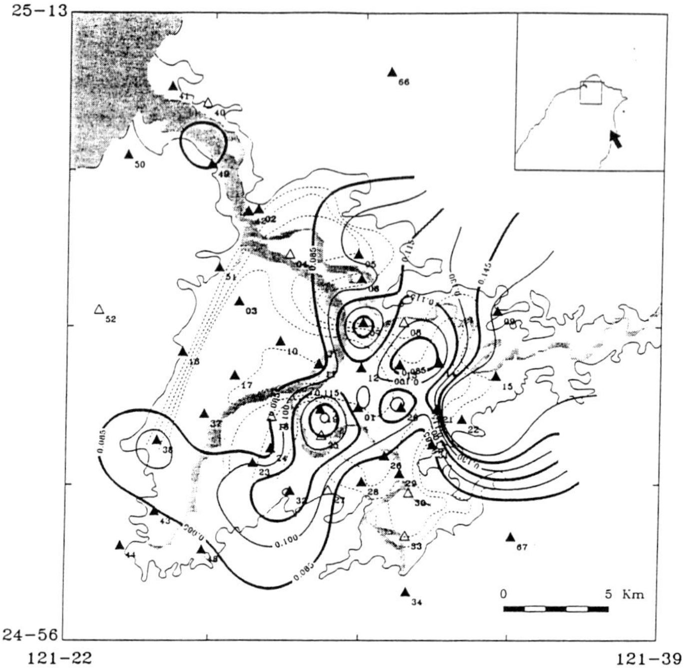

(b)

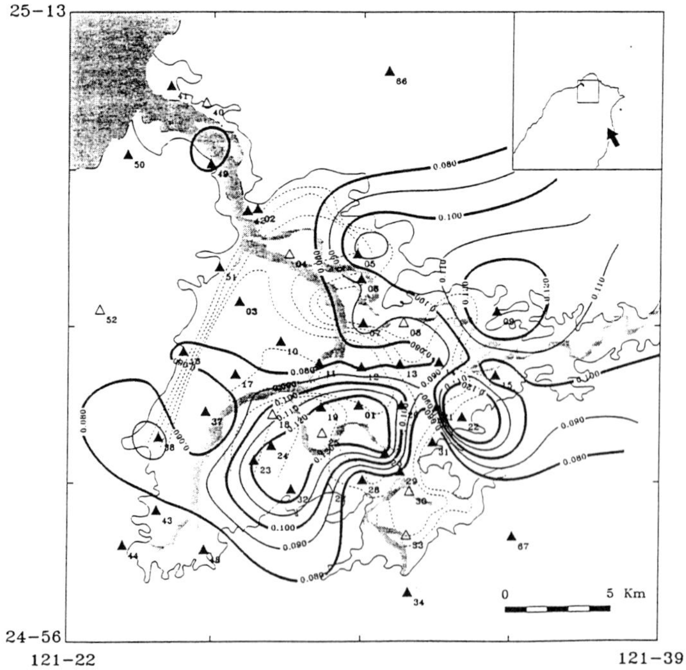

Fig. 8. Acceleration response spectra contours at 0.51 seconds with 5% damping ratio (a) EW component, (b) NS component.

(a)

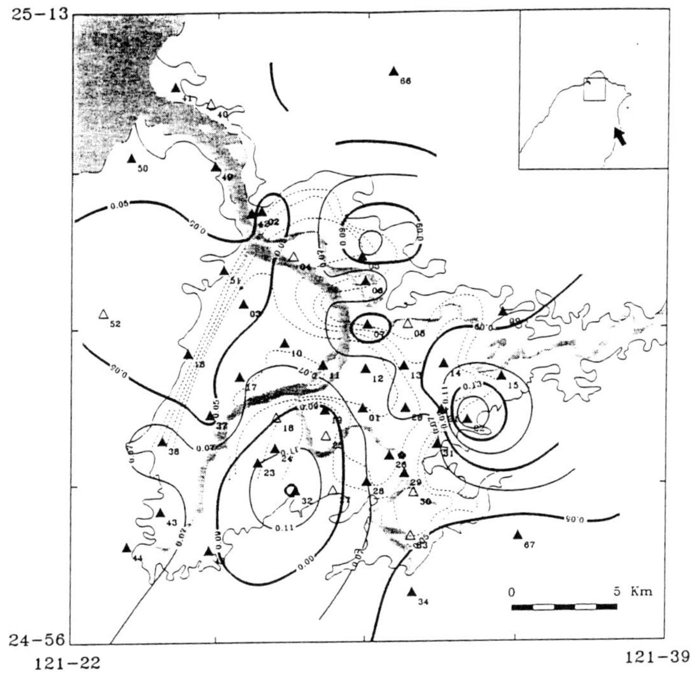

(b)

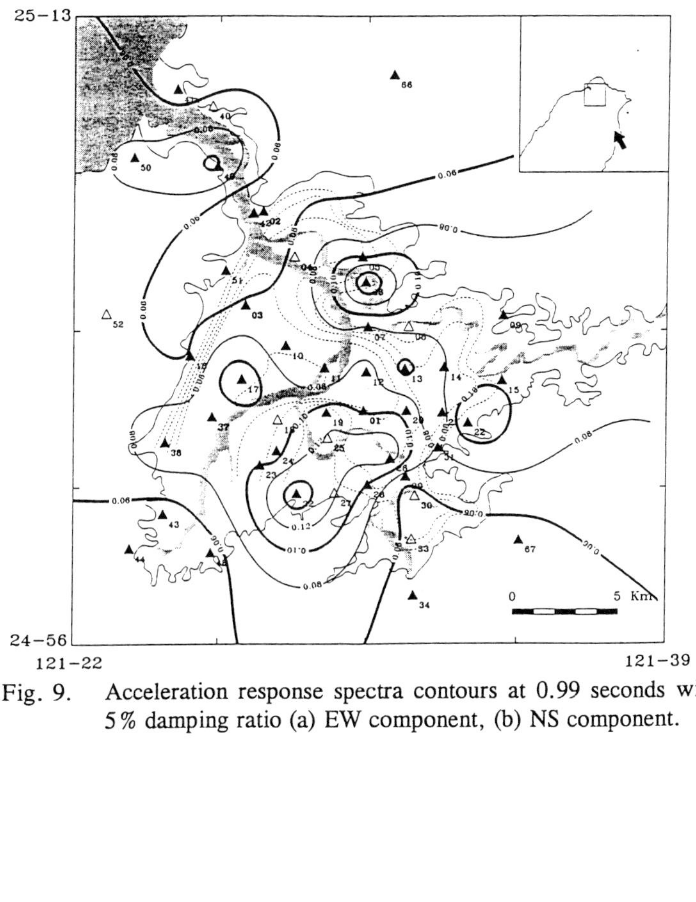

Fig. 9. Acceleration response spectra contours at 0.99 seconds with 5% damping ratio (a) EW component, (b) NS component.

(a)

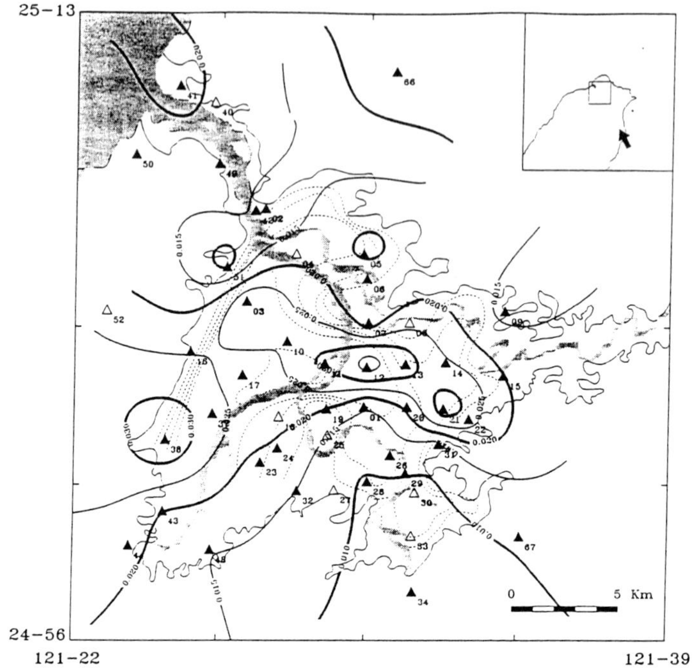

(b)

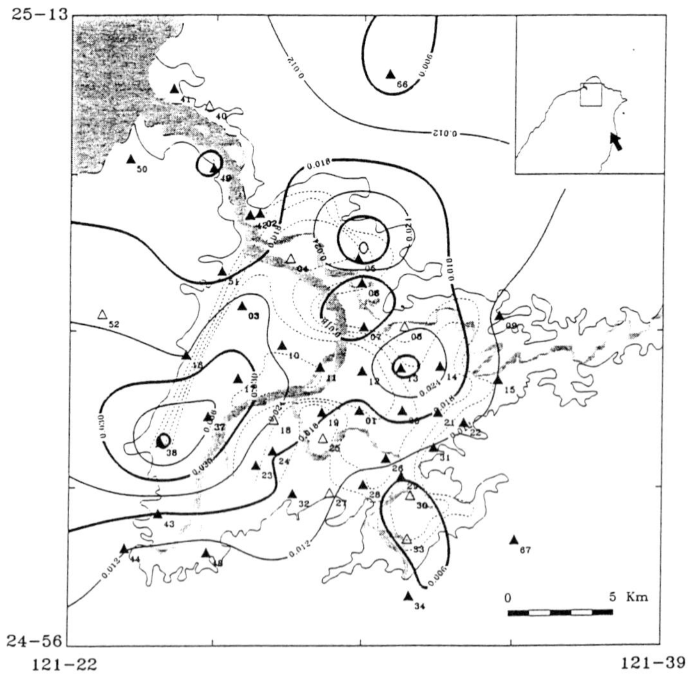

Fig. 10. Acceleration response spectra contours at 2.06 seconds with 5% damping ratio (a) EW component, (b) NS component.

(a)

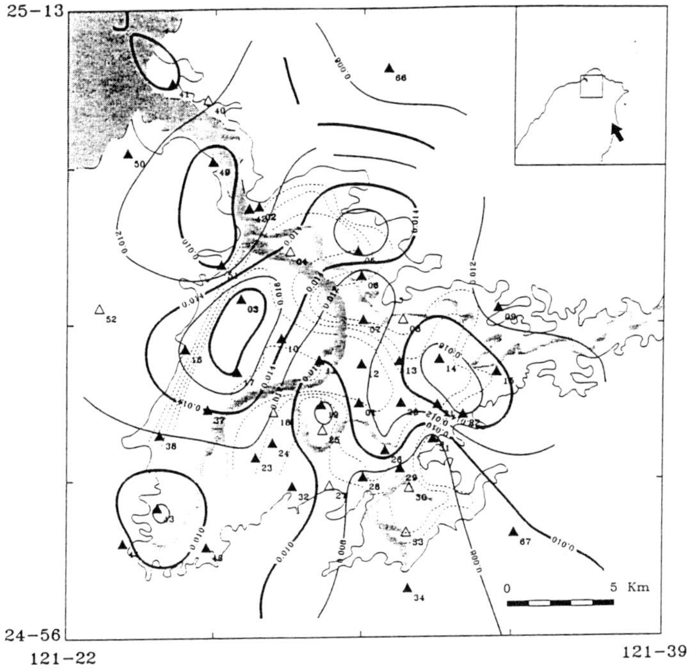

(b)

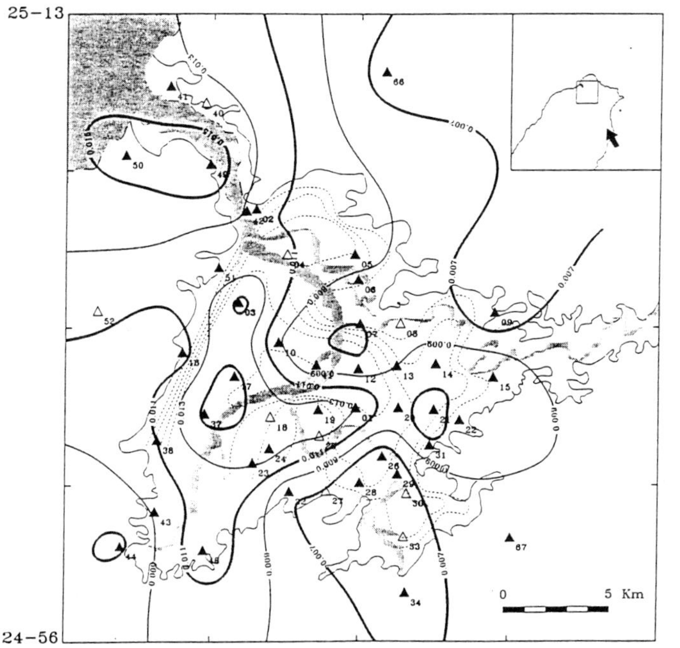

Fig. 11. Acceleration response spectra contours at 3.06 seconds with 5% damping ratio (a) EW component, (b) NS component.

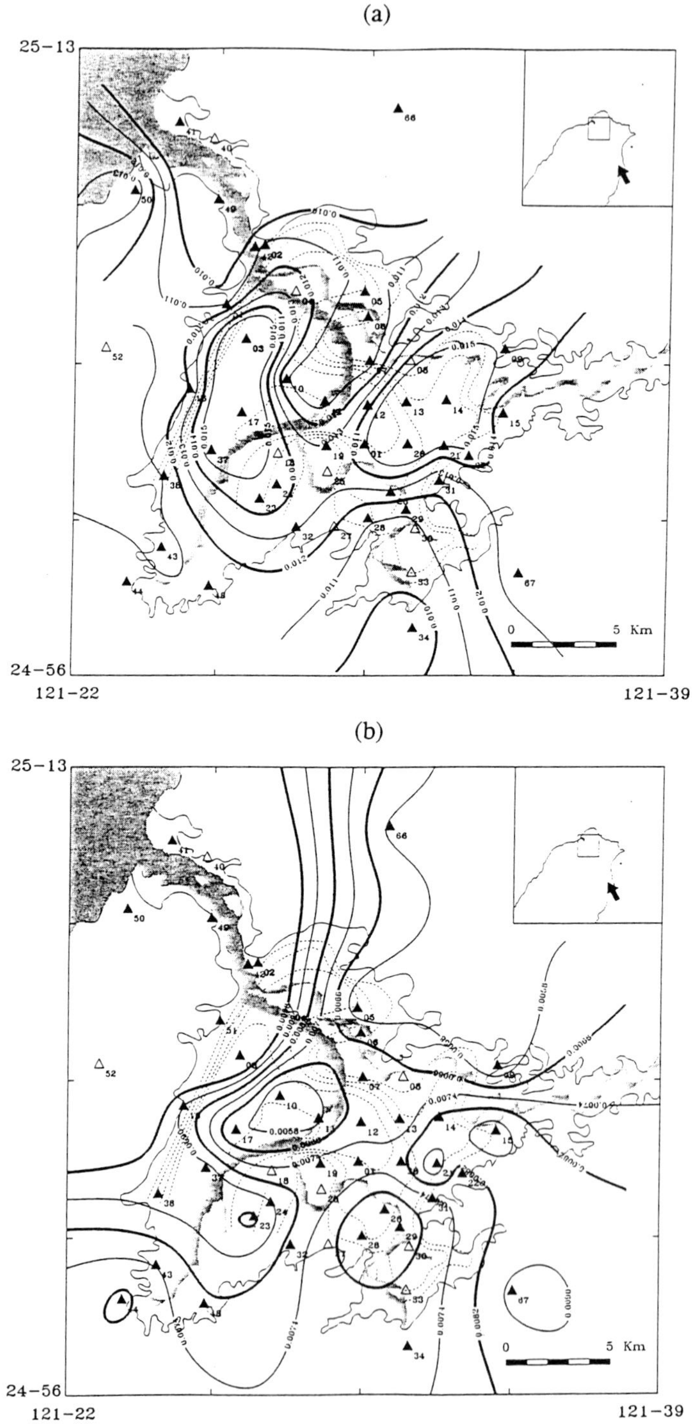

Fig. 12 Acceleration response spectra contours at 4.87 seconds with 5% damping ratio (a) EW component, (b) NS component.

From previous analysis we can realize that Taipei area can not just using one Building Code for whole basin area. Different area will have different local site effects. The variation in the same frequency can be about 5 times different. The records of an moderate earthquake are used to study the variation of spectral acceleration in Taipei Basin. 101 periods are selected from 0.02 to 15 seconds with equal logarithmic interval to calculate the 5% damping response spectra. From the microtremor survey and record of Nov. 14, 1986 show that at about 1.65 seconds have a rather high peak spectrum (Tsai, 1988). Fig. 7 shows the contour of acceleration response spectra with 5% damping ratio at 1.69 seconds calculated from this event. The pattern is the same as that in Fig. 6a of the 0.48 Hz spectra's ratio. Both components all show the contour high areas are along the Dahang Creek and the Keelung river. The contours of acceleration response spectra at some other periods of 0.51, 0.99, 2.06, 3.06, and 4.87 seconds which engineers may interested are shown in Figs. 8-12, respectively, for reference. These contours of acceleration response spectra can be seen as the responses of different storey structures in the Taipei Basin.

SUMMARY

From the records of the Taipei strong motion observation network we can study the local site effects and basin effects by compare the records at different site. Responses in the low frequency band are more correlated with the basin structure from both analyses in time and frequency domains. This observation project just beginning in this area and stations in different sites still will be added recently. Hopefully, the characteristics of the ground motions in the Taipei basin during earthquake can be realized through the study, and the seismic resistant design code of Taipei city can be improved accordingly. This, in turn, will mitigate the possible damages in the future earthquakes.

REFERENCES

Fong, C. C., T. L. Shaw, C. C. Liu, and Y. B. Tsai (1976). Geophysical survey in Taipei City, Technical Report of Institute of Earth Sciences, Academia Sinica.

Hsu, M. T. (1983). Evaluation of magnitude and intensity of the damage earthquakes in the Ming and Ching Dynasties, *Bull. of Weather Soc. of China*, 29, 4, 1-18.

Lee, S. M. (1988). Progress model fro shear wave velocity of Sandy and Silty soils for Taipei basin, Proc. CCNAA-AIT Joint Seminar on Research and Application for Multiple Hazards Mitigation, 489-494.

Lin, Y. I., C. Yang, M. C. Chang, and M. S. Chun (1980). Shear wave velocity measurement results of the top soil (Sungshan Formation) in Taipei basin and its preliminary study at the structural resistant design, Report of Dept. of Physics, National Taiwan Educational Univ.

Moh and Associates Inc. (1989). The geotechnical properties of subsurface layers in Taipei City, Taipei basin, Moh and Associates Inc., 19p., 14 maps.

Shin, T. C. (1993). Progress summary of the Taiwan Strong Motion Instrumentation Program, Abstract of the Symposium on Taiwan Strong Motion Instrumentation Program, 1-10.

Tsai, I. C. (1988). Strong ground motion characteristics of Taipei basin in November 15, 1986 earthquake and its impact on structural design, Proc. CCNAA-AIT Joint Seminar on Research and Application for Multiple Hazards Mitigation, 195-208.

Tsai, Y. L. (1994). Seismic shallow reflection survey in Sanchung-Hsinchuan-Shuhlin area of Taipei basin, *M.S. Thesis*, Inst. of Geophysics, National Central University.

Wang Lee, C. M., Y. M. Cheng, and Y. Wang (1978). Geology of the Taipei basin: *Taiwan Mining* 30, 4, 350-380.

Wang, C. Y., C. T. Sun, W. C. Shaw, and Y. L. Tsai (1993). Seismic strata of Taipei basin (I): Kuandu-Luchou area, Report of Institute of Geophysics, National Central University.

Wang, C. Y., W. C. Hsiao, and C. T. Sun (1994). Reflection seismic stratigraphy in the Taipei basin (I) - Northwestern Taipei, *J. Geo. Soc. China*, 37, 1, 69-95.

Yeh, Y. T., K. L. Wen, and Y. H. Yeh (1994). A integrated survey of subsurface geology and engineering environment of Taipei basin - Study of basin effects on seismic waves (I), Tech. Rept. of Central Geological Survey, No. 83-004.

Earthquake Engineering

SYSTEMATIC DETERMINATION OF SEISMIC SAFETY LEVELS FOR DESIGN OF R/C BUILDINGS

A. H. -S. Ang* and D. de Leon**

Department of Civil and Environmental Engineering, University of California, Irvine, CA, U.S.A.

ABSTRACT

An approach for the systematic determination of the appropriate target reliability (or acceptable risk) for the earthquake-resistant design of R/C buildings is proposed. Optimal reliabilities for damage control and life safety are determined on the basis of minimizing the respective life-cycle costs. This requires the proper integration of seismic hazard analysis, damage and reliability assessments, and formulation of appropriate life-cycle cost functions. The resulting target reliabilities are then translated to corresponding seismic base coefficients for damage control and life safety.

KEYWORDS

Earthquake engineering, design, target reliability, optimal design, acceptable risk, buildings, damage control, life safety.

INTRODUCTION

In engineering design, the determination of the proper level of safety or performance requires consideration of acceptable risk. The required acceptable risks for life safety and damage control may be determined on the basis of minimizing the respective expected life-cycle costs, from which the corresponding criteria for design may be derived.

Cost is invariably important in engineering design; in particular, the total life-cycle cost of the system is of special concern. Therefore, in the formulation or development of criteria for design, a balance between performance and life-cycle cost is an important objective. Traditionally, this balance is "achieved" or attempted through judgement and intuition. A systematic and quantitative approach for this purpose is suggested, which requires the proper integration of the following:

(i) The quantitative damage and reliability assessments of a structure subjected to an anticipated earthquake, or ensemble of potential earthquakes.

----

* Professor of Civil Engineering

** Graduate Student Researcher

(ii) Development of realistic cost functions that are related to or vary with the level of safety (or probability of failure), and the formulation of expected life-cycle costs.

(iii) Consideration of the seismic hazard at the site of a structure; e.g. as defined through a hazard curve.

(iv) Trade-off between safety (or performance) and life-cycle cost to determine the optimal target reliabilities for design.

The theoretical and analytical tools for the above purposes are now available, as described briefly below.

DAMAGE ASSESSMENT

Clearly, in design for damage control, a tolerable level of structural damage must be defined. For this purpose, a predictive model is needed in order to calculate the damage expected during a given or anticipated earthquake. The model, however, must be related to or consistent with observed seismic damages. This can be accomplished by calibrating the calculated damages of a selected group of structures with the corresponding observed damages of the same structures from past earthquakes.

On the basis of this calibration, and the subsequent disposition of the damaged structures, intermediate damage levels may be defined including the limit of reparable damage..

Damage Model for Reinforced Concrete Elements

A seismic damage model for reinforced concrete has been developed by Park and Ang (1985). Damage is assumed to be caused by the maximum structural deformation and the dissipated hysteretic energy, and the total damage is expressed in terms of a damage index

$$D = \frac{x_m}{\delta_u} + \frac{\beta}{Q_y \delta_u} \cdot E_h \qquad (1)$$

where:

x_m = maximum deformation;
E_h = dissipated hysteretic energy;
δ_u = deformation capacity;
Q_y = static yield strength;
β = constant.

The above damage index has been normalized, such that at collapse the value of D is 1.00. Based on extensive analyses of available test data for beam and beam column elements, the damage index at collapse has been determined (Park and Ang, 1985) to have a lognormal distribution with a median of 1.0 and a c.o.v. of around 50%.

Damage of Structure

The above model refers only to the damage of structural elements or components (e.g. beams and columns). The global damage of a complete structure is, of course, a function of the damages of its constituent components. The determination of the global damage may be based on the following:

$$(D_t > d) = \underset{i}{U}(D_i > d) \tag{2}$$

where,

D_i = the damage of critical component i ;
D_t = the global damage of the structure ; and.
U = the union of events.

Eq. (2) means that the global damage will exceed d when one or more of its critical components have damages exceeding d. For non-critical components, (D_i>d) may require several components with damages all exceeding d. It follows that the probability of the global damage can be assessed as,

$$P(D_t > d) = P[\underset{i}{U}(D_i > d)] \tag{3}$$

For a specified value of d, representing the tolerable damage level, the pertinent damage exceedance probability, Eq. (3), can be assessed, e.g. by fast Monte Carlo simulations (Wang and Ang, 1993). In the above damage assessment the response statistics (namely, the maximum deformation and dissipated hysteretic energy) of the components are required. This involves the nonlinear response analysis of the structure under a given earthquake. The model for this analysis is the following.

The equation of motion may be expressed in matrix form as,

$$M\ddot{x} + C\dot{x} + Kx + (1 - \alpha)Kz = -M\ddot{x}_g \tag{4}$$

where:

$\ddot{x}$, $\dot{x}$ and x = relative acceleration, velocity and displacement vectors;
M, C and K = mass, viscous damping and initial stiffness matrices;
α = ratio of post- to pre-yielding stiffnesses;
$\ddot{x}_g$ = ground acceleration.

And the components of the vector z is governed by a differential equation

$$\dot{z} = \frac{1}{\eta}[A\dot{x} - \nu(\beta|\dot{x}||z|z + \gamma\dot{x}|z|^2)] \tag{5}$$

where A, ν, β, γ and η are parameters that govern the shape and amplitude of the hysteresis loop.

FORMULATION OF COST FUNCTIONS

The optimal risk or target reliability may be determined on the basis of achieving the minimum expected life-cycle cost. Therefore, the cost items that are pertinent are those that would vary with the risk (probability of failure); any cost item that is independent of the risk will not affect the determination of the optimal risk and may be neglected.

For the purpose of formulating the expected life-cycle cost and determining the optimal target reliability for design, the cost items may be classified into three categories as follows:

1. Those that vary with risk; i.e. the cost will increase or decrease with the risk underlying the design of a structure, e.g. initial cost of a structure; damage/repair cost.

2. Those that are independent of risk, e.g. insurance premium; costs of excavation and site preparation.

3. Those that are consequences of damage or collapse of a structure; e.g. economic losses (business interruption, loss of revenue), cost of injury.

Ideally, insurance premium ought to vary with the acceptable risk underlying the design of a structure. Realistically, however, it is more likely to be independent of the underlying risk.

Cost items of the first category are explicitly functions of the underlying risk, whereas in the second category the costs are constants and thus will not influence the determination of the optimal risk. For the third category, the expected cost of each consequence is the actual cost multiplied by the risk (probability of failure).

For consistency, all costs may be expressed in terms of present value; besides discount and inflation rates, this requires also information on the future occurence of earthquake-induced damage or collapse. In approximation, such damage may be assumed to occur during the mid life of the structure.

Initial Cost Function

The initial cost of a structure will vary with the target reliability (or risk) underlying its design; in particular, this cost will decrease with risk (or increase with reliability), as shown in Fig. 1. To develop this initial cost function for a structure or a class of structures, the same structure must be designed repeatedly for varying levels of risk; e.g. using several values of the seismic base coefficients. The cost for each of the designs can then be estimated which may consist of material cost, construction cost, design cost, etc. Under the earthquake for a designated site (e.g. defined by a spectrum shape) with a given intensity, the probability of damage exceedance for each of the designs can be assessed through Eq. (3). These calculated probabilities may then be related to the respective costs for the different designs, thus obtaining the initial cost function for the given intensity. By varying the intensity of the same earthquake, a family of initial cost functions is developed for the structure as illustrated conceptually in Fig. 1.

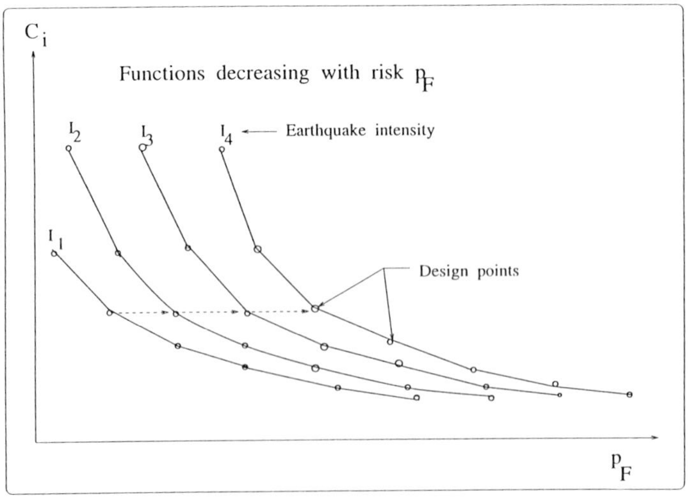

Fig. 1. Initital cost functions

Damage/Repair Cost Function

In the case of damage and subsequent repair, the cost would increase with the acceptable risk (i.e. decrease with reliability); a structure with low reliability level can be expected to sustain more severe damage than if it were designed with higher reliability. Therefore, the damage/repair cost will increase with risk as shown in Fig. 2.

The repair costs of specific structures that have been damaged from past earthquakes are essential for developing the damage/repair cost function. The global damages (in terms of the damage index) of the same structures can be calculated, from which a regression relation may be obtained between the damage/repair cost and the median global damage, as illustrated in Fig. 4 for the data from Mexico City.

Subjected to a given intensity of the earthquake for a designated site, the median damages of the structures designed above can be assessed, from which the damage/repair costs for the different designs are obtained from the pertinent cost relationship (e.g. Fig. 4). The damage exceedance probabilities for the same designs can be calculated, and thus the damage/repair cost as a function of the damage probability is developed for a given intensity of the earthquake. By varying the intensity, a family of similar damage/repair cost functions would be generated for the structure, as illustrated in Fig. 2.

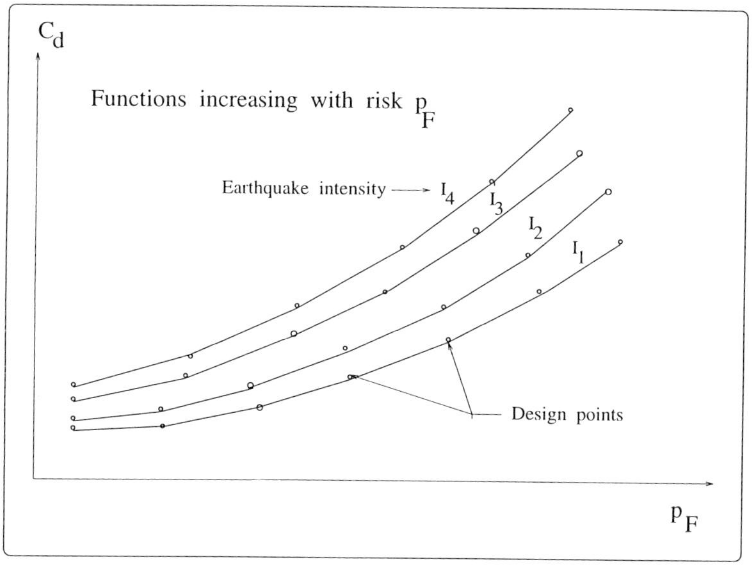

Fig. 2. Damage/repair cost functions

Other Cost Items

The other cost items that also constitute the total life-cycle cost of a structure would include the insurance premium, the economic loss caused by structural damage or collapse, and in the case of severe damage and collapse the cost of injury and life loss. The latter costs are direct or indirect consequences of structural damage or collapse.

There is increasing interests in the study of potential economic and other losses from seismic damage and collapse of structures (e.g. ATC-13, 1985; NRC, 1989; NRC, 1992). Data on building stocks or inventories that may be subject to risk are also being collected and examined (e.g. Jones, et al, 1987; Jones and Chang,

1993). This and similar information will be pertinent also to the development of improved criteria for design, particularly criteria for damage control in which the potential economic loss may be of principal importance.

Expected Life-Cycle Costs

For damage control or prevention, the total expected life-cycle cost function of a structure, therefore, would be

$$C_t = C_i + p_f C_d + p_f (C_c + C_e + C_{nd}) \quad \textbf{(6)}$$

where:

C_i	=	the initial cost function;
C_d	=	the damage/repair cost function;
C_c	=	loss of contents (partial);
C_e	=	the economic loss caused by a structural damage;
C_{nd}	=	the cost of non-disabling injury caused by a structural damage;
p_f	=	the probability of excessive damage.

Observe that C_c, C_e, and C_{nd} are, respectively, the loss of contents, economic loss and cost of injuries which are all consequences of a structural damage; therefore, the respective expected costs are obtained by multiplying the actual cost by p_f. All the cost items may be in terms of present value, based on a specified discount rate and an average inflation rate for a period equal to the half life of the structure.

In the case of life safety, the total expected life-cycle cost would be,

$$C_t = C_i + p_f C_r + p_f(C_c + C_e + C_{in} + C_f) \quad \textbf{(7)}$$

where:

C_i	=	the initial cost function
C_r	=	the replacement cost of the collapsed or demolished structure;
C_c	=	loss of contents (total);
C_e	=	the economic loss caused by structural collapse;
C_{in}	=	cost of all injuries (disabling and non-disabling);
C_f	=	cost of human fatality;
p_f	=	probability of structural collapse.

Eq. (7) is similar to Eq. (6), except that the damage/repair cost is C_d replaced by the replacement cost C_r, and the cost of human fatality C_f is added; moreover, the cost of injuries includes both disabling and non-disabling injuries, and C_c represents the total loss of the contents.

In either case, i.e. Eq. (6) or (7), the expected life-cycle cost function will be of the form shown in Fig. 3.

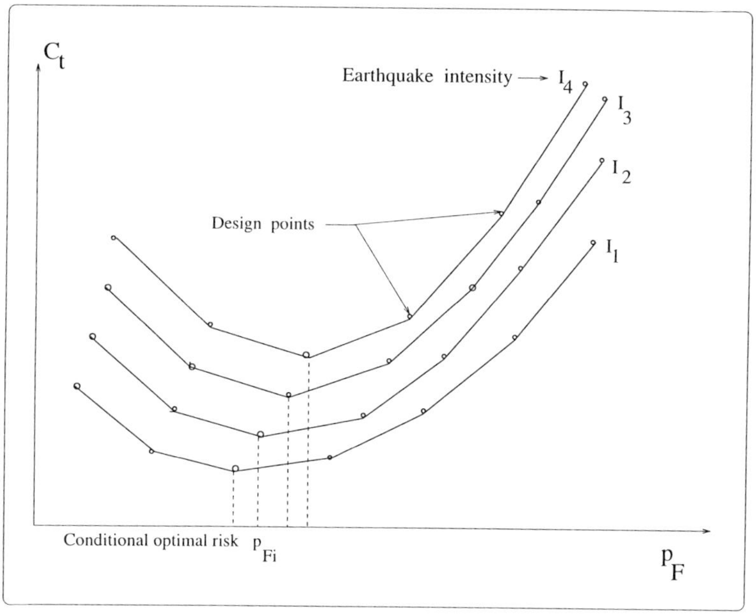

Fig. 3. Total expected life-cycle cost functions

DETERMINATION OF OPTIMAL RISKS

The optimal risks for damage control and life safety may be determined on the basis of the respective minimum expected life-cycle costs; i.e. minimizing C_t of Eq. (6) or (7) to obtain the corresponding p_f.

Damage Control

For a given intensity of earthquake, the expected life-cycle cost function can be determined according to Eq. (6); by varying the intensity, a family of expected life-cycle cost functions are generated as shown in Fig. 3. Observe that for a given intensity, there is an optimal probability of damage exceedance (optimal risk) corresponding to the minimum life-cycle cost; this conditional optimal risk will vary with the intensity as indicated in Fig. 3.

The seismic hazard at the site of a structure will naturally affect the safety and performance of the structure. Properly, therefore, the seismic hazard over the life of a structure should be taken into account. To include this factor, the conditional optimal risks determined above for the different intensities should be weighted by the respective probabilities of the different intensities, resulting in the expected optimal risk for damage control,

$$E(p_f) = \sum_{all\ i} (p_f | a_i) \cdot P(a_i) \tag{8}$$

where:

$(p_f | a_i)$ = conditional optimal risk at intensity a_i;
$P(a_i)$ = probability that the lifetime intensity is a_i.

Life Safety

Similarly, for different seismic intensities at the site, a family of expected life-cycle cost functions for life safety can be generated based on Eq. (7); the resulting cost functions would be similar to those shown in Fig. 3, except that the cost will be a function of the probability of collapse, $p_f=P(D_t> 1.0)$.

Observe that for a given intensity, there will also be a conditional optimal risk (probability of collapse) corresponding to the minimum life-cycle cost; this conditional optimal risk will also vary with the intensity, and thus by weighing the conditional optimal risks for the various intensities by the corresponding probabilities of the respective intensities, the expected optimal risk for life safety is obtained through Eq. (8), where $(p_f|a_i)$ is the optimal probability of collapse for given intensity a_i.

ILLUSTRATIVE APPLICATION

The approach is illustrated with the development of criteria for reinforced concrete buildings in Mexico City. Damage and cost data are based on reports available for the 1985 Mexico City earthquake. In particular, a class of framed R/C apartment buildings of five to ten stories is considered in this study.

Analyses of Damaged Buildings

A number of R/C buildings that were damaged to different degree of severity during the 1985 Mexico City earthquake were analyzed using the damage model described earlier. The respective median global damage indices were obtained; the calculated indices were then calibrated with the observed damages of the respective buildings. Also, based on the subsequent disposition of each of the buildings (some were repaired whereas others were demolished), the reparable damage level was determined to be at $D_t=0.5$.

For a limited number of buildings (five), the actual repair costs were reported (Guerrero, 1990). From these repair cost data and the calculated median global damage indices, D_m, for the corresponding buildings, the relationship between repair cost and the median global damage index is obtained as shown in Fig. 4.

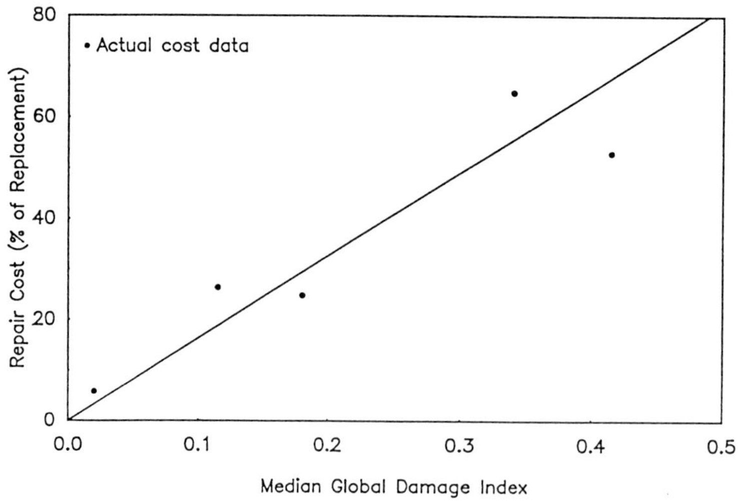

Fig. 4. Repair cost vs. damage relationship for Mexico City

Development of Cost Functions

For the initial cost function, a seven-story R/C building is designed in accordance with the 1987 Mexican Code (DDF, 1987). The resulting structure is assumed to represent a class of R/C frame buildings of 5-10 stories. The building plan and elevation are shown in Fig. 5. The same building is designed repeatedly using the several seismic base coefficients enumerated in Table 1.

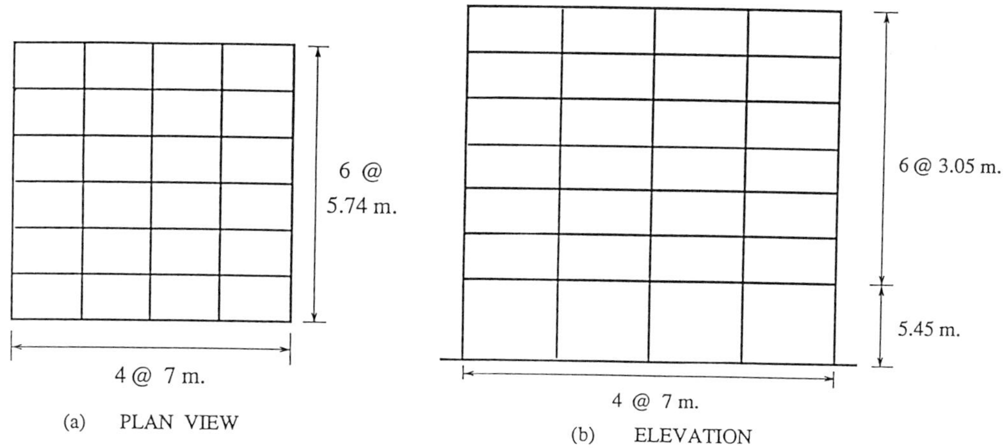

Fig. 5. Plan and elevation of 7-story R/C building

Table 1. Seismic base coefficients used in design

0.10	0.15	0.20	0.25	0.30	0.35	0.40
0.45	0.50	0.55	0.60	0.65	0.70	0.75

Based on the resulting design for each building, the cost for material is estimated, whereas the cost for construction is assumed to be the same as the material cost. The cost for design will most likely not change with the base coefficient, and therefore is a constant. The cost associated with the loss of contents is estimated as 0.5 C_i (for collapse) and $D_m C_i$ (for intermediate damage).

Each of the buildings is then subjected to the 1985 Mexico City earthquake (defined by the SCT spectrum) as shown in Fig. 6; the intensity (in terms of the peak ground acceleration, PGA) is varied. Under a given intensity, the probabilities of damage exceedance of the different designs are assessed, which can be related to the respective initial costs; thus yielding the initial cost function for the particular intensity, as indicated in Fig. 7. By varying the intensity, similar initial cost functions are obtained for other intensities, as illustrated in Figs. 7 and 8 for PGA of 0.15g and 0.25g, respectively.

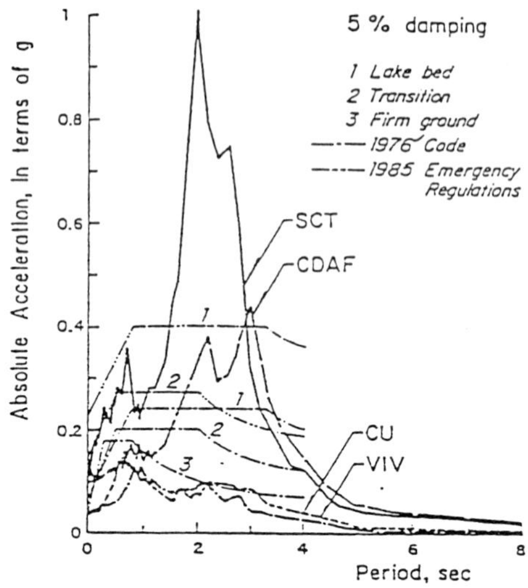

Fig. 6. Accleration response spectra for Mexico City

Under a given ground motion intensity, the median global damage indices of the various designs are also calculated as part of the above damage assessments. Then from the repair cost vs. damage relationship of Fig. 4, the required damage/repair costs for the respective designs are obtained. These costs are then related to the corresponding calculated damage exceedance probabilities, thus generating the required damage/repair cost function for the particular intensity. For other intensities, similar cost functions are obtained as illustrated also in Figs. 7 and 8 for two intensities.

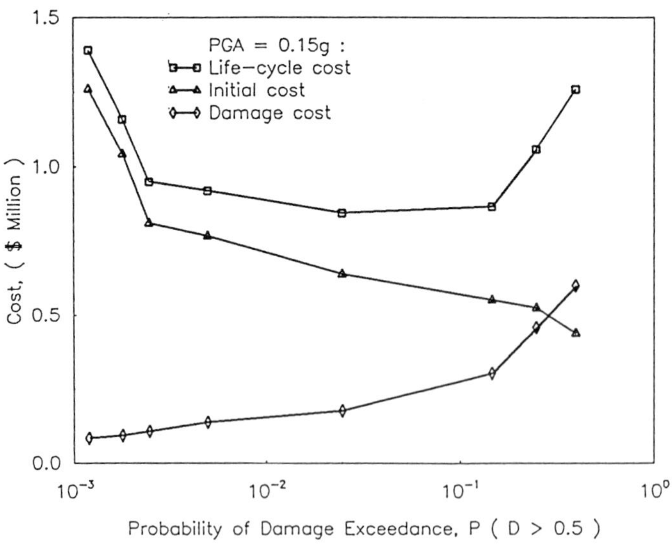

Fig. 7. Cost functions for damage control (PGA=0.15g)

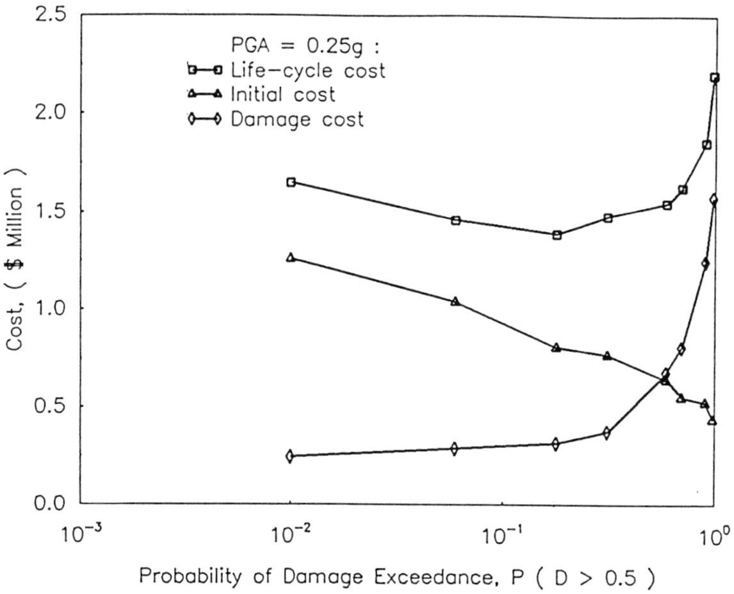

Fig. 8. Cost functions for damage control (PGA=0.25)

The economic loss caused by structural damage (business interruption, loss of revenue, etc) clearly will depend on the usage of the structure and the severity of the damage. This loss may be assumed to be a quadratic function of the median global damage, with the maximum loss corresponding to the damage at which the structure must be demolished and reconstructed; namely, at D_t=0.5. In general, the economic consequences of structural damage or collapse are difficult to determine (NRC, 1992), and available data are scarce; any loss estimation is, therefore, subject to high degree of uncertainty. In the case of a residential apartment building, the maximum economic loss may be estimated as the loss of rentals during the period of reconstruction. For this case, the maximum loss for Mexico City may be estimated based on the following assumptions:

(i) the reconstrucion period is 2 years;

(ii) the average monthly rental is $20 per square meter of floor area.

Thus, the maximum possible economic loss caused by structural damage for apartment buildings is,

$$C_e = 20 \ x \ 12 \ x \ 2A = 480A \tag{9}$$

Therefore, for a median global damage of D_m, the economic loss would be

$$C_e = 480A \left(\frac{D_m}{0.5} \right)^2 \tag{10}$$

Injury and Fatality

The cost associated with human injury and fatality may be estimated as follows.

For Mexico City, the cost of injuries may be estimated based on the following assumptions:

(i) the cost of each disabling injury is $117,000, whereas for each non-disabling injury the cost is $1667;

(ii) the average number of injuries per unit area of collapsed buildings is 0.0168 per sq. m (estimated from data reported in Tokyo Metropolitan Government, 1986; UNAM Inst. of Engineering, 1985);

(iii) 2/3 of all injuries are non-disabling, and 1/3 are disabling.

With these assumptions, the cost of non-disabling injury is,

$$C_{nd} = 0.0168 \; x \; \frac{2}{3} A \; x \; 1667 \left(\frac{D_m}{0.5} \right) = 37.33 \, A D_m \tag{11}$$

here A = the floor area of building in sq. m; whereas for total injuries, the cost would be,

$$C_{in} = 0.0168 \left(\frac{1}{3} \; x \; 117{,}000 \; + \; \frac{2}{3} \; x \; 1667 \right) A = 672 \, A \tag{12}$$

Finally, the cost of a human life may be estimated as the economic loss to society caused by the loss of a life; e.g., this may be estimated as the present value of the average contribution to the gross national product during the remaining productive life of an individual (Rosenblueth, 1976). For Mexico the average per capita income is $4680 (Europa Publ., 1993). The average number of fatality per unit area of a collapsed building is estimated to be 0.0122 per sq. m (Tokyo Metropolitan Government, 1986; UNAM Inst. of Engineering, 1985). On these bases, and assuming an average remaining productive life of 30 years, the cost associated with loss of life caused by structural collapse is,

$$C_f = 0.0112 A \; x \; 4680 \; x \; 30 = 1572 A \tag{13}$$

where A = floor area of a building in sq. m.

The insurance premium and legal cost are assumed to be independent of the acceptable risk, and thus will not effect the determination of the optimal risk.

The above cost items are current costs. In formulating the expected life-cycle cost functions, all the costs are converted into present value assuming a discount rate of 5% per annum and an average inflation rate of 7% over an expected duration of 25 years (one half of the building life of 50 years).

Determination of Optimal Risks

Damage Control -- For damage control, the life-cycle cost functions were developed in accordance with Eq. (6) for different potential intensities in Mexico City; the resulting cost functions are displayed in Fig. 9 in which the conditional optimal risk for each intensity is indicated at the corresponding minimum cost.

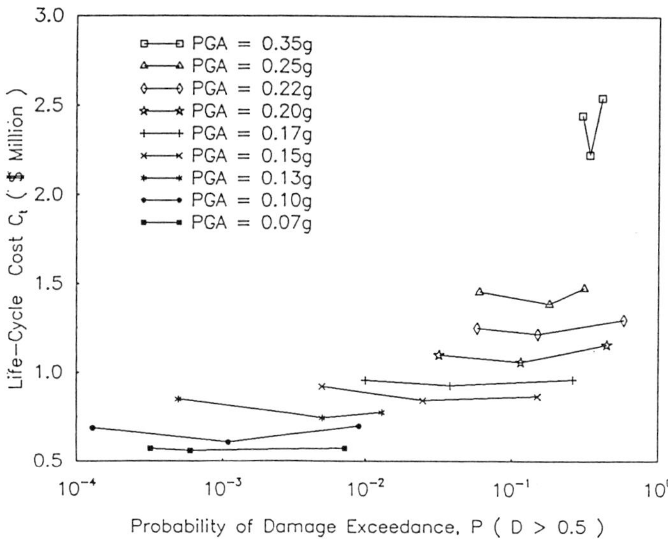

Fig. 9: Life-cycle cost functions with respective conditional optimal risks for damage control

The seismic hazard curve for Mexico City is shown in Fig. 10, presented for periods of one year and 50 years. This is generated following the model suggested by Esteva and Ruiz (1989) for the seismic hazard of Mexico City. By weighing or convolving the conditional optimal risks for the different intensities with the hazard curve for a life of 50 years, the expected optimal risk for damage control is obtained as,

$$E[P(D_t > 0.5)] = 0.105$$

By examining the family of life-cycle cost functions (see Fig. 9), and observing the design that corresponds to this expected optimal risk, the corresponding seismic base coefficient is determined to be 0.30. These results are for a tolerable damage of d=0.5 which is the limit of reparable damage. Obviously, other (lower) tolerable damage levels may be specified for which the corresponding expected optimal risks and base coefficients may similarly be obtained. If lower allowable damage is specified (e.g. d<0.5), the corresponding optimal base coefficient can be expected to increase (i.e. higher than 0.30).

Life Safety -- For life safety, the corresponding life-cycle cost functions were developed following Eq. (7) and using the cost components of Eqs. (9) through (13), again for various potential earthquake intensities in Mexico City; the resulting family of cost functions is shown in Fig. 11, in which the conditional optimal risks for the various intensities are indicated at the respective minimum costs. By convolving these optimal risks with the 50-year hazard curve for Mexico City, the expected optimal risk for life safety is obtained as

$$E[P(D_t > 1.0)] = 1.6 \times 10^{-3}$$

The associated seismic base coefficient is determined to be 0.45. Observe that in the new seismic code of Mexico City (revised in 1987), the base coefficient is 0.40 which is close to the optimal risk determined here.

The above results are summarized in Table 2.

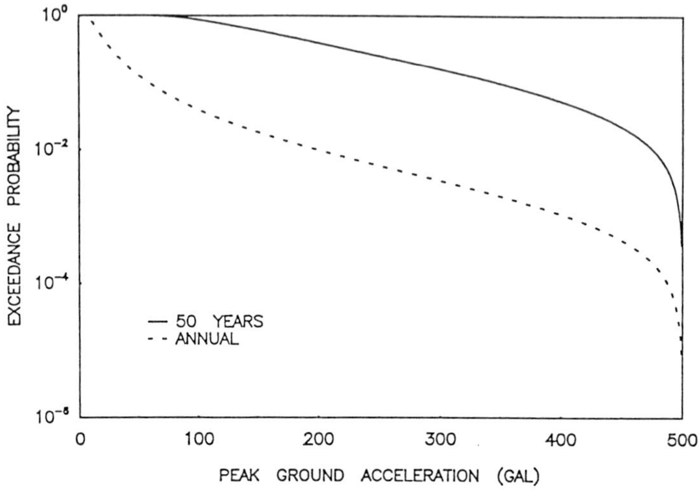

Fig. 10: Seismic hazard curve for Mexico City

Table 2. Summary of results

Limit State	Expected Optimal Risk*	Seismic Base Coefficient
d=0.5	0.105	0.3
d=1.0	1.6×10^{-3}	0.45

*For 50-year life

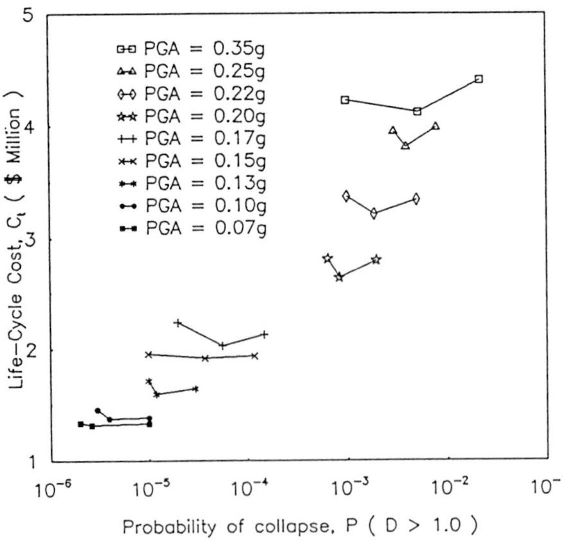

Fig. 11: Life-cycle cost functions with respective conditional optimal risks for life safety

SUMMARY AND CONCLUSIONS

The proper level of safety or reliability for earthquake-resistant design may be determined on the basis of minimum expected life-cycle cost. This is specially pertinent in the case of design for damage control. A systematic procedure for this purpose is proposed, involving the integration of technical and economic factors. The necessary models and tools are now available to allow the systematic determination of the optimal risks for life safety and damage control, from which the appropriate design criteria for safety and performance can be developed. The procedure is illustrated for R/C buildings in Mexico City.

ACKNOWLEDGEMENTS

This study was conducted under the partial support of the National Science Foundation under Grant CMS-9201812, and the fellowship support from the National Council for Research and Technology of Mexico. These supports are gratefully acknowledged.

REFERENCES

Applied Technology Council, ATC-13 (1985), "Earthquake Damage Evaluation Data for California".

Departamento del D.F. (1987), "Reglamento de construcciones para el Distrito Federal. Normas Tecnicas" (in Spanish), Mexico, D.F.

Esteva, L. and Ruiz, S. (1989), "Seismic Failure Rates of Multistory Frames", J. of Structural Engineering, ASCE, Vol. 115, No. 2, pp. 268-284.

Guerrero, J.J. (1990), "Vulnerabilidad Sismica de Edificios de Concreto Reforzado en la Delegacion Cuauhtemoc", Master thesis (in Spanish), DEPFI, UNAM, Mexico, D.F.

Institute of Engineering, UNAM (1985), "Efectos de los sismos de septiembre de 1985 en las construcciones de la Ciudad de Mexico, Aspectos estructurales", Report (in Spanish), November.

Jones, B.G and Chang, S.E. (1993), "Indirect Methods for Estimating the Built Physical Environment for Risk and Damage Assessment and Relief, Recovery, and Reconstruction Planning", Proc. National Earthquake Conference, Memphis, pp. 593-602.

Jones, B.G., Manson, D.M., Hotchkiss, C.M. and Savonis, M.J. (1987), "Determination of Building Stocks for Urban Vulnerability Analysis and Earthquake Loss Estimation", Working Paper, Cornell University.

National Research Council (1989), "Estimating Losses from Future Earthquakes", Panel Report, National Academy Press.

National Research Council (1992), "The Economic Consequences of a Catastrophic Earthquake", Proc. of a Forum, National Academy Press.

Park, Y.J. and Ang, A.H-S. (1985), "Mechanistic Seismic Damage Model for Reinforced Concrete", J. of Structural Engineering, ASCE, Vol. 111, No. 4, pp. 722-739.

Rosenblueth, E. (1976), "Towards Optimum Design Through Building Codes", J. of Structural Division, ASCE, Vol. 102, No. 3, pp. 591-607.

Europa Publications Limited (1993), "South America, Central America and the Caribbean, 1993", 4th Edition, England.

Tokyo Metropolitan Government (1986), "Report on the Investigation of the Earthquake in Mexico (September 19, 1985)".

Wang, G.S. and Ang, A.H-S. (1993), "Adaptive Kernel Method for Evaluating Structural System Reliability", Proc. ICOSSAR '93, Innsbruck, pp. 1495-1500.

INELASTIC BEHAVIOR AND LOAD DISPLACEMENT EQUATIONS OF LOW-RISE RC SOLID AND PERFORATED SHEAR WALLS

F. Y. Cheng* and K. Y. Lou**

Department of Civil Engineering, University of Missouri-Rolla, Rolla, MO 65401, U.S.A.

ABSTRACT

Inelastic behavior of low-rise reinforced concrete (RC) shear walls with or without openings is reported herein with analytical and experimental comparisons. The analytical studies are based on finite element formulation with N.S. Ottosen's failure criterion of concrete and nonlinear constitutive relationships. Isoparametric hexahedron elements with eight nodes are used for concrete while steel is assumed to be uniformly distributed over concrete elements. The walls studied have height to width ratios varying from 0.5 to 1.0 and an opening ratio (opening area to wall surface area) from 0 to 0.2, for which changes in horizontal and vertical steel ratio (from 0.25% to 1.0%) and the influence of diagonal steel are considered. Computer results compared reasonably well with experimental data, including failure mechanics, the relationship between shear and lateral displacement, the influence of openings on and the contribution of diagonal steel to a wall's resistance capacities at various loading stages. Equations based on regression are derived for calculating crack, yielding and ultimate loads and their associated displacements of various low-rise shear walls.

KEYWORDS

Earthquakes, shear walls, inelastic, ultimate, reinforced concrete, nonlinear finite elements, yielding

INTRODUCTION

Considerable interest has focused on low-rise shear walls (ratio of height to width less than 1) because they can effectively resist shear forces in building structures. As early as the 1950s, Galletly (1952) and Benjamin and Williams (1957) conducted experimental investigations to determine the strength of reinforced concrete (RC) shear walls under static monotonic loadings. Antebi et al. (1960) developed an experimental method to evaluate the dynamic strength of low-rise shear walls with boundary elements. All these studies led to design criteria for RC shear walls under blast loadings (Anderson et al., 1964).

Yamada et al. (1974) experimentally investigated RC shear walls with openings and boundary elements. Studies show that the occurrence of brittle fracture, or ductile fracture of shear wall, was mainly controlled by shape and size of opening. But the effect of diagonal steel on the wall's opening was not investigated.

*Curators' Professor of Civil Engineering; **Ph.D. Candidate

Barda et al. (1976) studied low-rise shear walls with flanged boundary elements under static monotonic or cyclic loadings. Test results indicate that the behavior of these walls was governed by shear.

Cardenas et al. (1980) tested shear walls with a height to width ratio of 1. Boundary columns were not considered; main variables were the amount and distribution of steel. From their results, low-rise rectangular walls were able to resist shear stress on the order of $10\sqrt{f_c'}$ psi. Watabe et al. (1989), along with others, quantitatively determined load-deflection for heavily reinforced concrete low-rise shear walls. Because specimens were column-bounded and mainly subjected to monotonically increasing load, equations could not be used in isolated walls with openings. ACI 318-89 (1989), based on Cardenas et al. (1973), Barda et al. (1977) and other research work, gives equations to calculate shear strength, shear reinforcement, minimum horizontal and vertical reinforcement ratio. The equation of shear strength cannot be applied to low-rise shear walls because the effect of the wall's height is not taken into account.

Since 1988, Sheu at National Cheng-Kung University (NCKU) in Taiwan and Cheng at University of Missouri-Rolla (UMR) have been engaged in joint analytical and experimental studies of isolated low-rise walls with or without openings. Their emphasis is on inelastic modeling for hysteresis rules and backbone curves as well as analysis or prediction of the capacity of shear wall structures (Cheng et al., 1989, 1991, 1992 and 1993).

The work presented herein is to extend the senior author's work on inelastic behavior of perforated isolated low-rise shear walls and load displacement equations at crack, yielding and ultimate loading stages. For designing a shear-wall structure, the aforementioned equations are needed to evaluate lateral loads and displacement of the constituent walls of the system. For inelastic analysis of a shear-wall structure, the load-displacement curve (backbone curve) of a shear wall must be obtained in order to employ hysteresis rules for that wall.

An algorithm of nonlinear finite element analysis is used to determine the behavior of low-rise shear walls with or without openings. Analytical results are first compared with experimental data, and load displacement equations are then formulated through regression procedures. Comparisons are based on test data by Vecchio (1990) for shear panels with or without openings and test results by Sheu (1988) with changes of height to width ratio, vertical and horizontal steel ratio, the amount of diagonal steel, and opening ratio. A large number of load-displacement curves are plotted and compare well with experimental data. Equations based on regression are then developed for determination of crack, yielding and ultimate loads and displacements, which can be conveniently used to predict the relationship between shear capacity and lateral displacement of various low-rise shear walls with or without openings.

IMPLEMENTATION OF NONLINEAR FINITE ELEMENTS

For inelastic analysis of RC walls with satisfactory prediction of concrete crack location, crack propagation, steel yielding, ultimate capacity and deformation, a good selection of failure criterion and constitutive relationship is essential and should have the following characteristics: (1) failure surface is smooth and convex; (2) meridians of failure surface in the principal stress coordinate system (see Appendix I) are parabolic and do not intersect the negative hydrostatic axis; (3) trace of failure surface in the deviatoric plane varies from nearly triangular to a circular shape with increase of hydrostatic pressure.

Failure criterion for concrete proposed by Ottosen (1977) has the aforementioned quality and is therefore adopted in this study. Constitutive model by Ottosen (1979) for determination of secant modulus of concrete is also adopted here (see Appendix I). The failure criterion and constitutive relationship for both concrete and reinforcement are incorporated into a nonlinear finite element model. Nonlinear material behavior is accommodated by using the secant stiffness method. An isoparametric hexahedral element with eight nodes and an elastoplastic model are employed for concrete and reinforcement, respectively. Reinforcement is assumed to be uniformly distributed over the concrete element with a perfect bond between them before concrete cracks. If concrete cracks, a smeared cracking model is applied to deal with the subsequent behavior of concrete. A FORTRAN computer program is developed for either personal computer or UNIX system. Figure 1 gives a flow chart to describe the nonlinear finite element program.

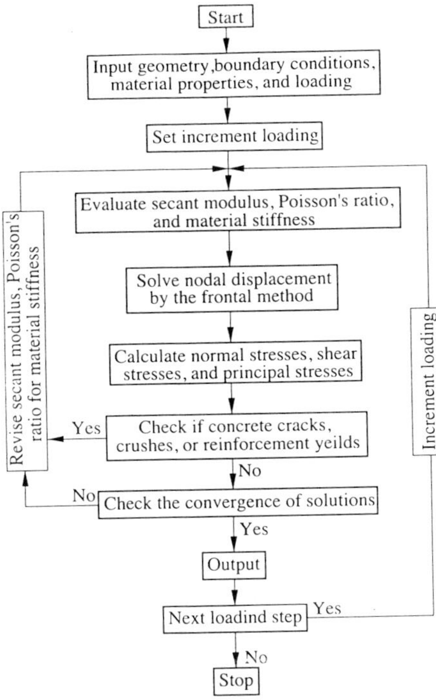

Fig. 1 Flow Chart for Nonlinear Finite Element Program

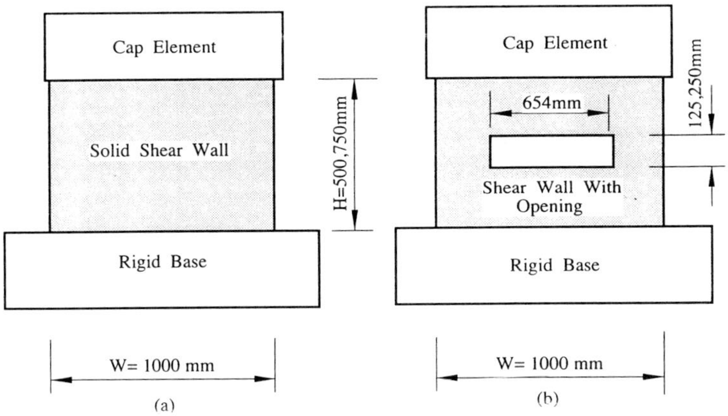

Fig. 2 NCKU Shear Wall: (a) Solid Shear Walls; (b) Shear Walls with Openings

EXPERIMENTAL DESCRIPTION

Low-rise shear walls without boundary elements tested by Sheu (1988) are rectangular in cross section, and have a depth of 100 mm and a width of 1000 mm. Their height varies from 500 mm to 750 mm, which gives a ratio of height to width from 0.50 to 0.75. A perforation is centered for shear walls with openings. Opening size is 654 mm x 125 mm for shear walls 500 mm high, and 250 mm x 654 mm for those 750 mm high. These openings approximate the size of windows and are thus common in shear wall structures. Evenly spaced horizontal and vertical reinforcements range from D10 (#3) to D13 (#4). Diagonal reinforcement for shear walls with openings is D13 (#4).

A cap element 300 mm x 300 mm in cross section is cast at the top of shear wall in order to apply uniform lateral force along the specimen's top surface. Two jacks are horizontally placed at each side of the cap element to apply monotonic or cyclic lateral force at the centerline of this element. A 300 mm x 550 mm rigid base is cast for the base of the shear wall while the base itself is bolted to the test bed. Both cap element and rigid base are heavily reinforced to prevent unexpected failure (see Fig. 2). Figure 3 shows the test setup.

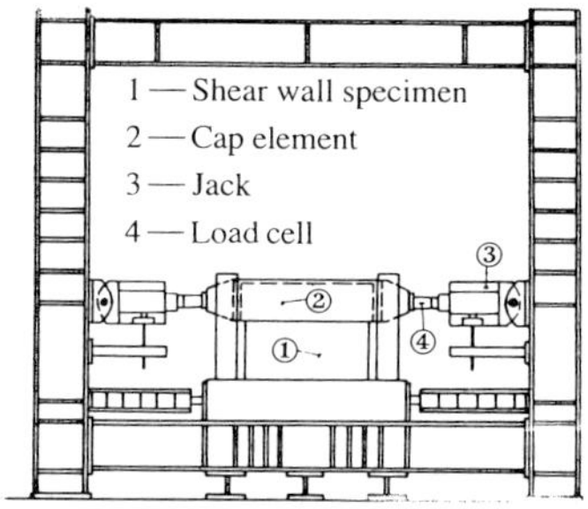

Fig. 3 Test Setup

VERIFICATION OF NONLINEAR FINITE ELEMENT RESULTS

As shown by the literature review, there are scant test data available to determine loads and displacements for shear walls with or without openings. To predict the behavior of shear walls, it is essential to adopt nonlinear finite elements because this approach is economic and reliable. For verifying the algorithm proposed by authors, a comparison is made with test data conducted by Vecchio et al. (1990). Vecchio tested nine panels divided into three groups. Each group contains a solid panel, a panel with a 150-mm square opening at the center, and a panel with the same opening additionally reinforced around the sides. Load condition varies from pure shear (group I), combined shear and biaxial compression (group II), and combined shear and biaxial tension (group III). Dimensions of all panels are 890 mm x 890 mm x 70 mm. Reinforcement is evenly placed in two layers and oriented parallel to the sides of the panel. Vertical and horizontal reinforcement ratios are 0.82% and 1.65%, respectively. Figure 4 presents these specimens in detail.

Figure 5 shows computer and experimental results from shear panels PC1A, PC2 and PC3. In general, computer results conform well with test data. These results depict the entire behavior of shear panels under uniform pure shear. Such behavior includes precracking stiffness, cracking load and deformation, post-cracking deformation, and ultimate load capacity. Before cracking, deformation of the shear panel is approximately proportional to the load. Cracking loads are close to experimental data. After cracking, the stiffness of panel PC2 is lower than that of test data. This is because concrete is assumed to be completely brittle after it cracks. Also, a perforation without additional reinforcement weakens the cross section and leads to reduction of stiffness after cracks appear. But cracks can attain stability. Force released by cracked concrete is carried by reinforcement, and there is less deformation under the same increment of

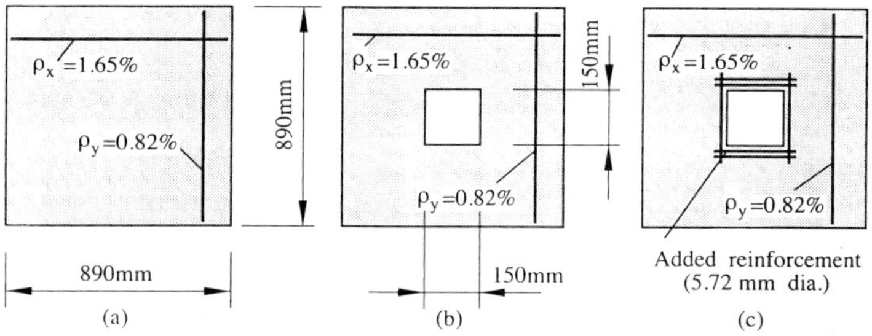

Fig. 4 Specimen Details by Vecchio: (a) Solid Shear Panel; (b) Shear Panel with Opening; (c) Shear Panel with Opening and Added Reinforcement

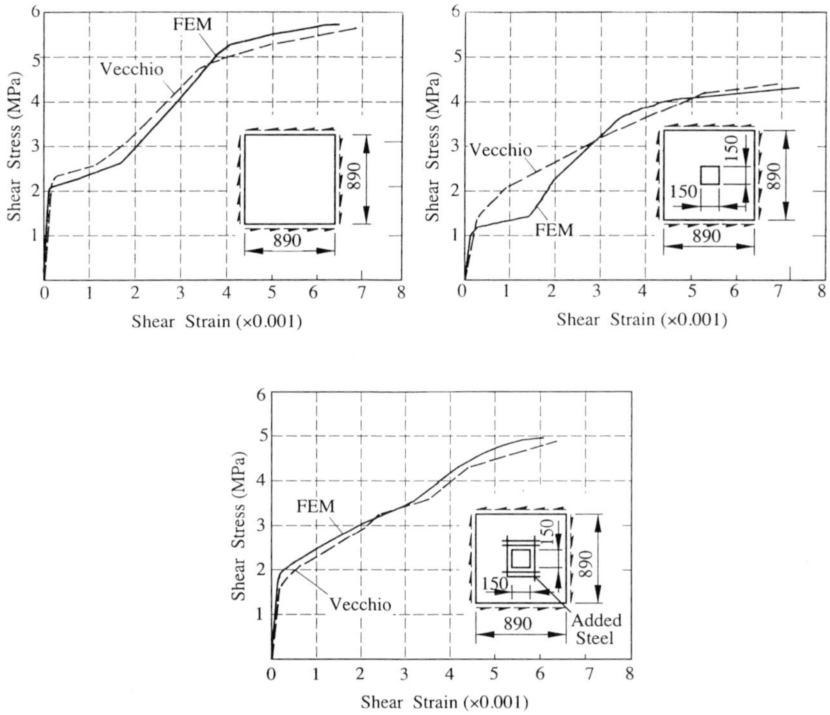

Fig. 5 Comparison of Test and Computer Results: (a) PC1A; (b) PC2; (c) PC3

pure shear. Thus, near the ultimate stage, the computer curve converges with experimental data. A good approximation is also observed in other panels under combined shear and compression as well as combined shear and tension.

THEORETICAL SOLUTIONS FOR LOW-RISE SHEAR WALLS WITH OR WITHOUT OPENINGS

Shear walls chosen for nonlinear finite element analysis are identified as SW-OE, SWO-7E, SW-9E and SWO-15E, among a number of NCKU test specimens. Material properties of these walls are summarized in Table 1. D10 and D13 are equivalent to No. 3 and No. 4 bars, respectively. Specimen details of SW-0E, SW0-7E, SW-9E and SW0-15E are shown in Fig. 6.

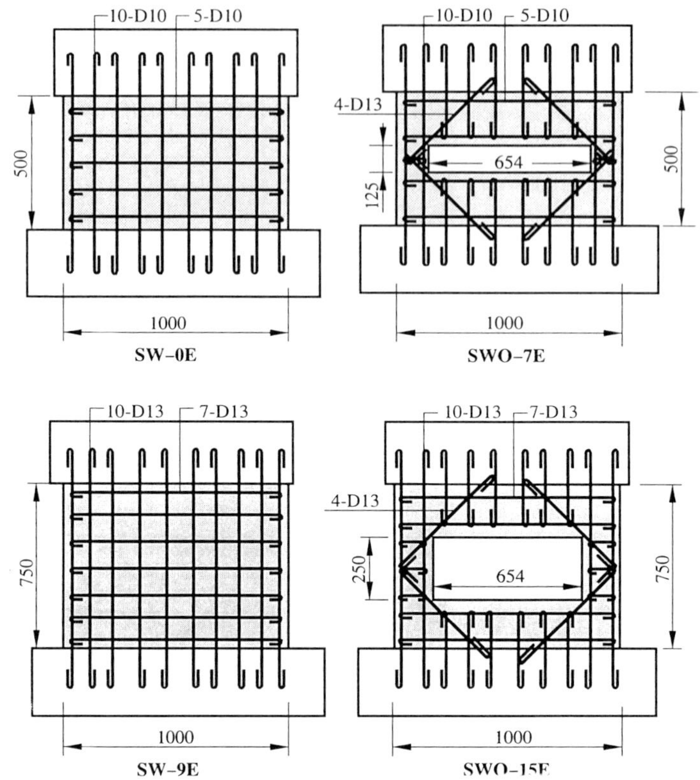

Fig. 6 NCKU Specimen Details: (a) SW-OE; (b) SWO-7E; (c) SW-9E; (d) SWO-15E

Table 1. Selected NCKU Shear Walls

Wall	WxHxB (mm)	f_c' (N/mm^2)	Vertical		Horizontal		Diagonal Bars	f_y (N/mm^2)	$\frac{H}{W}$
			Bars	ρ_v (%)	Bars	ρ_h (%)			
SW-OE	1000x500x100	25.4	10D10	0.7133	5D10	0.7133	None	510	0.5
SWO-7E	1000x500x100	32.8	10D10	0.7133	5D10	0.7133	D13	510	0.5
SW-9E	1000x750x100	29.4	10D13	1.267	7D13	1.183	None	461.7	0.75
SW0-15E	1000x750x100	27.9	10D13	1.267	7D13	1.183	D13	461.7	0.75

Figure 7 shows the layout of finite element meshes for shear walls SWO-7E and SW-9E, respectively. For SWO-7E, there are 52 concrete and 52 steel elements uniformly distributed over concrete elements. For SW-9E, the specimen is divided into 80 concrete and 80 steel elements. In order to simulate experimental conditions, displacements at the bottom of specimens are restrained. Lateral force is uniformly distributed at each node on top of the shear walls. Test and computer results for load-displacement curves of specimens SWO-7E and SW-9E are compared in Fig. 8.

At the beginning of loading, the increment of displacement for shear wall SWO-7E is proportional to the increment of loading. During this stage, both finite element analysis and test results present linear behavior. When loading increases to 38.4 KN, the concrete elements at the upper right and lower left corner of the opening begin to crack. At the same time, a crack at the right bottom corner appears.

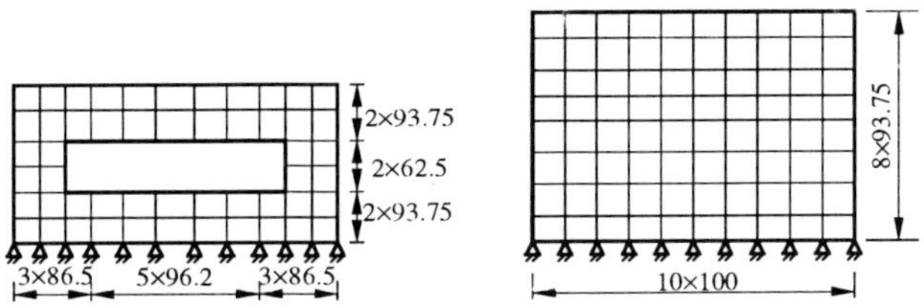

Fig. 7 Finite Element Mesh: (a) SWO-7E; (b) SW-9E

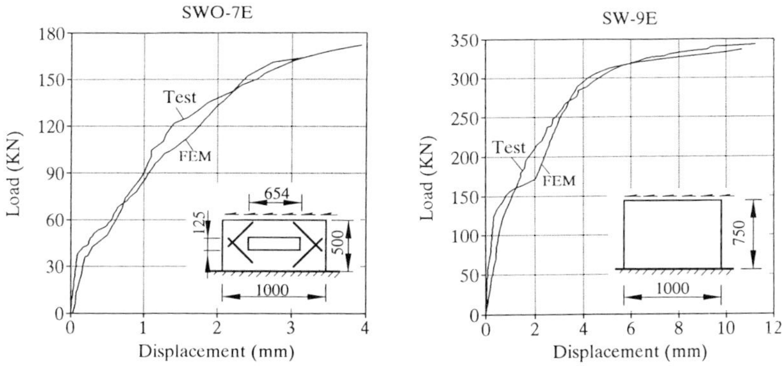

Fig. 8 Load and Displacement Curves: (a) SWO-7E; (b) SW-9E

Cracking load for the test is 36.3 KN. The computer's crack pattern is very similar to that of tests as shown in Fig. 9.

As lateral load increases, cracks propagate. Stiffness of the shear wall decreases and the load-displacement curve deviates from a straight line. Consequently, test and analytical curves exhibit nonlinear behavior. Vertical steel begins to yield around 142.2 KN while test value is 121.7 KN. After this stage, adding a small increment of load can lead to a large increment of displacement. Ultimate load for finite element analysis is 172 KN, which is close to that of test data (163.9 KN). Theoretical crack patterns under the ultimate load are similar to the test crack pattern (see Fig. 9). Figure 10 shows crack patterns for specimen SW-9E. Again, computer results conform well to test results.

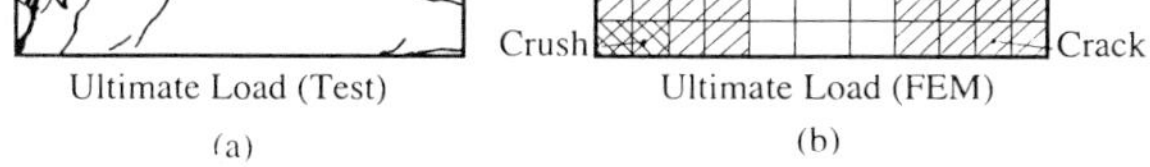

Fig. 9 Crack Patterns of SWO-7E under Cracking and Ultimate Load

To prove that the load-displacement curve of shear walls with openings will converge to that of solid shear walls, three shear walls with different openings are considered. Wall dimensions are 1000 mm wide, 500 mm high, and 100 mm thick. One is a solid wall (SW-OE); the other two have centered openings of 100 x 200 mm and 100 x 100 mm. Horizontal steel ratio ρ_h and vertical steel ratio ρ_v are 0.7133%, the same

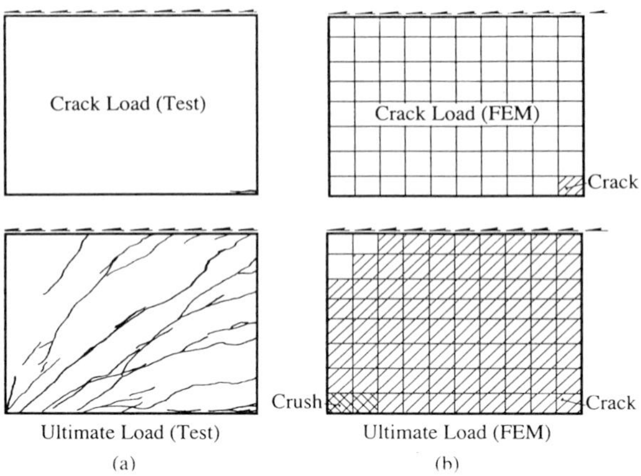

Fig. 10 Crack Patterns of SW-9E under Cracking and Ultimate Load

as for all walls. Other material properties of SW-OE are shown in Table 1. Figure 11 represents a comparison among the three cases. One may observe that their load-displacement curves are similar. If the size of the opening decreases, then ultimate load increases and ductility decreases. Note that the load-displacement curves of perforated and solid shear walls converge. Also note that curve ‘a’ compares well with experimental data as shown later in Fig. 16.

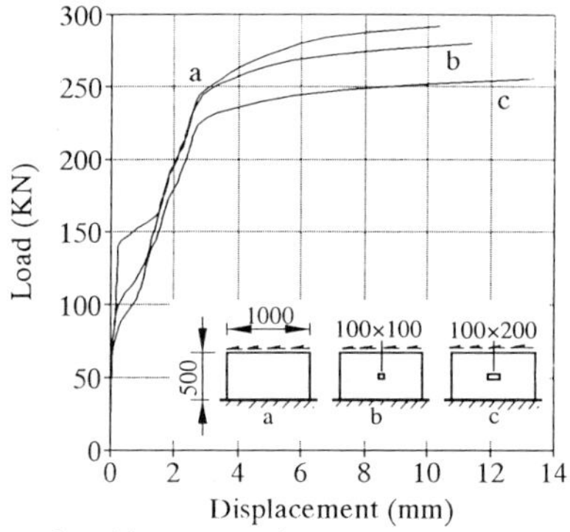

Fig. 11 Load Displacement Curves with Change of Opening

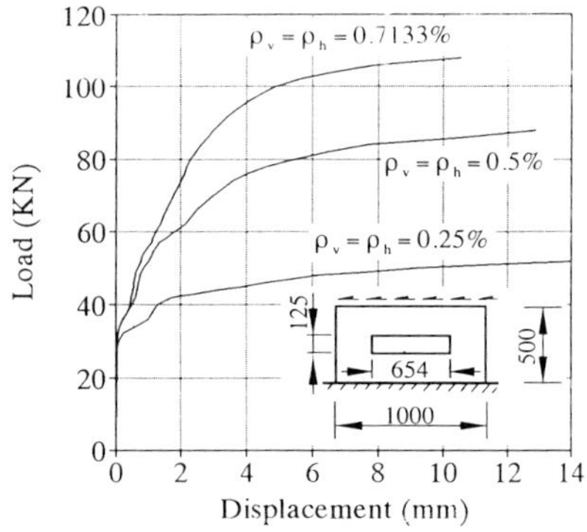

Fig. 12 Load Displacement Curves with Change of Steel Ratio

Figure 12 shows a group of load-displacement curves of shear walls with changes in reinforcement. Opening size is 654 x 125 mm. Vertical and horizontal steel ratios are the same for each case; they are 0.25%, 0.5% and 0.7133%. Compressive strength of concrete is 25.4 N/mm^2 and tensile strength of reinforcement is 510 N/mm^2.

Before concrete cracks, the stiffness of walls is almost the same (see Fig. 12). Reinforcement thus plays a small role in resisting shear forces at the pre-cracking stage. After cracking, the stiffness of load-displacement curves increases with an increase in vertical and horizontal reinforcement. This means that an increase in reinforcement can effectively reduce the propagation of cracking and can increase the wall's capacity to resist lateral force. A decrease in vertical and horizontal reinforcement leads to an increase in ductility and a decrease in shear capacity.

A comparison of changing diagonal reinforcement for shear walls with openings is presented in Fig. 13. Overall dimensions are 1000 mm wide, 500 mm high and 100 mm thick with an opening of 400 x 125 mm. Vertical and horizontal steel ratios are 0.50% for all cases. Diagonal reinforcement varies from 0, D10

(71.33 mm^2), D13 (126.7 mm^2) to D16 (197.93 mm^2), and has a 45° angle with a horizontal line. Compressive strength of concrete is 25.4 N/mm^2, and tensile strength of reinforcement is 510 N/mm^2. At the pre-cracking stage, an increase in diagonal reinforcement hardly changes the shear wall's stiffness. But at the post-cracking stage, diagonal reinforcement has a significant effect on stiffness. A reduction in diagonal reinforcement leads to a decrease in stiffness. Ultimate load increases with an increase in diagonal reinforcement.

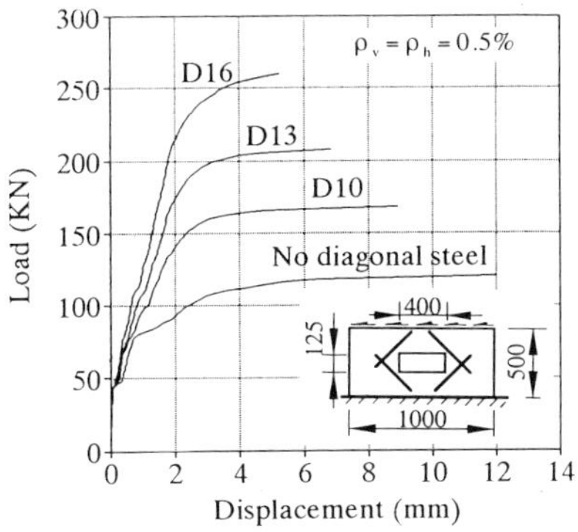

Fig. 13 Load Displacement Curves with Change of Diagonal Steel

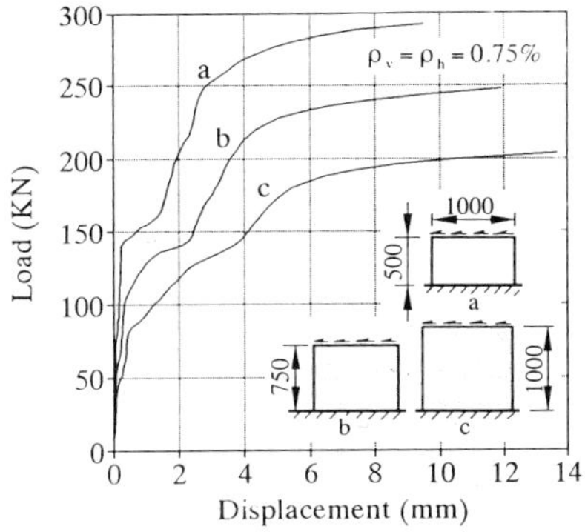

Fig. 14 Load Displacement Curves with Change of Height of Walls

Figure 14 compares the analytical results by changing the wall's height. Depth and width of the three shear walls are identical at 100 mm, 1000 mm, respectively. Height varies from 500 mm, 750 mm to 1000 mm. Vertical and horizontal steel ratios are 0.75%. Compressive strength of concrete is 25.4 N/mm^2 while tensile strength of reinforcement is 510 N/mm^2. An increase in ultimate loading and a decrease in ductility due to the reduction in height are observed.

EQUATIONS OF LOAD-DISPLACEMENT CURVES FOR SHEAR WALLS WITH OR WITHOUT OPENINGS

Further computer results are developed for the determination of load-displacement curves of shear walls with or without openings. For solid shear walls, height, vertical and horizontal steel ratio, compressive strength of concrete and tensile strength of reinforcement are considered as parameters shown in Table 2. For perforated walls, Table 3 summarizes the dimensions of walls and openings and material properties as well as vertical, horizontal and diagonal steels.

Table 2. Variation of Parameters' Material Properties for Solid Walls

WxHxB (mm)	Vertical and Horizontal Steel Ratio $\rho_v = \rho_h$ (%)	f_c' (N/mm^2)	f_y (N/mm^2)	$\frac{H}{W}$
1000x500x100	0.25, 0.5, 0.75, 1.0	25.4	510	0.5
1000x750x100	0.25, 0.5, 0.75, 1.0	25.4, 29.4	510, 461.7	0.75
1000x1000x100	0.25, 0.5, 0.75, 1.0	25.4	510	1

There are 16 cases for solid and 132 cases for perforated shear walls based on test data and computer results. For instance, in a solid wall of 1000 x 500 x 100 mm having compressive strength of concrete, 25.4 N/mm^2, and steel strength, 510 N/mm^2, there are four different vertical and horizontal steel ratios from 0.25% to 1.0%. Consequently, this yields four cases. When compressive strength of concrete, 25.4 N/mm^2, and steel strength, 510 N/mm^2, are used as one group while compressive strength of concrete, 29.4 N/mm^2, and steel strength, 461.7 N/mm^2, are used as another group for the 1000 x 750 x 100 mm wall, we then have eight cases. For perforated shear walls with a size of 1000 x 500 x 100 mm, opening

Table 3. Variation of Parameters' Material Properties for Perforated Walls

WxHxB (mm)	Opening Ratio ρ_o	Vertical and Horizonal Steel Ratio $\rho_v = \rho_h$ (%)	Diagonal Reinforcement	f_c' (N/mm^2)	f_y, f_{yd} (N/mm^2)	$\frac{H}{W}$
1000x500x100	0.1, 0.1635, 0.2	0.25, 0.5, 0.7133	0, D10(#3), D13(#4), D16(#5)	25.4	510	0.5
1000x750x100	0.1, 0.15, 0.218	0.25, 0.5, 0.75, 1.0	0, D10(#13), D13(#4), D16(#5)	27.9	461.7	0.75
1000x1000x100	0.1, 0.15, 0.2	0.25, 0.5, 0.75, 1.0	0, D10(#3), D13(#4), D16 (#5)	27.9	461.7	1.0

ratio is 0.1, vertical and horizontal steel ratio is 0.25, and we have four cases due to four different diagonal reinforcements varying from zero to D16. With changes in dimensions, opening ratio, vertical and horizontal steel ratio, and diagonal reinforcement, there are 132 cases. Cracking load and displacement, yielding load and displacement, and ultimate load and displacement are obtained by regression and given as follows.

Cracking load P_{cr} in unit N

$$P_{cr} = \left\{ 6.233+10.827\rho_o-101.78\rho_o^2+(6.398+9.834\rho_o)\left(\frac{W}{H}\right)+(1.542-24.618\rho_o)\left(\frac{W}{H}\right)^2\right\}10^{-3}WBf_c' \tag{1}$$

Cracking displacement Δ_{cr} in unit mm

$$\Delta_{cr} = \left[1+4.4705\rho_o+125.8122\rho_o+(6.493-99.8875\rho_o+174.0799\rho_o^2)\left(\frac{H}{W}\right)-(6.5325-98.0783\rho_o + 273.529\rho_o^2)\left(\frac{H}{W}\right)^2\right]\frac{\left(1+\dfrac{3E_tIK}{GA_cH^2}\right)}{3E_tI}H^3P_{cr} \tag{2}$$

Yielding load P_y in N

$$\begin{aligned} P_y = & \left\{1 - 44.366\rho_o - 249.074\rho_o^2 + (15.433 - 129.775\rho_o + 1103.385\rho_o^2)\left(\frac{W}{H}\right)\right. \\ & \left. + (1.650 + 26.705\rho_o - 461.869\rho_o^2)\left(\frac{W}{H}\right)^2\right\}10^{-3}WBf_c' + \left\{0.272-8.151\rho_o + 34.180\rho_o^2\right. \\ & \left. + (0.092 + 11.279\rho_o - 52.171\rho_o^2)\left(\frac{W}{H}\right) + (0.001 - 4.106 + 17.960\rho_o^2)\left(\frac{W}{H}\right)^2\right\}\rho WBf_y \\ & + \left[0.758 - 0.667\left(\frac{W}{H}\right) + 0.169\left(\frac{W}{H}\right)^3\right] nA_d f_{yd}\cos\phi \end{aligned} \tag{3}$$

Yielding displacement Δ_y in mm

$$\Delta_y = \Big\{ 0.0024 - 0.0408\rho_o + 0.2070\rho_o^2 + \left(0.7442 + 0.0085\rho_o - 21.5337\rho_o^2\right)\rho + \Big[0.4639 - 6.0695\rho_o + 19.1698\rho_o^2 - \left(9.537 - 1.9278\rho_o - 339.1297\rho_o^2 \right)\rho - \left(0.459 - 6.8013\rho_o + 22.7969\rho_o^2\right) \left(\frac{H}{W} - 0.1\right)^{10\rho} \Big] \frac{f_c'}{f_y} + \left(6.2330 + 32.8707\rho_o - 120.6860\rho_o^2 \right)\frac{f_c'}{f_y}\rho_d \Big\} \sqrt{HW} \quad (4)$$

Ultimate load P_u in N

$$P_u = \Big\{ 1 + 505.033\rho_o - 2118.820\rho_o^2 + \left(11.843 - 601.298\rho_o + 2564.563\rho_o^2\right)\left(\frac{W}{H}\right) + \left(6.177 + 106.819\rho_o - 660.897\rho_o^2\right)\left(\frac{W}{H}\right)^2 \Big\} 10^{-3} WBf_c' + \Big\{0.388 - 10.383\rho_o + 45.341\rho_o^2 + \left(0.024 + 13.269\rho_o - 62.696\rho_o^2\right)\left(\frac{W}{H}\right) + \left(0.001 - 4.529\rho_o + 20.497\rho_o^2\right)\left(\frac{W}{H}\right)^2 \Big\} \rho WBf_y + \left[0.024 \left(\frac{W}{H}\right) + 0.094 \left(\frac{W}{H}\right)^3 \right] nA_d f_{yd} \cos\phi \quad (5)$$

Ultimate displacement Δ_u in mm

$$\Delta_u = \Big\{ 0.0178 - 0.0564\rho_o + 0.2050\rho_o^2 - \left(0.9243 + 0.0173\rho_o - 52.3631\rho_o^2\right) \rho + \Big[0.3008 - 0.8848\rho_o - 24.1924\rho_o^2 + \left(4.6108 + 62.9711\rho_o - 1209.9000\rho_o^2\right) \rho - \left(0.2844 - 1.3419\rho_o - 25.9565\rho_o^2\right) \left(\frac{H}{W} - 0.1\right)^{10\rho} \Big] \frac{f_c'}{f_y} - \left(0.8176 - 4.9281\rho_o + 20.429\rho_o^2\right) \frac{f_c'}{f_{yd}} \rho_d^{\frac{1}{4}} \Big\} \sqrt{HW} \quad (6)$$

where

ρ_0 = opening ratio, $\frac{A_o}{A_c}$; A_0 = opening area (mm^2); A_c = area of shear wall H x W (mm^2); H = height of shear wall (mm); W = width of shear wall (mm); f_c' = compressive strength of concrete (N/mm^2); ρ = smaller value of horizontal or vertical reinforcement ratio; B = depth or thickness of shear wall (mm); f_y = tensile strength of reinforcement (N/mm^2); A_d = area of one diagonal reinforcement (mm^2); F_{yd} = tensile strength of diagonal reinforcement (N/mm^2); n = number of diagonal reinforcements at cross section W x B; ϕ = angle between diagonal reinforcement and horizontal direction; E_i = initial modulus of concrete (N/mm^2); G = shear modulus of concrete $G = \frac{E_i}{2(1+\nu_i)}$ (N/mm^2); ν_i = Poisson's ratio; I = moment of inertia (mm^4); K = numerical factor = 1.2; ρ_d = equivalent diagonal reinforcement ratio

$$\rho_d = \frac{nA_d}{WH} \cos\phi$$

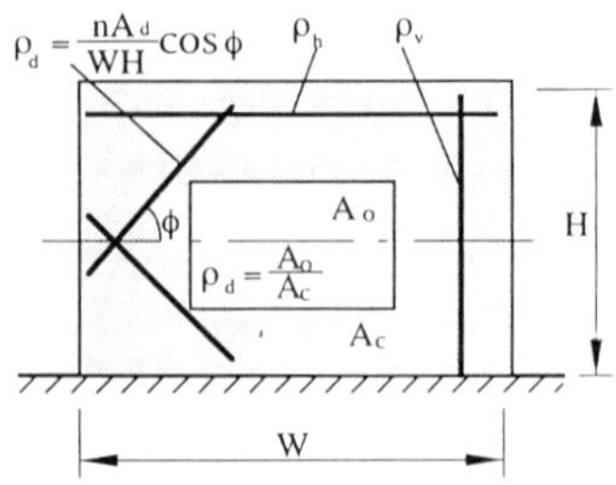

Fig. 15 Definition of Sympols for Shear Wall

As shown by experimental data (Sheu, 1988) and computer results, when vertical reinforcement begins to yield, the slope of the shear wall's load-displacement curve is not greatly reduced. This is because uniform distribution of vertical reinforcement and a relatively large cross section W x B lead to a slow rotation of the shear wall. Yielding load is thus defined as the beginning of significant change in the slope of the load-displacement curve.

Note the load-displacement curve. A straight line is used to depict the relationship between load and displacement before cracking, and between yielding and ultimate load. For the range between cracking load and yielding load, a second-order polynomial is adopted to determine the relationship between load and displacement.

$$P = \alpha\Delta^2 + \beta\Delta + \gamma \tag{7}$$

In Eq. 7, displacement, Δ, equals cracking displacement, Δ_{cr}, and the load P is cracking load P_{cr}. When Δ is equal to yielding displacement, Δ_y, P becomes yielding load, P_y, and the slope at Δ_y equals that of a straight line for yielding and ultimate load. Parameters α, β, and γ can be derived as

$$\alpha = \frac{\dfrac{\Delta_y - \Delta_{cr}}{\Delta_u - \Delta_y}(P_u - P_y) + P_{cr} - P_y}{(\Delta_y - \Delta_{cr})^2}$$

$$\beta = \frac{2\Delta_y(P_y - P_{cr}) - \dfrac{\Delta_y^2 - \Delta_{cr}^2}{\Delta_u - \Delta_y}(P_u - P_y)}{(\Delta_y - \Delta_{cr})^2}$$

$$\gamma = P_{cr} - \frac{\Delta_{cr}}{(\Delta_y - \Delta_{cr})^2}\left\{(P_y - P_{cr})(2\Delta_y - \Delta_{cr}) - \frac{\Delta_y - \Delta_{cr}}{\Delta_u - \Delta_y}(P_u - P_y)\Delta_y\right\}$$

Proposed equations for load-displacement of low-rise shear walls with or without openings have the following characteristics: (1) Shear walls with larger openings have lower ultimate shear capacity and appear more ductile; (2) As the height of shear walls increases, a decrease in ultimate load and an increase in ductility are observed; and (3) A reduction in vertical, horizontal or diagonal reinforcement leads to an increase in ductility and a decrease in ultimate shear capacity. These characteristics hold true for shear walls with or without openings.

Figure 16 compares test data, computer results, and proposed equations. These equations conform well to test data and computer results. A prediction of load displacement for low-rise shear walls is also given. Four shear walls with or without openings, and of different heights, are used for comparison. Their material properties are shown in Table 4. Good conformance is also observed in Fig. 17 where the initial elastic modulus of concrete is 25,166 N/mm^2, and Poisson's ratio for concrete is 0.2.

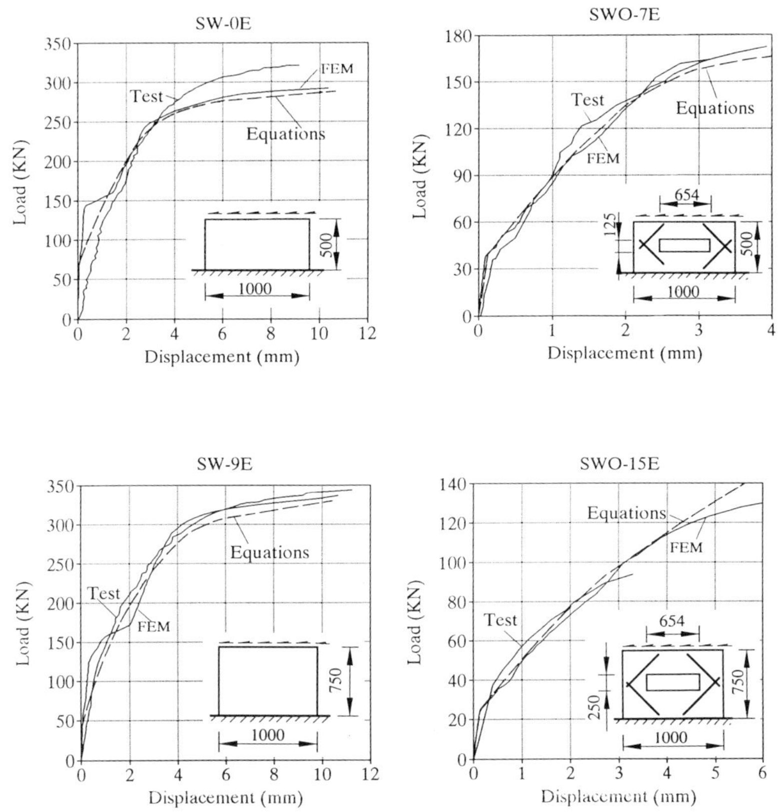

Fig. 16 Comparison between Test, Computer Results and Proposed Equations: (a) SW-OE; (b) SWO-7E; (c) SW-9E; (d) SWO-15E

Table 4. Properties of Solid and Perforated Shear Walls

WxHxB (mm)	Opening Area (mm^2)	Opening Ratio	f_c' (N/mm^2)	Vertical and Horizontal Reinforcement Ratio (%)	Area of Diagonal Steel (mm^2)	f_y, f_{yd} (N/mm^2)	$\frac{H}{W}$
5486.4x3657.6x152.4	None	0	28.148	0.55	None	422.221	0.67
5486.4x3657.6x152.4	2194.56x914.4	0.1	28.148	0.55	4x129	422.221	0.67
5486.4x4876.8x152.4	None	0	28.148	0.55	None	422.221	0.89
5486.4x4876.8x152.4	2194.56x1219.2	0.1	28.148	0.55	4x129	422.221	0.89

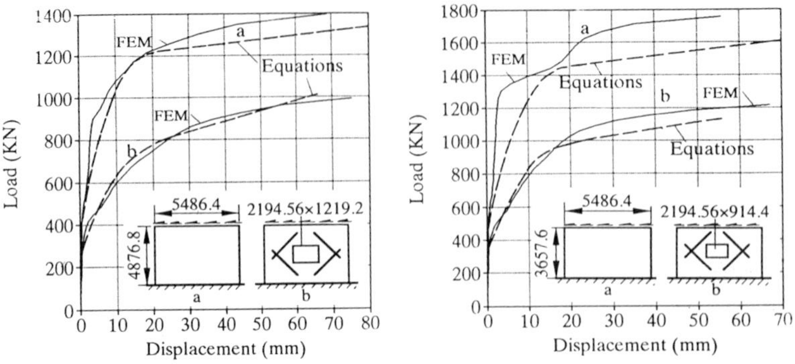

Fig. 17 Prediction of Load Displacement Curves of Shear Walls

CONCLUSIONS

A computer program based on nonlinear finite element method is developed. This program can predict the failure mechanics of the relationship between lateral load and displacement for shear walls with or without openings. It can also determine the influence of openings as well as the contribution of vertical, horizontal, and diagonal reinforcement on the stiffness of shear wall and shear capacity. Computer results conform well to test data. Regression formulas are generalized to evaluate cracking load, cracking displacement, yielding load, yielding displacement, ultimate load and ultimate displacement. These formulas reflect the characteristics of shear walls: (1) shear walls with larger openings have lower ultimate shear capacity and appear more ductile; (2) as the height of shear walls increases, a decrease in ultimate load and an increase in ductility are observed; (3) a reduction in vertical, horizontal or diagonal reinforcement leads to an increase in ductility and a decrease in ultimate shear capacity. The empirical equations display a good approximation to computer and test results. Thus they can be used to predict lateral load displacement of shear walls with or without openings to aid design as a supplement to ACI specifications.

ACKNOWLEDGMENTS

This research is partially supported by the National Science Foundation under the grants NSF CMS 9416463 and NSF MSS 9214664. The support is gratefully acknowledged.

REFERENCES

ACI Committee 318 (1989). Building Code requirements for reinforced concrete (ACI 318-89). American Concrete Institute (ACI), Detroit.

Anderson, F.E., Hanson, R.J., Murphy, H.L., Newmark, N.M., and Merit, P.W. (1964). *Design of Structures to Resist Nuclear Weapons Effects*. American Society of Civil Engineers (ASCE), New York, N.Y.

Antebi, J., Utku, S., and Hansen, J.R. (1960). The response of shear walls to dynamic loads. Report, Dept. of Civil and Sanitary Engrg., Massachusetts Inst. of Tech. (MIT), Cambridge, Mass.

Barda, F., Hanson, J.M., and Corley, W.G. (1976). An investigation of the design and repair of low-rise shear walls. *PCA Research and Development Bulletin, RD035D*. Portland Cement Association, Skokie, Ill.

Barda, F., Hanson, J.M., and Corley, W.G. (1977). Shear strength of low-rise walls with boundary elements. *Reinforced Concrete Structures in Seismic Zones*, SP53-8, ACI. Detroit, Mich., 149-202.

Benjamin, J.R., and Williams, H.A. (1957). The behavior of one-story reinforced concrete shear walls. *J. Struct. Engrg. Div.*, ASCE, *83*. 1254/1-49.

Cardenas, A.E., Hanson, J.M., Corley, W.G., and Hognestad, E. (1973). Design provisions for shear walls. *ACI J. 70*, 221-230.

Cardenas, A.E., Russell, H.G., and Corley, W.G. (1980). Strength of low-rise structural walls. *Reinforced Concrete Structures Subject to Wind and Earthquake Forces*, SP63-*10*. ACI, Detroit, Mich., 221-224.

Cheng, F.Y. (1989). Coupling bending and shear hysteretic models of low-rise RC walls. *Earthquake concrete in shear*. Elsevier Science Ltd., 276-288.

Cheng, F.Y. and Mertz, G.E. (1991). Macro-model for inelastic low-rise RC shear walls. Proc. 10th Conference on Electronic Computation, ASCE, 155-164.

Cheng, F.Y., Mertz, G.E., Sheu, M.S., and Ger, J.F. (1993). Computed versus observed inelastic seismic low-rise RC shear walls. *J. Struct. Engrg. 119*, 3255-3275.

Coladant, C., and Foure, B. (1989). Experimental behavior of RC shear walls for NPP. Trans., 10th Int. Conf. on Struct. Mech. in Reactor Tech., *H*, AASMiRT, 245-249.

Galletly, G.D. (1952). Behavior of reinforced concrete shear walls under static load. Report, Dept. of Civ. and Sanitary Engrg., MIT, Cambridge, Mass.

Ottosen, N.S. (1977). A failure criterion for concrete. *J. Engrg. Mech. Div.*, ASCE, *103*, 527-235.

Ottosen, N.S. (1979). Constitutive model for short-time loading of concrete. *J. Engrg. Mech. Div.*, ASCE, *105*, 127-141.

Sheu, M.S. (1988). Behavior of low-rise RC shear walls subjected to reversed cyclic loading. Technical Report to National Science Council. Arch. Engrg. Dept., National Cheng Kung University, Tainan, Taiwan.

Vecchio, F.J., and Chan, C.C.L. (1990). Reinforced concrete membrane elements with perforations. *J. Struct. Engrg.*, ASCE, *116*, 2344-2360.

Wataba, M., Fukuzawa, R., Chiba, O., and Hatori, T. (1989). Study on load-deflection characteristics of heavily reinforced concrete shear walls. Trans., 10th Int. Conf. on Struct. Mech. in Reactor Tech. *H*, AASMiRT, 209-214.

Yamada, M., Kawamura, H., and Katagihara, K. (1974). Reinforced concrete shear walls with opening. *Shear in Reinforced Concrete*, *2*; SP42-25. ACI, Detroit, Mich., 559-578.

Yang, J.S., Cheng, F.Y. and Sheu, M.S. (1992). Inelastic modeling and response of low-rise RC shear walls and systems. Proc. 10th World Conf. on Earthquake Engrg., Madrid, A.A. Balkema, *8*, 4313-4319.

APPENDIX I

Ottosen's concrete failure criterion under triaxial loading (Ottosen, 1977) expresses invariants as

$$f(I_1, J_2, \cos 3\theta) - a\frac{J_2}{f_c'^2} + \lambda\frac{\sqrt{J_2}}{f_c'} + b\frac{I_1}{f_c'} - 1 = 0 \tag{8}$$

where

$I_1 = \sigma_{ii}$, first invariant of stress tensor

$J_2 = \frac{1}{2} S_{ij} S_{ij}$, second invariant of stress deviator tensor

$$\cos 3\theta = \frac{3\sqrt{3}}{2}\cdot\frac{J_3}{J_2^{3/2}}$$

$J_3 = \frac{1}{3} S_{ij} S_{jk} S_{ki}$, third invariant of stress deviator tensor

$$\lambda = \begin{cases} K_1\cos\left(\frac{1}{3}\cos^{-1}(K_2\cos 3\theta)\right), & \cos 3\theta \geq 0 \\ K_1\cos\left(\frac{\pi}{3} - \frac{1}{3}\cos^{-1}(-K_2\cos 3\theta)\right), & \cos 3\theta < 0 \end{cases}$$

Parameters a, b, K_1 and K_2 are determined by the ratio of uniaxial tensile strength to uniaxial compressive strength. In a specific stress state, if $f(I_1, J_2, \cos 3\theta) \leq 0$, then concrete does not fail; otherwise, failure occurs.

θ is the angle between point P and positive principal axis σ_1 in a deviationic plane as shown in Fig. 18.

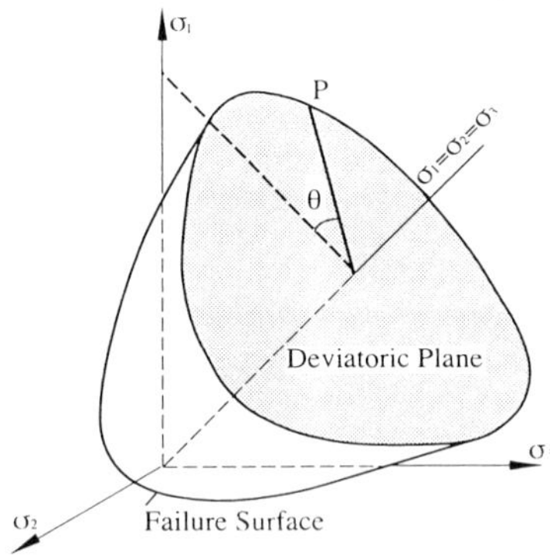

Fig. 18 Failure Surface for Concrete under Triaxial Load

Plastic index β' was developed by Ottosen (1979) to describe the nonlinear behavior of concrete under triaxial compressive loading. Based on equivalent stress-strain relationship, the following secant modulus E_s in triaxial compression is adopted.

$$E_s = \frac{1}{2}E_i - \beta'\left(\frac{1}{2}E_i - E_f\right) \pm \sqrt{\left[\frac{1}{2}E_i - \beta'\left(\frac{1}{2}E_i - E_f\right)\right]^2 + E_f^2\beta'\left[D(1-\beta) - 1\right]} \tag{9}$$

where

$E_c = \dfrac{f_c'}{\epsilon_c}$, secant modulus at uniaxial compression failure

ϵ_c = peak strain at uniaxial compression

$\beta' = \dfrac{\sigma_3}{f_c'}$, plastic index

σ_3 = maximum compressive principal stress. If tensile principal stress σ_1 appears, then $\sigma_3' = \sigma_3 - \sigma_1$ is used to replace σ_3 in β'

$E_f = \dfrac{E_c}{1 + 4(A-1)x}$, secant modulus at triaxial failure

$$x = \left(\frac{\sqrt{J_2}}{f_c'}\right)_f - \frac{1}{\sqrt{3}}$$

$$A = \frac{E_i}{E_c}$$

D = parameter to determine post-failure behavior of concrete
E_i = initial elastic modulus

Positive or negative signs refer, respectively, to the ascending or descending part of the equivalent stress-strain curve.

Concrete is assumed to be completely brittle after cracking. Note what happens at the post-failure stage. If concrete is subjected to both tension and compression, then the secant modulus, E_s, can be determined by a parallel translation of the original descending part of the stress-strain curve as

$$E_s = \frac{\beta' E_{MN} E_A E_M}{\beta' E_A E_M + \beta_f E_{MN} (E_M - E_A)} \tag{10}$$

in which E_A, E_M, E_{MN} and β_f are shown in Fig. 19.

The secant Poisson's ratio, ν_s, is modified according to plastic index β' as

$$\nu_s = \begin{cases} \nu_i & \beta' \le \beta_a \\ \nu_i - (\nu_f - \nu_i)\sqrt{1 - \left(\dfrac{\beta' - \beta_a}{1 - \beta_a}\right)^2} & \beta' > \beta_a \end{cases} \tag{11}$$

where

ν_i = initial Poisson's ratio
ν_f = secant Poisson's ratio at failure
β_a = around 0.8

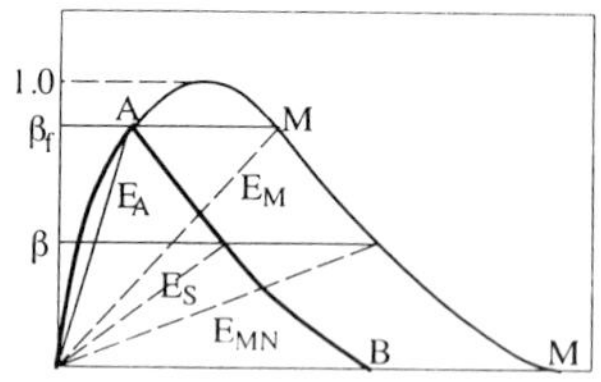

Fig. 19 Constitutive Relation of Concrete

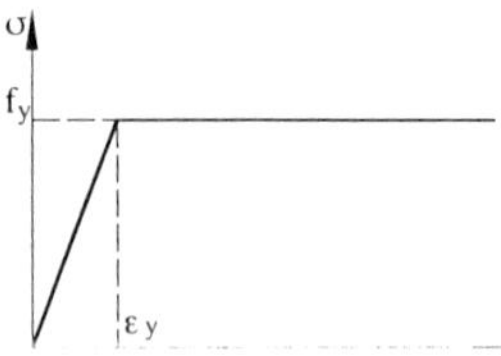

Fig. 20 Elastoplastic Model for Reinforcement

As shown in Fig. 20, the elastoplastic model is adopted for reinforcement.

APPENDIX II. NOTATIONS

The following symbols are used in this paper:

$A = E_i/E_c$; A_o = opening area; A_t = area of shear wall HxW; A_d = area of one diagonal reinforcement; a,b = parameters in failure criterion; B = width of shear wall; D = parameter for determination of post-failure behavior of concrete; E_A = Young's modulus in Eq. 10; E_c = secant modulus at uniaxial compression failure; E_f = secant modulus at triaxial failure; E_i = initial elastic modulus of concrete; E_M, E_{MN} = Young's moduli in Eq. 10; E_s = secant modulus of concrete; f_c' = compressive strength of concrete; f_y = tensile strength of reinforcement; f_{yd} = tensile strength of diagonal reinforcement; G = shear modulus of concrete; I = moment of inertia; I_1 = first invariant of stress deviator tensor; J_2, J_3 = second and third invariant of stress deviator tensor, respectively; K = numerical factor; K_1, K_2 = parameters for determination of parameter λ; n = number of diagonal reinforcements; P, P_{cr}, P_y, P_u = load, cracking load, yielding load and ultimate load, respectively; S_{ij} = stress deviator tensor; x = invariant for determination of E_f; W = width of shear wall; α, β = parameters in Eq. 7; β_a = parameter in Eq. 11; β', β_f = plastic index and plastic index at failure, respectively; γ = parameter in Eq. 7; $\Delta, \Delta_{cr}, \Delta_y, \Delta_u$ = displacement, cracking, yielding, and ultimate displacement, respectively; ϵ_c = peak strain at uniaxial compression; λ = parameter in failure criterion; θ = angle in Fig. 18; ρ = smaller value of horizontal or vertical reinforcement ratio; ρ_o = opening ratio; ρ_d = diagonal reinforcement ratio; ρ_h, ρ_v = horizontal and vertical reinforcement ratio, respectively; σ_{ij} = stress tensor; $\sigma_1, \sigma_2, \sigma_3$ = principal stresses; ν_i = initial Poisson's ratio; ν_f, ν_s = secant Poisson's ratio at failure and secant Poisson's ratio, respectively; ϕ = angle between diagonal reinforcement and horizontal direction.

RESEARCH ON INFILLED HOLLOW CLAY TILE WALLS

J. E. Beavers,* R. M. Bennett** and
R. D. Flanagan†

** Natural Hazards Services Division, MS Technology, Inc., Oak Ridge, TN, U.S.A.*
*** Department of Civil Engineering, University of Tennessee, Knoxville, TN, U.S.A.*
† Technical Programs and Services Department, Martin Marietta Energy Systems, Inc., Oak Ridge, TN, U.S.A.

ABSTRACT

This paper reports some of the results of an unreinforced hollow clay tile wall (HCTW) research program conducted by Martin Marietta Energy Systems, Inc., for the U.S. Department of Energy. The purpose of the program was to assist in determining the response of infilled HCTW buildings to earthquake forces.

KEYWORDS

Earthquakes, masonry, hollow clay tile, existing buildings

INTRODUCTION

During the early to middle 1900s, hollow clay tile (HCT) was used extensively as a building construction material throughout the United States. Many buildings were constructed of steel and/or concrete frames with HCT infilled walls. In the central and eastern United States, seismic building codes did not exist at the time and the buildings were not designed for seismic loads.

This research program as originally described (Beavers *et al.*, 1991) focuses on hollow clay tile wall (HCTW) buildings located in zones of low-to-moderate seismicity, although the results are directly applicable to high seismicity zones. The results of this program have shown that infilled, unreinforced HCTW building construction has inherent strength above that normally allowed in design by codes, and therefore, retrofit may not be required in low-to-moderate seismicity zones if a risk-based approach is adopted. A risk-based approach is based on the principle that while the building may not meet today's codes and standards requirements, the expected seismic performance is reasonably understood and has sufficient safety margins.

Another principle that has an impact on the issue of "to retrofit or not to retrofit" in low-to-moderate seismicity zones is the cost increment of retrofit. In many cases, the cost increment to retrofit a building to a higher seismic capacity rather than a lower seismic capacity is much less than the cost to retrofit to the lower seismic capacity. For example, the cost to install a one-inch diameter bolt required for the retrofit of a building in a moderate or high seismic zone is basically the same cost to install a one-half inch diameter bolt in the same building for retrofit in a low seismic zone. Thus, the real cost impact for retrofit of buildings in low-to-moderate seismic zones is not the level of retrofit but the decision, or requirement, to retrofit.

Building codes in the United States generally consider the limit state for unreinforced masonry to be cracking due to its brittle nature. However, the overall seismic resistance of an infilled masonry building panel can be an order of magnitude higher than that when the first crack occurs. This leads to a key proposition that must be adopted in the risk-based approach to a "no retrofit philosophy" for existing construction in low-to-moderate seismic zones. That proposition is that the HCTWs will crack when the earthquakes occur, but overall stability must be maintained. Thus, significant cracking of infilled walls is being described by the authors as acceptable damage for buildings in low-to-moderate seismic zones. Therefore, the principal basis of the research and development program on HCTWs described in this paper is to investigate and develop computational methods for determining post-cracking behavior, including ultimate load, of infilled HCTW building construction.

PREVIOUS HOLLOW CLAY TILE RESEARCH

Most of the work completed during the last decade in the United States has focused on the seismic performance of reinforced concrete block and brick masonry. The U.S. Technical Coordinating Committee for Masonry Research program as discussed by Noland (1990) and funded by the National Science Foundation has done some work on unreinforced masonry, but focused on reinforced masonry. More recently, as a result of the unreinforced masonry retrofit ordinance established in the Los Angeles Building Code (1985), and the reauthorization of the United States National Earthquake Hazards Reduction Program (1990) increased activity for testing and development of computational methods for unreinforced masonry construction is occurring in the United States. As a result of the National Earthquake Hazards Reduction Program, the President of the United States signed Executive Order 12941 on December 1, 1994, for assessing and enhancing the seismic safety of all federally owned and leased buildings not designed for seismic loads.

For the HCTW configurations that are being investigated, no previous research work was found other than studies of material properties, such as Plummer (1962) and Johnson and Matthys (1973). However, unreinforced masonry research on concrete block and brick have application to the HCTW program. Experimental and analytical work on the out-of-plane response of unreinforced concrete block infilled walls was performed (Dawe and Seah, 1988; Dawe and Seah, 1989). An extensive annotated bibliography was published (Lawrence, 1979) on the lateral loading of masonry infilled panels. The findings of a major test program on unreinforced infilled concrete block walls subjected to in-plane loads were summarized by Dawe and Seah (1989) and Dawe and Hatzinikolas (1988). Unreinforced masonry infilled walls on a shake table were tested by Dawe *et al.* (1989). The work of Page *et al.* (1985) in developing a nonlinear constitutive relationship and biaxial failure surface for unreinforced masonry subjected to in-plane loads is also applicable to the HCTW program. This model was used in the evaluation of test results of infilled frames (Dhanasekar and Page, 1986).

BUILDING TYPES

The building types studied in this program were steel frame with infilled HCTWs and were constructed in 1940s and 1950s. Typically the buildings were industrial type, one-to-five-stories in height, and had steel framing with simple framing connections and very little or no cross-bracing. This type construction resulted in the infilled HCTWs becoming the major lateral load resisting elements. Although the buildings were designed for the HCTWs to be nonload bearing, significant load bearing can exist due to the method of construction and building additions that had occurred. The walls were constructed with HCT cores running horizontally, resulting in a full bedding joint, and were of either eight or thirteen inches in thickness. The eight-inch walls were constructed with a single eight-inch tile. The thirteen-inch walls have an eight-inch and a four-inch tile laid in a staggered fashion, Fig. 1. All exterior walls are thirteen inches thick.

Fig. 1. Existing Hollow Clay Tile Wall

The potential collapse mechanism's of these buildings are loss of lateral load capacity by (1) out-of-plane failure of the infilled HCTWs as a result of cracking caused by out-of-plane inertia loads, (2) out-of-plane failure as a result of cracking caused by in-plane shear loads, (3) in-plane corner crushing or diagonal cracking causing loss of shear capacity and (4) out-of-plane failure as a result of cracking caused by building drift.

The seismic performance of this type construction is highly dependent upon the true boundary or interface conditions. Typical HCTWs being investigated rest on a running bond supported by a concrete floor. The vertical interface condition is the HCT inserted between the flanges of the framing steel. Several variations exist including the wall being slightly offset from the steel centerline. The top is grouted against the bottom flange of the beam.

HCTW TESTING PROGRAM

Elements of the HCTW test program were published (Fricke and Jones, 1989; and Fricke and Jones, 1991) as it was being developed and included some early test results. Since that time, the program was expanded into a total of approximately 18 types of tests with multiple test of each type, resulting in 200 major in situ and laboratory tests. A flow diagram of the HCTW research program is shown in Fig. 2. The major in situ test were out-of-plane air bag tests, while major laboratory tests were out-of-plane deformational and in-plane tests. Shake table testing has also been conducted but the results have not been fully analyzed at the time of this writing. The HCTW test program was quite extensive, took about twelve months in planning, and the testing program lasted for about four years.

The purpose of the static testing was threefold. At first the tests were designed to determine existing conditions. These include deformability tests, push tests, and bond wrench tests. Some of the results of these tests were correlated to assemblage tests. For example, comparison of laboratory prism compression tests and the in situ deformability test helped determine the feasibility of using the deformability test to assess the strength of a wall.

The second purpose for static testing was to develop an appropriate constitutive model. Both the elastic pre-cracking behavior and the post-cracking performance were examined. These tests include prism compression tests, flexural bond tests, prism split tensile tests, and biaxial tests.

The third purpose of the tests was to identify system behavior, and thus appropriate idealizations for computational analysis. The emphasis was on post-cracking behavior. These tests include air bag tests, in-plane racking tests and various other laboratory tests. This paper focuses on the results of these tests and development of the constitutive model.

One of the principal purposes of the laboratory static in-plane and out-of-plane tests was to gain knowledge of the in-plane performance following post-cracking caused by out-of-plane deformations. The out-of-plane loads were supplied by point loads on the steel frame to allow for the study of the impact of interstory drift. Frame flexibilty and interface conditions were also examined.

RESULTS OF STATIC TESTING PROGRAM

As shown in Fig. 2, approximately 18 different types of tests were conducted. The tests to be discussed in this paper will be those that had the largest impact on the development for computational modeling. These tests were the laboratory prism compression strength and in-plane and out-of-plane behavior tests. Some discussion is also presented on the results of full scale in situ tests.

The Prism Compression Strength Tests

Masonry prisms consisting of 2′ by 4′ assemblages were tested under uniaxial compressive loading. The stress-strain relationships were somewhat linear prior to peak capacity. Prisms with the compressive force parallel to the cores of the tile exhibited significant postpeak capacity. Conversely, prisms with the compressive force normal to the cores had little postpeak strength. Fortunately for the buildings under study, the HCTW cores ran horizontal (Fig. 1) or parallel to the lateral compressive load, which is believed to have increased the postpeak capacity of the infilled frames. Average gross compressive strengths parallel to the tile cores were 436 psi and 372 psi for 8" and 13" prisms respectively. Average gross compressive strengths normal to the tile cores were 810 psi and 332 psi for 8" and 13" prisms respectively.

The In-Plane Behavior

Thirteen cyclic in-plane racking laboratory tests were performed on large-scale clay tile infill specimens to determine the behavior. The tests were designed to evaluate a variety of engineering conditions including varying frame stiffness, single and double wythe construction, varying panel aspect ratio, offset panels, weak column/infill interface, and corner openings.

In-plane stiffness and strength of the steel framing is increased dramatically by the addition of the clay tile infill. The rigidity of the infill stiffens the steel framing while the framing confines the masonry panel allowing greater strength and ductility. For many specimens, the ultimate strength of the infilled panel was an order of magnitude higher than the bare frame. The behavior was characterized first by diagonal panel cracking followed by corner crushing at ultimate capacity. Significant postpeak strength was observed indicating continued energy absorption capability. Details of the in-plane testing are contained in Flanagan *et al.,* (1993b).

The Out-of-Plane Behavior

A total of eight laboratory infill panels were tested normal to the plane of the wall under either simulated drift or uniform lateral pressure. One large in situ infill wall was also tested under uniform lateral pressure.

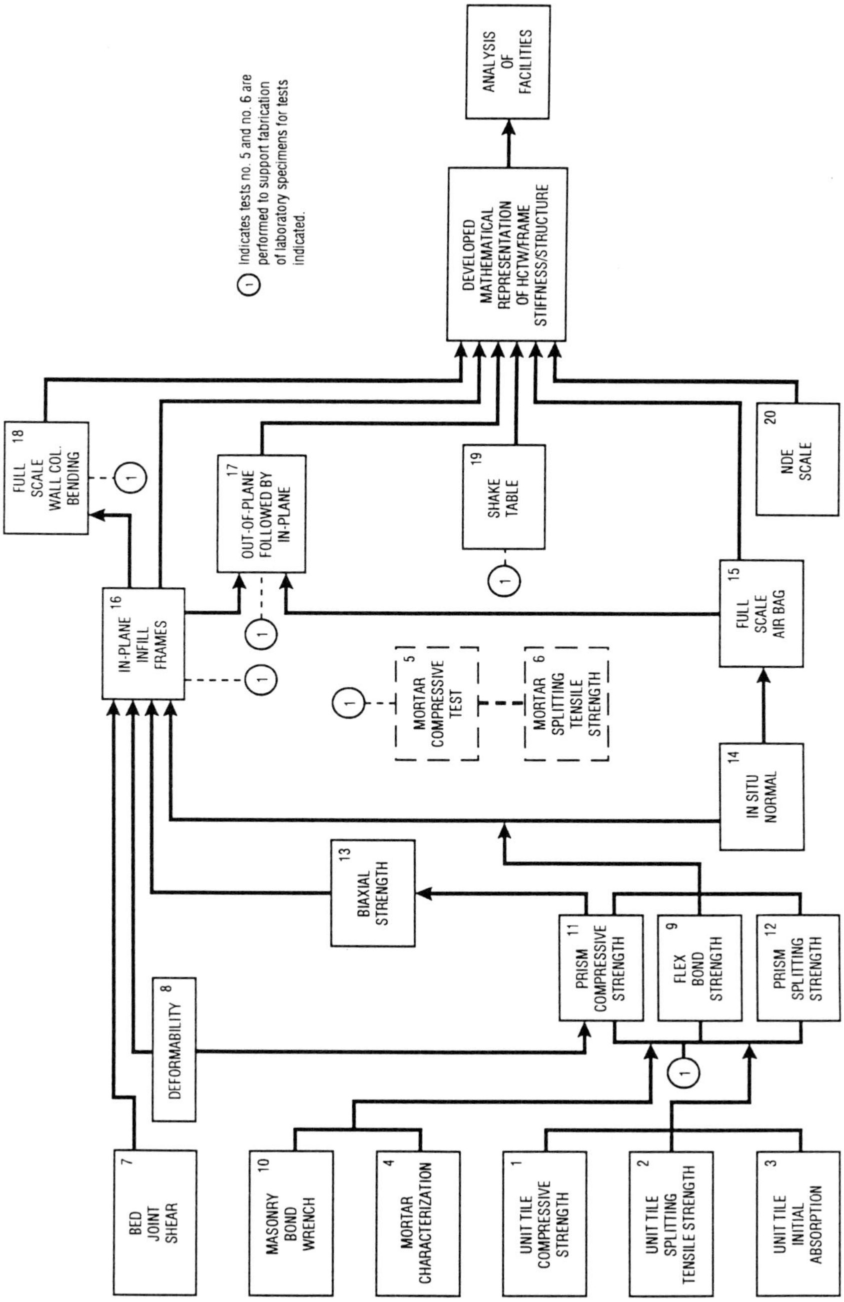

Fig. 2. HCTW Research Program Flow Diagram

Laboratory Inter-Story Drift

The stability of infill panels under out-of-plane inter-story drift was evaluated by testing two specimens. Details of this study are contained in Flanagan and Bennett (1992). The specimens were subjected to cyclic out-of-plane displacements imposed at either the top or midheight of the columns. There was little relative movement of the infill panels with respect to the steel framing as the out-of-plane drift loading was applied. Infill panels constructed snugly to the framing, but without ties or other reinforcing, were shown to remain stable when subjected to cyclic out-of-plane drift displacements within the range that a typical infill might experience.

The specimens were then loaded in-plane to failure to determine the effect of the out-of-plane drift on in-plane stiffness and strength. The results of the in-plane tests were compared to a control specimen which was similar except having been loaded only in-plane to failure. The in-plane failure modes were similar to that of the control specimen. Only minimal degradation of in-plane stiffness and strength resulted from the out-of-plane drift displacements.

Laboratory Uniform Lateral Pressure

Three infill panels were tested out-of-plane with a uniform lateral loading supplied by an air bag. The tests were designed to simulate the inertial effects of the panel normal to its plane. Static load-unload cycles were applied to the specimens until ultimate capacities were reached and overall panel stability began to deteriorate. The test specimens were designed to envelope the range of typical infill height to thickness ratios. Three different wall thickness were tested; two single wythe panels of 4" and 8" and one 13" double wythe panel. The ultimate capacities appeared to vary with the square of the panel thickness.

The behavior was dominated by arching action of the masonry. With the presence of panel edge restraint, membrane forces developed in the infill as fracture lines divided the panel into separate segments. As the segments moved out of the plane of the wall and rotated about their boundaries, arching forces continued to develop until failure in the masonry occurred along the boundaries. Details of the uniform lateral pressure tests are in Flanagan and Bennett (1994).

Two additional specimens were tested using sequential in-plane racking and out-of-plane pressure loadings. One specimen was first loaded to approximately 75% of ultimate in-plane capacity and then subjected to uniform lateral pressure loadings. The ultimate lateral capacity was approximately 85% of a similar undamaged specimen. The other specimen was first loaded with an uniform out-of-plane pressure of approximately 75% of ultimate. The infill was then subjected to in-plane racking forces and achieved approximately the same ultimate capacity as a similar undamaged specimen.

Finally, one specimen was loaded with combined in-plane and out-of-plane forces. After a series of cyclic racking loads were completed, the in-plane drift was held as uniform lateral pressure load-unload cycles were applied. This sequence was repeated until ultimate in-plane capacity of the infill was reached. Significant out-of-plane capacity was observed at high levels of in-plane drift. Although some interaction of in-plane and out-of-plane forces exists, the interaction did not appear to be significant at low to moderate levels of loading.

In Situ Uniform Lateral Pressure

One full-scale in situ air bag test was conducted on the first floor of a five-story steel-framed building. The infilled panel tested was an 8-inch HCTW 29 feet long and 12 feet high with the bottom of the panel grouted to a concrete floor, the top grouted to a W30 x 108 beam, and the ends encased partially between W14 x 142 columns. The results of this test showed that significant arching action occurred as uniform lateral load was applied to the wall. This arching action resulted in the wall being able to withstand an out-of-plane load equivalent to 2.1g as noted in Huff *et al.* (1992). The response of this in situ infilled wall performed in the same manner as the laboratory walls.

COMPUTATIONAL MECHANICS

Seismic analyses of complex industrial facilities are typically performed using three dimensional equivalent static or response spectra techniques. The methodology presented for seismic evaluation of clay tile infills is also applicable for simpler conditions where two dimensional analysis or cruder approximate techniques are employed.

The process of evaluating infilled frames is as follows. The masonry infill is idealized as an equivalent compression strut in the plane of the wall. Because the stiffness of the infill is a function of the load magnitude, the strut model is necessarily nonlinear. After seismic loads are determined, the in-plane capacity of the infill is assessed based on allowable strain in the masonry. The resulting out-of-plane drift stability is assessed by comparison with allowable ratios. Out-of-plane inertial strength is evaluated by comparing maximum panel accelerations with equivalent acceleration levels computed from uniform lateral load equations incorporating arching action. Frame members are evaluated based on the controlling in-plane drift and any significant thrust forces due to arching.

The conditions required for a compression strut to develop must be present. Openings, interface gaps, and other discontinuities may affect development of a compression diagonal. Equivalent strut methods are typically formulated without consideration of vertical forces. Thus, vertical loads should be omitted during lateral load analysis and superimposed later.

Attention to detail in modeling the compression strut is also important. For instance, the use of a nonsymmetric strut may result in erroneous axial forces in members at reentrant corners of buildings. It is recommended that a compression only strut be used in each direction. If a compression only element is not available, a tension-compression truss member may be used with half the strut area in each diagonal direction.

The compression strut formulation suggested is based on work by (Stafford-Smith and Carter, 1969). The method has been extended to include nonlinear behavior of the infill. The contribution to lateral stiffness may be computed using Equations 1-4, see Fig. 3.

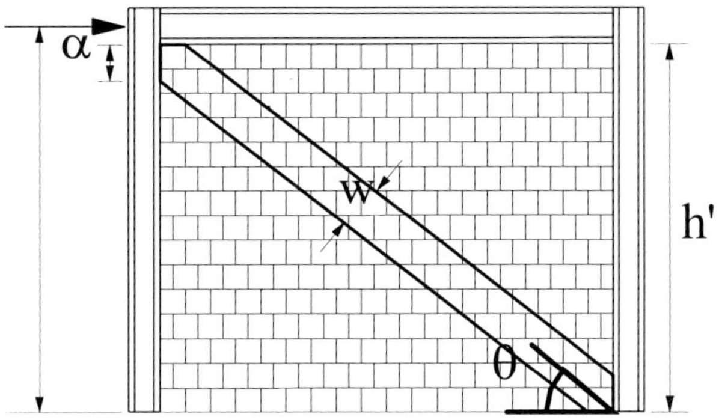

Fig. 3. Compressive Strut

The length of column bearing on the infill, α (in), is given by

$$\alpha = \frac{\pi}{C\lambda} \tag{1}$$

where

$$\lambda = \sqrt[4]{\frac{E_m\ t\ \sin 2\theta}{4\ E_s\ I_c\ h'}} \tag{2}$$

in which E_m is the gross elastic modulus of the masonry (psi), $E_s\ I_c$ the flexural rigidity of the column (psi, in^4), θ the slope of the infill diagonal to the horizontal, t the infill gross thickness (in), and h′ the infill height (in). The width of the equivalent strut, w (in), is

$$w = \frac{\alpha}{\cos\theta} \tag{3}$$

and the area of the equivalent strut, a (in^2) is

$$a = wt \tag{4}$$

It is suggested that the length of contact be limited to 20% of the panel height (h'). The parameter C varies with the in-plane drift displacement as shown in Table 1.

Table 1 Typical values of C for varying in-plane drift displacements

C	Displacement (in)	Typical Infill Damage
5	0.0-0.05	None
7	0.05-0.2	Diagonal Mortar Joint Cracking
11	0.2-0.4	Off Diagonal Mortar Joint Cracking
14	0.4-0.6	Banded Diagonal Mortar Joint Cracking
16	0.6-0.8	Corner Mortar Crushing and Tile Cracking
18	0.8-1.0	Tile Faceshell Splitting (Corner Regions)
-	1.5 x Disp @ Ultimate	Tile Faceshell Spalling

For specimens in the range of height tested, ultimate in-plane capacity of the clay tile infills was limited by a 1" in-plane drift displacement. This corresponds to an in-plane drift of approximately 1%. An upper limit of 50 kips for single wythe panels and 60 kips for double wythe panels should also be used for the in-plane capacity of the infills (maximum horizontal component of equivalent strut force). Stiff frames with a greater length of contact along the infill will be limited by this criteria and will reach ultimate capacity at a lower in-plane displacement.

Beyond peak capacity the in-plane strength of the infill is assumed to reduce to 75% of peak at an in-plane drift of 1.5 times the displacement at peak. Postpeak testing of the infilled frames indicates significant capacity well beyond a displacement of 1.5 times that reached at peak capacity. However, this displacement serves as a practical limit in predicting repairable damage levels of the masonry as well as an indicator of potential damage to the columns.

As indicated in Fig. 4, the results of this equivalent strut method compare well with the envelope of observed in-plane hysteretic behavior. The test specimen consisted of partially restrained connections, so a fixed base and pinned base calculation with the compressive strut formulation is given. The masonry elastic modulus used in the calculations is from the prism tests with the compressive load parallel to the tile cores (330 ksi for 8" and 410 ksi for 13" panels). Although the state of stress in the corners of the masonry panels is biaxial at low precracking load levels, the behavior of the panel appears to be more influenced by the compressive behavior in the lateral direction at higher load levels.

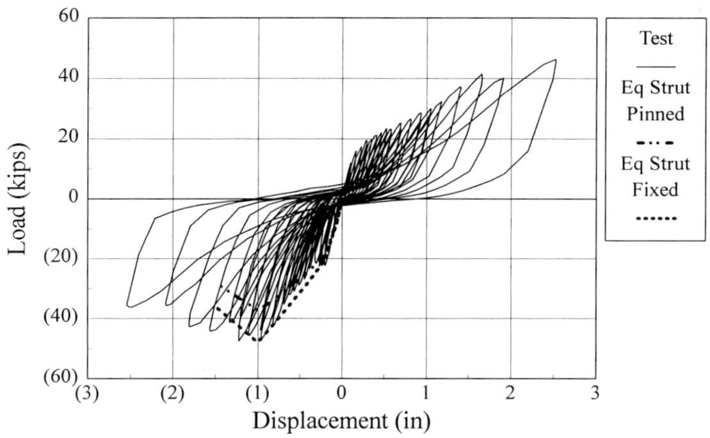

Fig. 4. In-Plane Hysteresis

Out-of-plane story drifts due to the relative top-bottom displacements of orthogonol walls may be evaluated based on an allowable drift ratio of 1%. Panels drifts exceeding 1% may be stable but require more detailed evaluation.

Out-of-plane inertial loads are evaluated for individual panels based on the maximum acceleration computed for a given elevation. The uniform lateral load capacity, q (psi), of an individual infill may be determined from

$$q = 0.8\, f'^{\,0.75}_{m}\, t^2 \left(\frac{\alpha}{l'^{\,2.5}} + \frac{\beta}{h'^{\,2.5}}\right) \tag{5}$$

in which f'_m is the masonry gross compressive strength (psi) and l' the infill length (in). Originally developed by Dawe and Seah (1989), this relationship incorporates arching action for infills supported on all four boundaries.

The parameter α is a measure of the relative stiffness of the bounding columns

$$\alpha = \frac{1}{h'} \left(E_s I_c h'^2 + G_s J_c t h' \right)^{0.25} < 35 \tag{6}$$

in which G_s is the column shear modulus, and J_c the column torsional constant. The parameter β is a measure of the relative stiffness of the bounding beam

$$\beta = \frac{1}{l'} \left(E_s I_b l'^2 + G_s J_b t l' \right)^{0.25} < 35 \tag{7}$$

in which I_b is the beam moment of inertia and J_b is the beam torsion constant. Effective acceleration limits may be computed from the ultimate static pressures and compared with the results of the dynamic analysis.

Fig. 5 presents a comparison of this uniform lateral capacity formulation for various h/t ratios versus the results of three static pressure tests. Good correlation of test data is made with these arching equations. The masonry strength used in the calculations is from the prism tests with the compressive load normal to the cores (810 psi for 8" and 332 for 13" panels). This is appropriate as the primary behavior mode of the panels was vertical arching (horizontal cracking) prior to peak capacity.

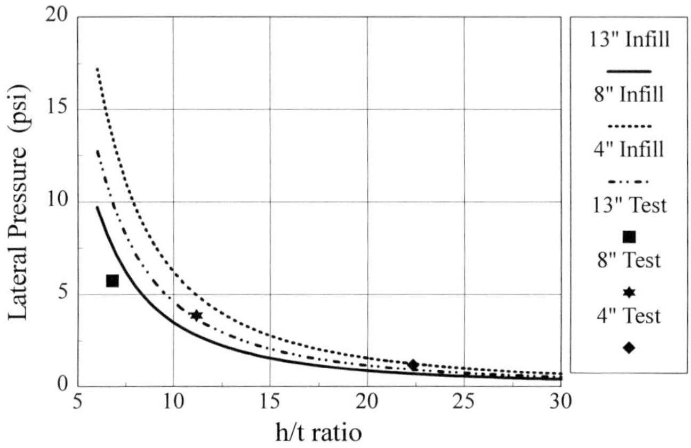

Fig. 5. Uniform Lateral Load Analysis

Frame member forces may be computed from the in-plane and out-of-plane behavioral conditions. Forces in the columns due to in-plane loading may be estimated based on the in-plane drift of the frame and the contact forces computed along the column infill boundary. Similarly, uplift forces in the beam may be computed based on the contact along the beam/infill boundary. Depending on the construction, the bending forces in the beam resulting from lateral loading may be negated by those resulting from vertical loads.

For out-of-plane drift, the frame forces are estimated directly from the level of frame drift. In the case of out-of-plane inertial loads, the frame thrust forces due to arching action require more detailed evaluation. Depending on the vertical loading in the beam and the presence of adjacent infills along the columns, these thrust forces may also be negated.

SUMMARY

Significant progress of the HCTW program has occurred to date. Tests have been conducted and computational methods have been developed that show reasonable comparison with other work. A large portion of the test program has been completed and tests included static and dynamic (shake table). The shake table test results are now under analysis and review.

Results to date have shown that significant inherent strength exists in the infilled unreinforced HCTWs and they can be considered as a structural element when evaluating buildings for seismic loads. Results also show that out-of-plane response is not a weak link in systems response and wall panels will not fail out-of-plane at the onset of earthquake loading, resulting in a loss of lateral load building stability. In fact, it was shown that out-of-plane failure does not occur until significant in-place crushing of the HCTW occurs and the postpeak in-plane strength begins to deterioate rapidly.

Computational methods were developed to better predict both building and component response. Due to the complexity of large structures, the development approach was toward more simplified models using an equivalent strut formulation.

The HCTW program has indeed shown that in low-to-moderate seismic zones, there is inherent lateral strength capacity such that major structural retrofit of the HCTW buildings studied is not required.

REFERENCES

101st Congress (1990). "National Earthquake Hazards Reduction Program Reauthorization Act," Public Law 101-614.

Beavers, J. E., R. M. Bennett and R. D. Flanagan (1992). Research on Unreinforced Hollow Clay Tile Walls and Developments of Computational Mechanics. In; Computation Mechanics in Structural Engineering - Recent Developments and Future Trends [eds. F. Y. Cheng and F. Zizhi]; Elsevier Applied Science

Dawe, J. L., and C. K. Seah (1989). "Out-of-Plane Resistance of Concrete Masonry Infilled Panels," Canadian Journal of Civil Engineering, 16, 854-864.

Dawe, J. L. and M. A. Hatzinikolas (1988). Behavior of Masonry Infilled Panels in Steel Frames. Proceedings, 8th International Brick and Block Masonry Conference, Dublin, Ireland, 516-27.

Dawe, J. L. and C. K. Seah (1989). Behavior of Masonry Infilled Steel Frames. *Canadian Journal of Civil Engineering*, Vol. 16, 865-76.

Dawe, J. L., A. B. Schriver and C. Sofocleous (1989). Masonry Infilled Steel Frames Subjected to Dynamic Load. *Canadian Journal of Civil Engineering,* Vol. 16, 877-85.

Dawe, J. L. and C. K. Seah (1988). Lateral Load Resistance of Masonry Panels in Flexible Steel Frames. Proceedings, 8th International Brick and Block Masonry Conference, Dublin, Ireland, 606-16.

Dawe, J. L. and C. K. Seah (1989). Out-of-Plane Resistance of Concrete Masonry Infilled Panels. *Canadian Journal of Civil Engineering*, Vol. 16, 854-64.

Dhanasekar, M. and A. W. Page (1986). Influence of Brick Masonry Infilled Properties on the Behavior of Infilled Frames. Proceedings, Institution of Civil Engineers, Vol. 81, 593-605.

Flanagan, R. D., R. M. Bennett and G. A. Barclay (1993b). "In-Plane Behavior and Strength of Structural Clay Tile Infilled Frames," Proceedings of the Sixth North American Masonry Conference, Philadelphia, PA, 371-382.
Flanagan, R. D., and R. M. Bennett (1994). "Uniform Lateral Load Capacity of Infilled Frames," Proceedings of the ASCE Structures Congress, Atlanta, GA.

Flanagan, R. D., and R. M. Bennett (1992). "Inter-Story Drift Effects on the In-Plane Capacity of Infilled Frames," Proceedings of the Third International Masonry Conference, London, England.

Flanagan, R. D., M. B. Butala and R. M. Bennett (1993a). "Properties of Structural Clay Load-Bearing Wall Tile," Proceedings of the Sixth North American Masonry Conference, Philadelphia, PA, 87-98.

Fricke K. E. and W. D. Jones (1991). A Test Program to Determine the Structural Properties of Unreinforced Hollow Clay Tile Masonry Walls of the DOE Oak Ridge Plants. *The Masonry Society Journal*, Vol. 9, 56-63.

Fricke, K. E. and W. D. Jones (1989). Seismic Analysis and Testing of Clay Tile Walls at the Oak Ridge Y-12 Plant. Proceedings, Second DOE Natural Phenomena Hazards Mitigation Conference, Knoxville, Tennessee. 162-71.

Johnson, F. B. and J. H. Matthys (1973). Structural Strength of Hollow Clay Tile Assemblages. *Journal of the Structural Division*, ASCE. Vol. 99, 259-75.

Lawrence, S.J. (1979). Lateral Loading of Masonry Infilled Panels: a Literature Review. Technical Record 454, Experimental Building Station, Sydney, Australia.

Los Angeles Building Code (1985). Division 88.

Noland. J. L. (1990). Status report: U.S. coordinated program for masonry building research. Proceedings, Fifth North American Masonry Conference, University of Illinois at Urbana - Champaign.

Page, A.W., P. W. Kleeman and M. Dhanasekar (1985). An In-Plane Finite Element Model for Brick Masonry. New Analysis Techniques for Structural Masonry, ASCE, pp. 1-18.

Plummer, H. C. (1962). Brick and Tile Engineering, Brick Institute of America.

Stafford-Smith, B., and C. Carter (1969). "A Method for the Analysis of Infilled Frames." Proceedings of the Institution of Civil Engineers, 44, 31-48.

VULNERABILITY STUDY OF SCHOOL BUILDINGS IN TAIWAN BEFORE AND AFTER RETROFITTING

M. -S. Sheu and C. -C. Hsu

Department of Architecture, National Cheng-Kung University, Tainan, Taiwan

ABSTRACT

This paper tries to compare the changes of earthquake vulnerability for RC school buildings before and after retrofitting through nonlinear spectra analysis. School buildings located in Taiwan are investigated. Before and after retrofitting, peak ground acceleration of elastic design spectra is triggered from zero up to collapse of the structure. At each shaking stage, member stiffness, damping ratio and $P-\Delta$ effects are modified gradually. This paper proposes analytically the vulnerability diagram showing correlations among peak ground acceleration, tangential period, base shear and roof deflection. From vulnerability diagram, collapse PGA for each structure can be read clearly. Retrofitting method includes insert of separation slits between windowsills and columns, adding plate hoops for short columns, adding shear walls. Combinations for different single retrofitting method are tried for example building until the collapse PGA meet the requirement of the risk curve at site of Taiwan.

KEYWORDS

Dynamics analysis, acceleration spectra, damping, retrofit, yield peak ground acceleration, collapse peak ground acceleration.

NONLINEAR SPECTRA ANALYSIS

Nonlinear spectra analysis has been shown to be one of the practical technique for vulnerability assessment before earthquake as well as damage assessment after earthquake (Sheu et al., 1994) In that analysis, the whole structure is considered as a 3 dimensional space structure and subjected to dynamic loadings. Site acceleration spectrum are used as external dynamic loads and triggered gradually from zero up to collapse of the structure. During triggering of acceleration, member stiffness and damping ratio are modified from step to step. The analytical result clearly shows the softening of the building, including changes of tangential period, base shear and roof deflection. It is a good reference for aseismic diagnosis.

Base on Sheu et al.(1994) and SAP IV software, parameters of nonlinear incremental spectra analysis for 3 dimensional RC structures are described as follows:

1. Load Combination: Since earthquake loads are reversed cyclically, the stress combination for major and minor axes of each member section should take consideration for both positive and negative earthquake loads.

2. Dynamic Strength of Materials: The concrete compressive strength is $1.18f_c'$; the yield stress of rebars under earthquake loading is considered as $1.14f_y$.

3. Effective Dynamic Flexural Stiffness for Beams and Columns: Effective dynamic flexural stiffness, Eq.(1), (Sukhvarsh et al., 1981), is employed and transferred into tangential stiffness.

$$I_e = [0.7M_c/M_m]I_g + [1 - 0.7M_c/M_m]I_c \quad (1)$$

where M_c and M_m are crack and max. moments; I_g and I_c are gross and cracked moment of inertias.

4. Effective Shear Rigidity for Beams and Columns: Before diagonal crack, tangential shear rigidity GA is taken as $EA_g/2.6$; After diagonal cracks it is $EA_g/7.8$.

5. Effect of Axial Force: Axial Force is considered for cracking moment, ultimate moment, cracking shear, ultimate shear and cracked moment of inertia at every incremental loading step. However, axial stiffness for beam, column and wall remains constant, EA_g, to compensate reversed tension and compression axial stresses.

6. Strength and Stiffness Change for RC Shear Walls: The empirical equations of Wang's paper (1980) are used. For example,

$$V_u = V_c + \oint \left\{ C_s A_v f_y \left(\frac{W - W_0}{S_v}\right)^{1.2} + A_h f_y \left(\frac{H - H_0}{S_h}\right) \right\} \quad (2)$$

where W and W_0 are widths of wall and opening; H and H_0 are heights of wall and opening; S_v and S_h are spacings of vertical and horizontal rebars. The reduction factor $\oint$ is 0.7 for wall with 2 boundary elements; 0.75 for wall with 3 boundary elements; 0.80 for wall with 4 boundary elements. C_s is function of aspect ratio H_s/W and opening ratio W_0/W of wall as follows:

$$C_s = 0.94 - 0.67H_s/W - (1.01 - 0.6H_s/W)(W_0/W) + (5.33 - 5.47H_s/W)(W_0/W)^2 \geq 0.2 \quad (3)$$

The secant shear rigidity at crack point, yield point and ultimate point are:

$$GA_c = \{0.005 - 0.005H_0/H + 0.038(H_0/H)^2\}EA_{gt}\sqrt{f_c'/280} \quad (4)$$

$$GA_y = \{0.024 + 0.014H_0/H - 0.023(H_0/H)^2\}EA_{gt}\sqrt{f_c'/280} \quad (5)$$

$$GA_u = \{0.01 + 0.034H_0/H - 0.035(H_0/H)^2\}EA_{gt}\sqrt{f_c'/280} \quad (6)$$

7. Damping Ratio of RC Structure: If the peak ground acceleration (PGA) is less than 40% of collapse ground acceleration (CGA), damping ratio is 4%; For PGA in between 40% to 75% of CGA, damping ratio is 7%; For PGA larger than 75% of CGA, damping ratio is 10%.

8. Incremental Ground Acceleration: For 4% damping ratio, incremental ground acceleration is 0.04 g; For 7% damping ratio, incremental acceleration is 0.02 g; For 10% damping, it drops down to 0.01 g.

9. $P - \Delta$ Effect of Floor Sideways: $P - \Delta$ effect due to vertical loads are considered as equivalent lateral forces acting on each floor.

10. Rigid Zone of Beam-Column Joints: The bending moment at beam-column joints is reduced to the moment at outer face of rigid zone. Using this reduced bending moment, the effective flexural stiffness is calculated at every load step.

In order to verify the accuracy of the dynamic analytical results, one shaking table specimens (Bertero et al., 1984), 1/5 scale 7-story frame with RC shear walls subjected to Taft and Miyagi-Oki earthquakes, is checked. Fig.1 shows the comparisons of the load-deflection envelopes. It is in good match. Four actual buildings seriously damaged after 1986 and 1990 earthquakes in Taiwan are also used to check the vulnerability assessment. They are predicted reasonably (Sheu et al., 1993).

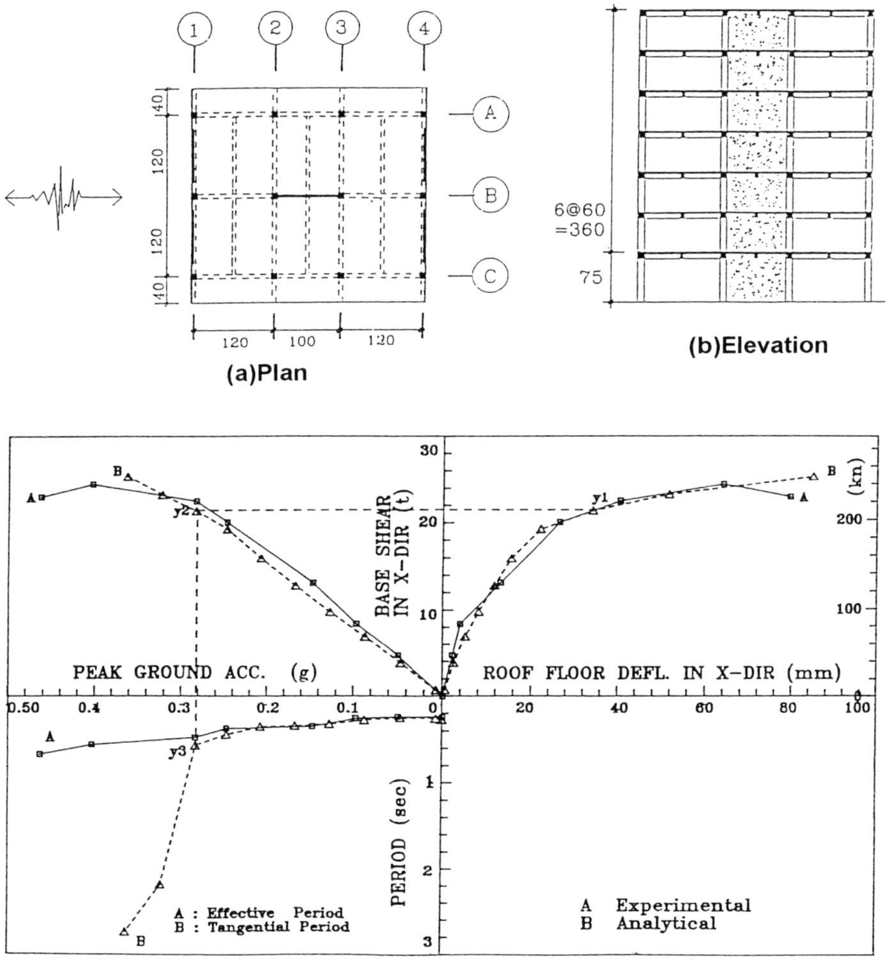

(a)Plan

(b)Elevation

(C) Diagnosis Chart

Fig.1 Vulnerability Diagram of Tested Specimen.

RETROFITTING STRATAGEMS

Fig.2 is a typical plan view for school buildings in Taiwan, especially for fundamental schools and middle schools. In general, it is two to four stories. Story height is about 400 cm. So RC shear walls or brick masonry walls are very useful for such low rise buildings to resist earthquake. However, there are tow deficiencies in the typical plan from earthquake resistant point of view:

1. Very few walls in X-direction (longitudinal direction),
2. Windowsill walls in X-direction reduce the net height and increase the stiffness of adjacent columns. These columns might absorb much shear forces and become brittle failure.

Retrofitting strategies proposed in this paper are shown in Fig.5:

1. To cut separation slits between windowsill wall and column face so as to normalize the height of column.

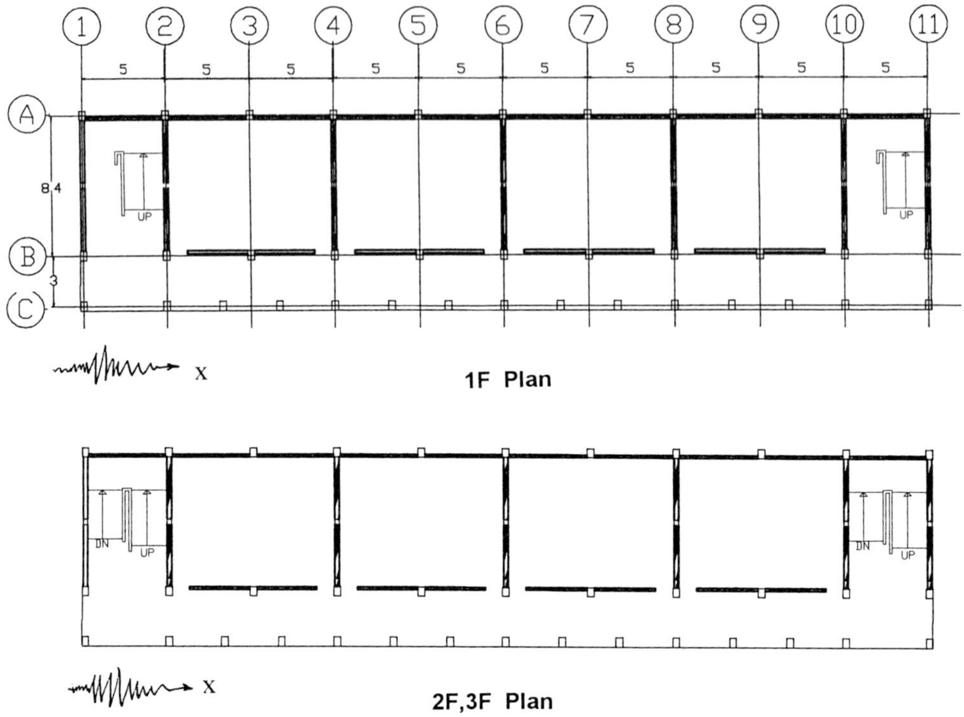

Fig.2 Example of School Building

2. To add plate hoops for all short columns so as to increase the capacity of column shear resistance.
3. To add RC wing walls in X-direction so as to increase the capacity of story shear resistance.
4. To cancel some wall openings by infilled shear walls so as to increase the capacity of story shear resistance.
5. Different combinations of the strategies listed above.

EXAMPLES

School building shown in Fig.2 will be employed as examples for retrofitting. The details of member size and reinforcing bar are shown in Hsu's paper (1994). Fig.3 is the earthquake risk curve of the site. Peak ground acceleration corresponding to 0.2% annual exceedance probability is 0.28 g. Since the important factor for school buildings specified by Taiwanese Building Code is 1.25, so 0.28 g should be multiplied by 1.25. In another words, the school building shown in Fig.2, with site risk curve shown as Fig.3, should not collapse at 0.35 g under earthquake. Otherwise, retrofitting is necessary.

Fig.4 is the site normalized elastic acceleration response spectra with different damping ratios. Using the technique of nonlinear spectra analysis described previously, the spectra shown in Fig.4 is input as external load in longitudinal direction (X-direction). The vulnerability diagram of the school building in X-direction is shown as "Original" in Fig.6. In which collapse PGA is 0.15 g. The collapse base

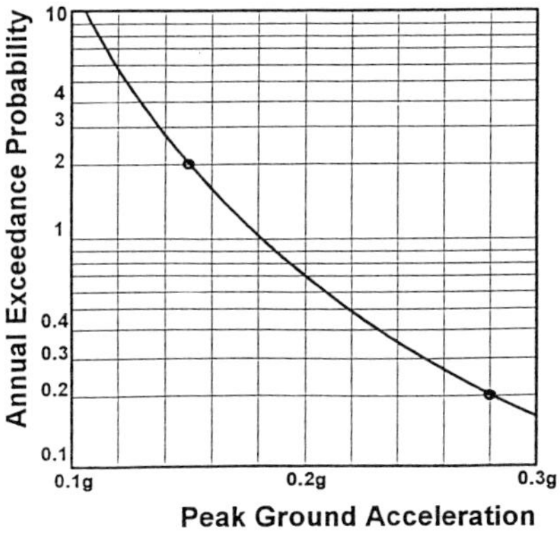

Fig.3 Earthquake Risk Curve at Site

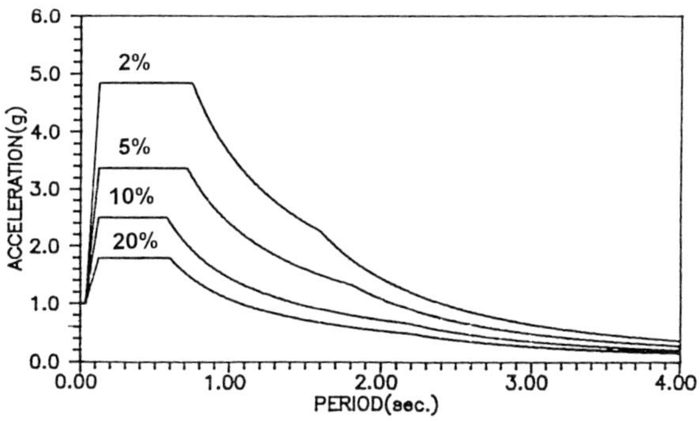

Fig.4 Normalized Elastic Acceleration Design Spectra at Site

shear is 9170 Kn. Roof deflection at collapse is. 20.1 cm in X-direction. Collapse PGA is less than 50% of the required 0.35 g. Failure mechanism of Frame A, B and C are shown in Fig.7. The natations in Fig.7 are:

C : flexural crack
y : flexural yield
U : flexural or shear ultimate
X : shear crack

From Fig.7, it is obvious that most of the columns in first story reach flexural ultimate. And the short columns on both ends of first story fail by shear.
Now retrofitting of the structure is necessary to increase the collapse PGA from 0.15 g to 0.35 g. 5 different retrofitting strategies are used to see how the collapse PGA and failure mechanism change.

1. Retrofitting S1: In Fig.2, Frame A, columns A1, A2, A10 and A11 in first and second stories are strengthened with plat hoops, SM50, 1cm thick, 10 cm wide and 25 cm spacing. Using nonlinear spectra analysis technique again, the vulnerability diagram after retrofitting S1 is shown as "S1" in Fig.6. Now, collapse PGA is 0.24 g, collapse base shear in X-direction is 11400 Kn, collapse roof

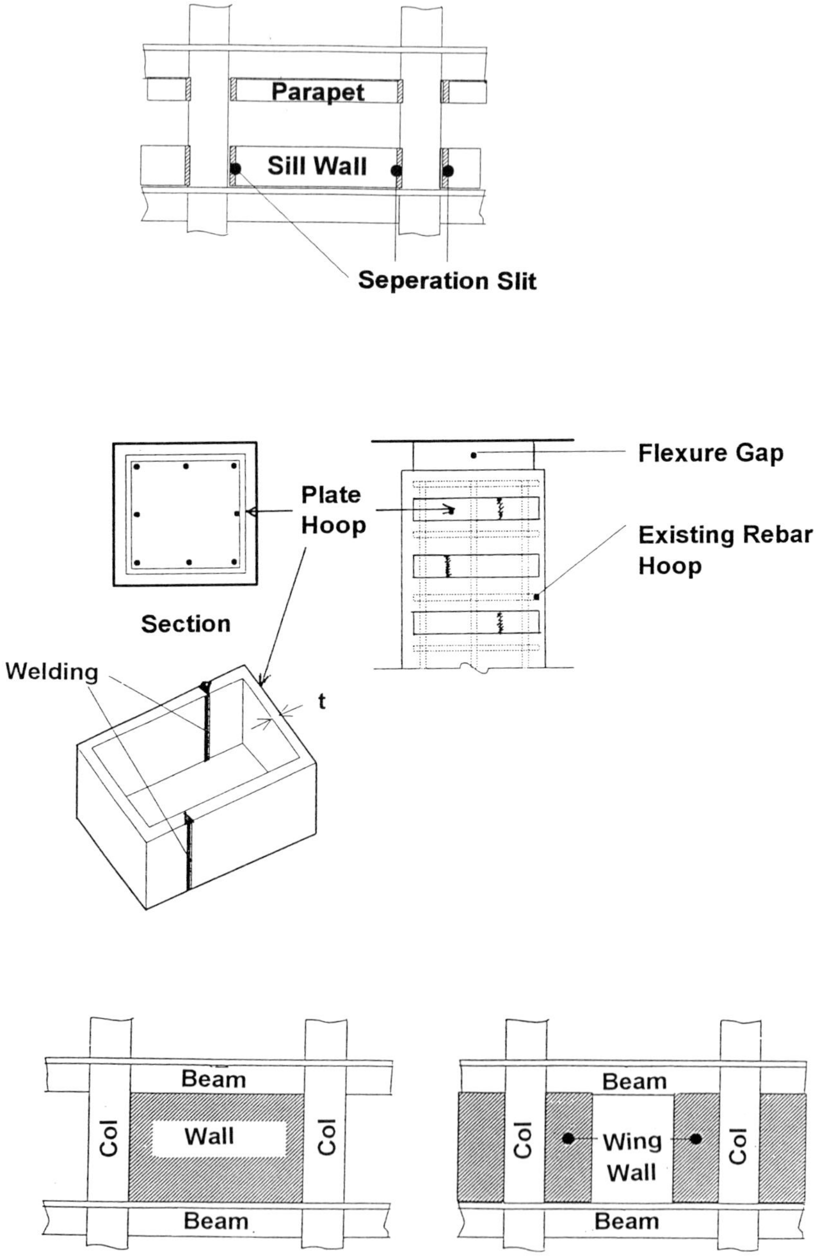

Fig.5 Retrofitting Strategies

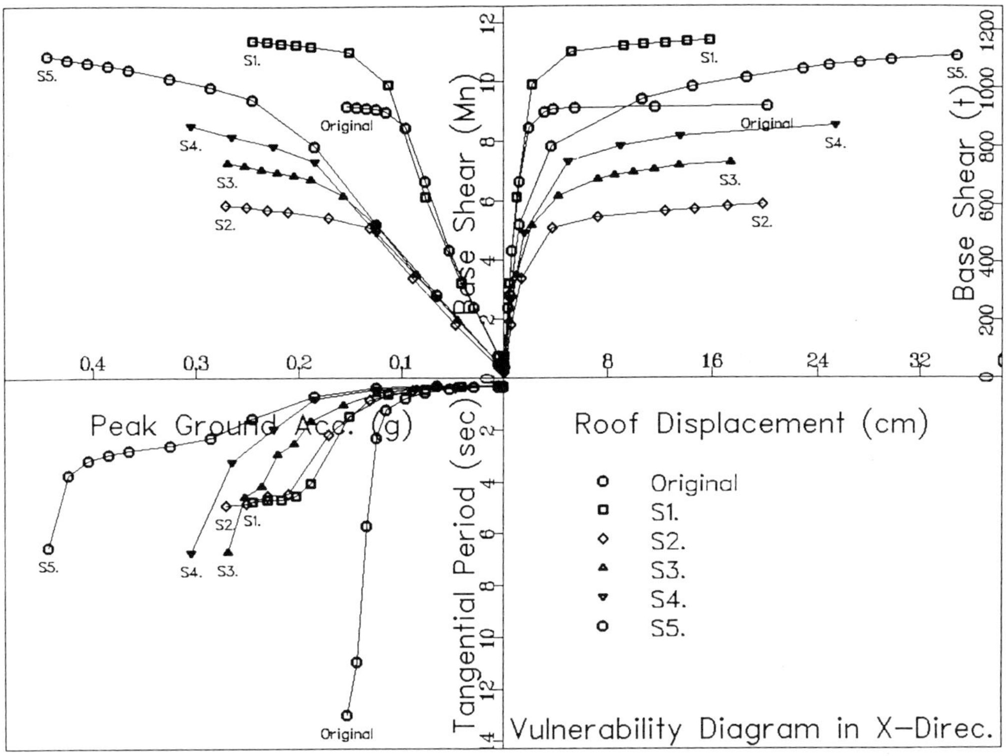

Fig.6 Vulnerability Diagram before and after Retrofittings

deflection in X-direction is 15.7 cm. Failure mechanism after retrofitting is shown in Fig.8. All the short columns beside windowsills in first story are flexural failure. Some more strengthening is necessary to increase collapse PGA up to 0.35 g.

2. Retrofitting S2: In addition to S1, short columns in Frame A and Frame B in first story will be separated from windowsill walls by 2 cm wide gaps. So that all the column height, except those in stair cases, in first story become normalized. After retrofitting, the whole structure is more flexible than before. It might absorb less earthquake force. “S2” in Fig.6 is the vulnerability diagram after retrofitting S2. Collapse PGA is 0.266 g. Collapse base shear in X-direction is 5900 Kn. Collapse roof deflection in X-direction is 19.7 cm. Failure mechanism after retrofitting S2 is shown in Fig.9. Since there are no separation slits for columns in second story, so 2 interior columns in Frame A and four interior columns in Frame B are failed by shear in second story.

3. Retrofitting S3: All the interior columns in Frame A and B will have separation slits in first and second story. The opening on right-hand side stair case of Frame A will be cancelled by infilled brick masonry wall in first story. The short columns in first and second stories of stair case will be strengthened by plate hoops, SM50, 1 cm thick, 15 cm wide and 25 cm spacing. “S3” in Fig.6 is the vulnerability diagram after retrofitting S3. Collapse PGA in X-direction is 0.264 g. collapse base shear in X-direction is 7300 Kn. Roof deflection at collapse in X-direction is 17.3 cm. Failure mechanism is shown in Fig.10. One interior column in first story of left-hand side stair case is shear failure because the net height of that column is very small.

4. Retrofitting S4: Retrofitting are the same as those for S3, except that the short columns in both stair cases in first and second stories are strengthened by plate hoops, SM50, 1 cm thick, 20 cm wide, 25 cm spacing. "S4" in Fig.6 is the vulnerability diagram. Collapse PGA increases to 0.299 g. Base shear at collapse in X-direction is 8500 Kn. Roof deflection at collapse in X-direction is 25.3 cm. Failure mechanism is shown in Fig.11. Almost all the columns in first and second stories get plastic hinges at one or two ends. And two interior columns in second story of Frame A are failed by shear.

5. Retrofitting S5: Retrofitting are the same as those for S4, except that the openings on right-hand side stair case of Frame A are infilled with brick masonry walls in first and second stories. "S5" in Fig.6 is the vulnerability diagram. Collapse PGA is 0.43 g. which is larger than required 0.35 g. The collapse base shear in X-direction is 10900 Kn. Roof deflection in X-direction at collapse is 34.9 cm, which causes 2.9% roof drift. Failure mechanism is shown in Fig.12. It is clear that almost all the columns become plastic hinges, and all brick masonry walls in both stair cases are failed by shear at final stage. It is also clear that after retrofitting "S5", Frame A and Frame B do not meet "strong column weak beam" requirement.

From the process of retrofitting S1 to S5, it is believe that the school buildings in Taiwan might be retrofitted by cutting separation slits between windowsill walls and column faces, by infilling brick masonry or RC shear walls in some stair cases and by adding plat hoops for short columns. The construction fee for these retrofitting is pretty cheap even for such poor structure as this example.

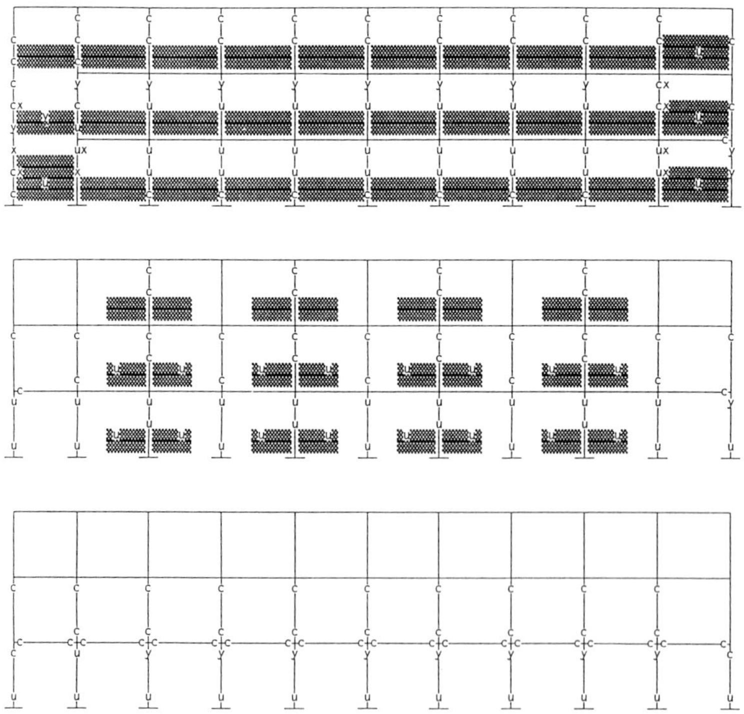

Fig.7 Failure Mechanism of Original Structure

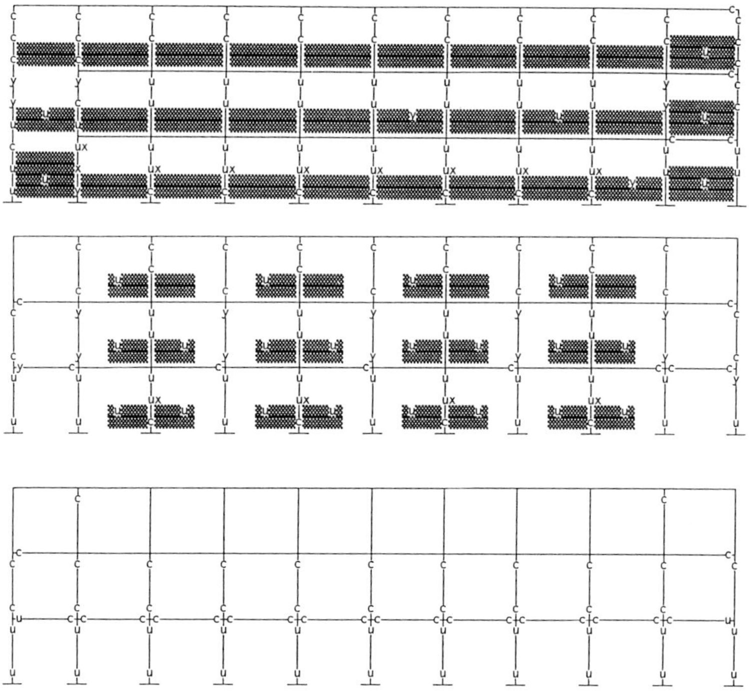

Fig.8 Failure Mechanism after Retrofitting S1

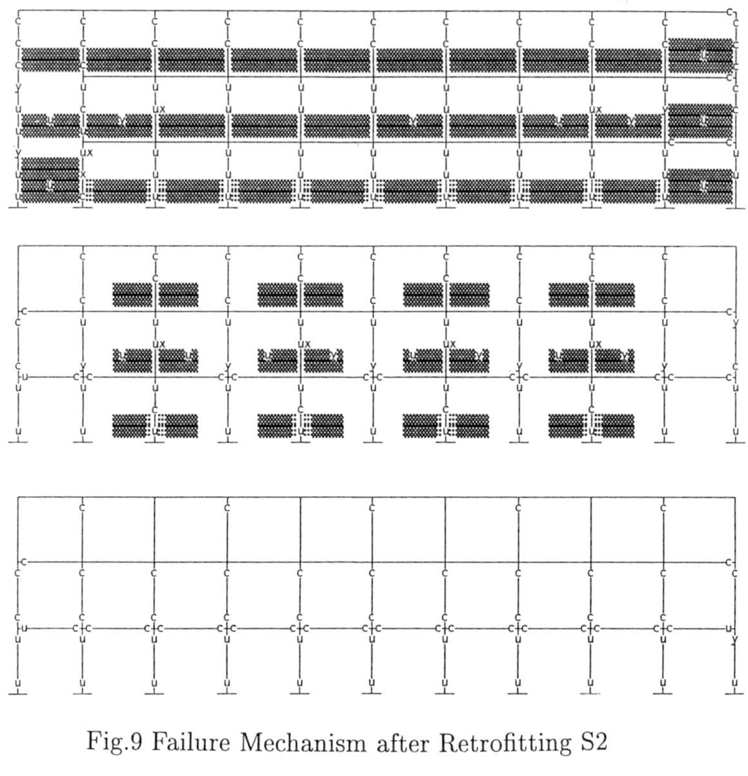

Fig.9 Failure Mechanism after Retrofitting S2

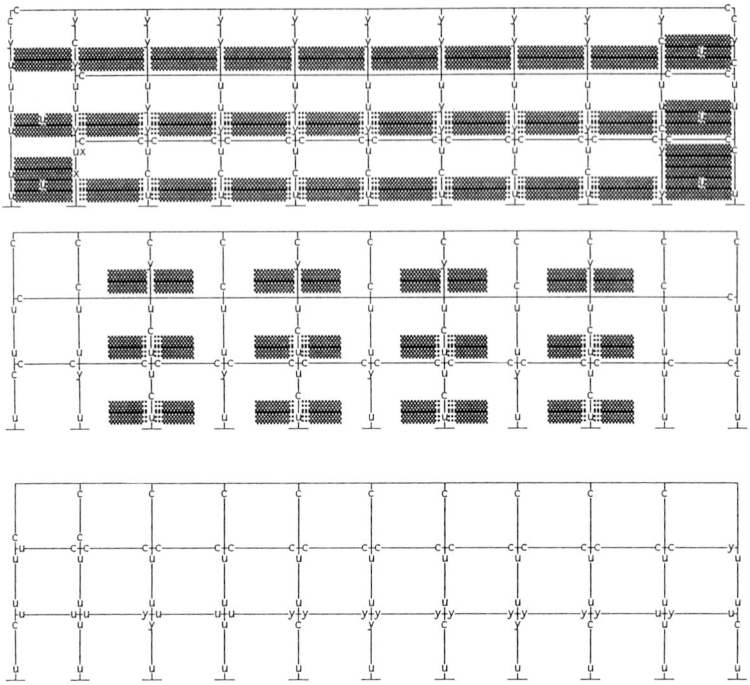

Fig.10 Failure Mechanism after Retrofitting S3

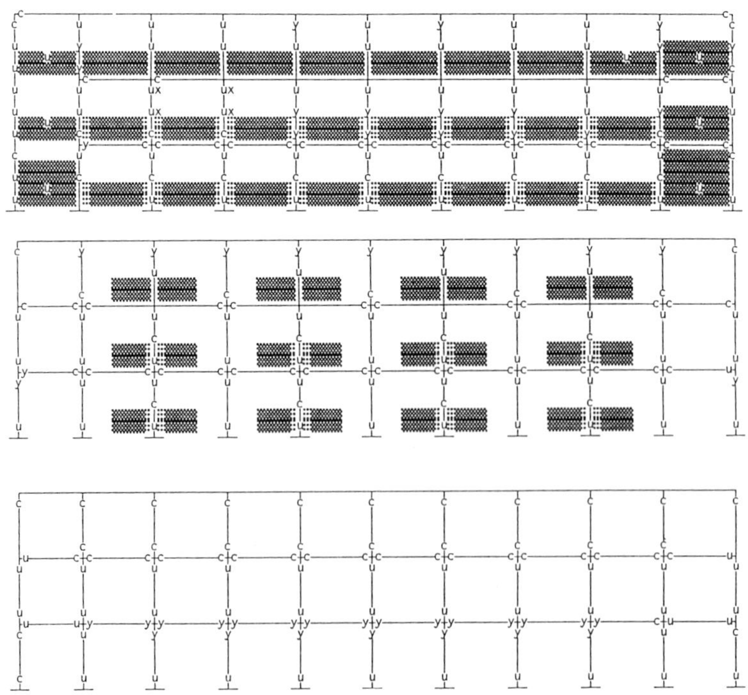

Fig.11 Failure Mechanism after Retrofitting S4

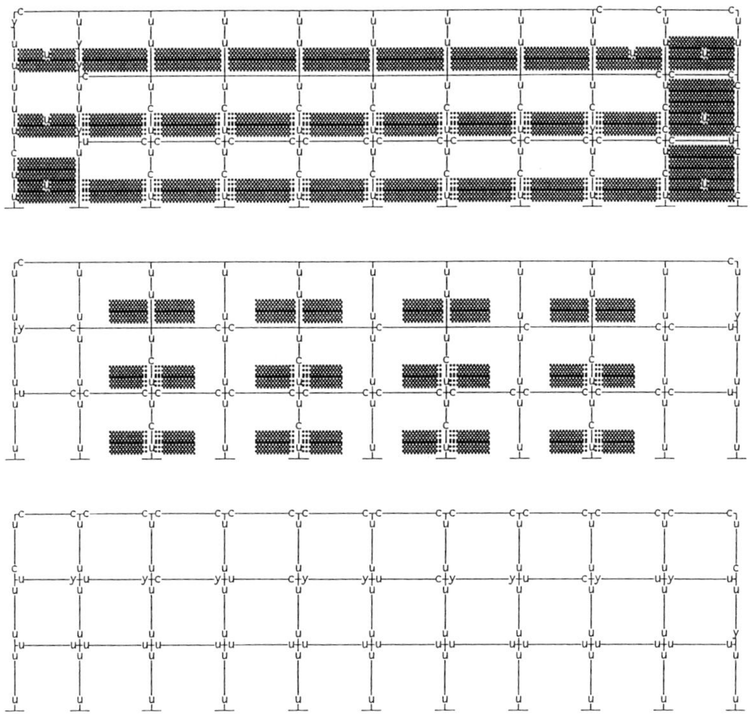

Fig.12 Failure Mechanism after Retrofitting S5

CONCLUSIONS

1. Nonlinear spectra analysis, proposed by this paper, is practical for vulnerability assessment for building structures before and after earthquake.

2. Most of the school buildings in Taiwan might be effectively retrofitted by cutting separation slits between windowsill walls and adjacent column faces, by infilling brick masonry or RC shear walls for some openings in stair cases and by adding plate hoops for short columns. The construction fee for these retrofittings is very cheap. However, they can increase at least 100% the collapse PGA after retrofittings.

REFERENCES

1. Bertero, O.V., Khan, A. E., Charney, F. A. and Sause, R. (1984). US-Jpan Cooperative Earthquake Research Program: Earthquake Simulation Tests and Associated Studies of A 7-Story RC Test Structure. Report No. NCB/EERC 84/05, Univ. of Calif., Berkely.

2. Hsu, C. C. (1994). Aseismic Diagnosis and Strengthening of RC School Structures, Master Thesis of Dept. of Architecture, National Cheng-Kung Univ., Taiwan, p. 159.

3. Sheu, M. S. and Chang, W. T. (1993). Damage Assessment for R C Buildings with Shear Walls, Structural Congress, *ASCE,* Irvine, Calif., USA.

4. Sheu, M. S. and Liu, P. M. (1994). Diagnosis of R.C. Structures after Man-Made Disasters — Fire Damge, 5th National Conference on Earthquake Engineering, Chicago, USA.

5. Sukhvarsh, J. and Milad M. (1981). Dynamic Stiffness of Concrete Beams, *ACI Journal.*

6. Wang, C. T. (1980). Aseismic Behaviours of RC Shear Walls with Openings but without Boundary Elements, Master Thesis of Dept. of Architecture, National Cheng-Kung Univ., Taiwan.

SYSTEMS INTEGRATED APPROACH IN NATURAL HAZARD MITIGATION

G. C. Lee and A. S. Dargush

*National Center for Earthquake Engineering Research,
SUNY/Buffalo, Red Jacket Quadrangle, Box 610025,
Buffalo, NY 14261-0025, U.S.A.*

ABSTRACT

Experiences of developing a systems integrated approach in earthquake hazard mitigation by the National Center for Earthquake Engineering Research will be presented, and its utilization and significance in natural hazards reduction will be discussed. Particular attention will be given to lessons learned from recent earthquakes, which both serve to validate and direct ongoing research for progress in hazard mitigation.

INTRODUCTION

The National Earthquake Hazards Reduction Program (NEHRP) was established in 1977, the product of increasing national concerns about seismic safety that were raised after the 1964 Alaska and 1971 earthquakes and compounded by scientific identification of the Palmdale bulge in Southern California. NEHRP was introduced as United States Congressional legislation to provide a comprehensive, integrated national program to reduce losses of life and property resulting from earthquakes. The Congress recognized that losses and disruption to the individual, the community, the State, and the Nation caused by earthquakes could be substantially reduced through the development and implementation of earthquake hazards reduction measures. Through the enactment of NEHRP, the Congress directed the Federal Government to provide a central focus for leading, coordinating, and conducting earthquake research hazard mitigation, and disaster preparedness. The Federal Emergency Management Agency (FEMA), the United States Geological Survey (USGS), the National Institute for Standards and Technology (NIST) and the National Science Foundation (NSF) have been charged with carrying out the principal Federal responsibilities outlined within NEHRP for the reduction of losses caused by earthquakes.

In 1990, NEHRP was amended by the NEHRP Reauthorization Act. Updated goals and objectives of the Act emphasize the improved understanding, characterization, and prediction of hazards and vulnerabilities; improved model building codes and land use practices; reduction of earthquake risks through education; development and improvement of design and construction techniques; improved mitigation capacity; accelerated application of research results; and exploration of mechanisms to improve the effectiveness of earthquake insurance.

Elements of NEHRP include:

- Fundamental Earthquake Studies
- Earthquake Hazard Potential
- Earthquake Effects and Engineering Research
- Planning for and Mitigating Earthquakes
- Information Systems and Dissemination
- Post-earthquake studies
- International Cooperation

Research conducted under NEHRP has led to significant advances in the knowledge and characterization of earthquake source mechanisms, estimation of recurrence intervals for historic earthquakes, more thorough understanding of ground motion and ground failure, and the effects of such phenomena on man-made facilities such as buildings, bridges, pipelines, dams, and critical facilities such as nuclear power plants, hospitals and emergency systems.

The National Center for Earthquake Engineering Research (NCEER) was established in 1986, by the National Science Foundation through its Earthquake Hazard Mitigation Program (EHM) in the Division of Civil and Mechanical Systems. The EHM program is one of NSF's contributors to NEHRP and is intended to assist the Foundation in meeting the following NEHRP-prescribed objectives:

- Support activities that improve the understanding of the tectonic framework that cause earthquakes and processes leading up to an earthquake
- Develop instrumentation and analytical methods for geotechnical site characterization
- Develop effective methods for determining the response of soil and earth masses to earthquakes and the interaction between soils and structures
- Develop analytical and experimental methods to predict the seismic response of buildings, lifeline systems and other major installations; develop engineering tools and advanced techniques to improve the seismic safety of structures; develop knowledge base for updating seismic design standards, codes and criteria
- Identify effective methods to improve seismic performance of architectural, mechanical, and nonstructural building components in both new and existing facilities; identify viable architectural methods to improve seismic-resistant design of new buildings as well as the repair and rehabilitation of existing buildings
- Develop and upgrade the nation's earthquake engineering and geotechnical experimental research facilities
- Understand societal and institutional issues related to earthquake hazard reduction which influence implementation of hazard reduction measures
- Support activities that promote the effective dissemination of earthquake engineering research findings
- Work with other agencies to conduct post-earthquake investigations
- Develop and encourage international research cooperation

ORGANIZATION	MECHANISM	GENERAL NEHRP OBJECTIVES
FEMA NIST USGS	• In-house Projects • Sponsored Projects	• Enhance knowledge and understanding of earth-quakes and their effects
NSF	• Individual Sponsored Projects	• Increase availability of information
	• Coordinated Sponsored programs (TCCMAR, RRRep, etc.)	• Target outreach and implementation efforts
	• NCEER: Systematically integrated interdisciplinary approach	• Transfer technology • Conduct problem-focused research
State and Local Government Other Mission Agencies	• In-house Projects • Sponsored Projects	• Encourage community adoption of codes

Fig. 1. NCEER's role in NEHRP

As shown in Figure 1, NSF uses a number of different mechanisms to conduct earthquake research. In addition to funding individual proposals which may be either unsolicited or invited in response to an NSF-defined initiative, NSF allows research to be conducted through coordinated sponsored programs such as TCCMAR (Technical Coordinating Committee for Masonry Research), the Repair and Rehabilitation Research Program for the Seismic Resistance of Structures, and PRESSS (Precast Seismic Structural Systems). With the USGS and FEMA, NSF also supports the Southern California Earthquake Center (SCEC), a cooperative research program to quantify the regional seismic hazard in Southern California and establish an integrated, relational model that can be used by scientists, engineers, and government officials to reduce risk.

The establishment of NCEER offers NSF an additional, unique mechanism to address earthquake engineering problems by using a systematically-integrated and interdisciplinary approach. Teams of researchers from institutions and organizations across the country are charged with advancing knowledge to minimize loss of life and property caused by earthquakes, with particular attention to the lesser-understood earthquake risk to the Central and Eastern US.

NCEER: A SYSTEMATICALLY INTEGRATED APPROACH TO EARTHQUAKE HAZARDS RESEARCH

NCEER Mission and Management

Since its establishment, the charge of the National Center for Earthquake Engineering Research has been to conduct mission-oriented research to advance the state of the art in earthquake engineering through both curiosity-driven and problem-focused research and a parallel program to promote knowledge utilization. To most effectively achieve the goals, the NCEER research program recognizes and reflects the interdisciplinary nature of disaster research and management, involving multidisciplinary, cross-professional teams to identify knowledge gaps, develop potential solutions, evaluate applications and recommend improvements in seismic methodologies and policies. Experts are drawn from fields of seismology, geotechnical and structural engineering as well as economics and the social sciences and represent academia, industry and the public sector. Teams contribute to four primary research concentrations which focus on the seismic performance, retrofit and enhanced design of buildings, nonstructural systems, lifelines (including gas, oil, water, electric power and telecommunications), highways and bridges. Studies in these problem-focused areas incorporate integral fundamental knowledge on ground motion and the seismic hazard, geotechnical influences, risk assessment, social systems and economics which is contributed by experts in these fields. Research on smart materials and intelligent systems for structural vibration control is conducted, as well, to develop innovative approaches to seismic design.

NCEER research is funded by the National Science Foundation, which requires that an additional and equal amount of matching funding be obtained from external non-federal sources. NCEER's leveraging of NSF research dollars has allowed the development of a cost-effective program, which can conduct fundamental research while using additional funding to support NCEER's efforts in knowledge utilization and implementation. The multidisciplinary, cross-professional nature of NCEER's research teams, which represent more than 80 individuals at institutions in 20 states, also provides additional opportunity for involvement in and contribution to other hazard-related research. In addition to this resource of technical expertise, the involvement of multiple institutions provides access to diversity of experimental and testing facilities, including shaking table and other structural testing devices, soil testing laboratories and centrifuge facilities.

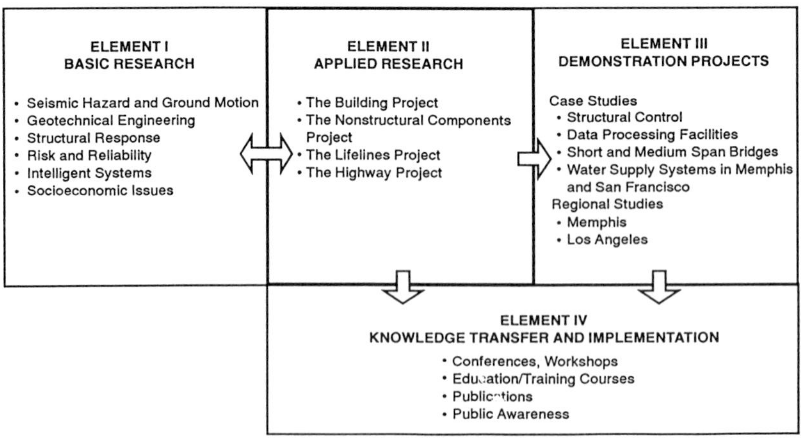

Fig. 2. NCEER Research and Implementation Plan

The Systems Approach to Earthquake Mitigation Research

The NCEER Program is driven by four problem-focused projects: Buildings, Nonstructural Components, Lifelines and Highways and Bridges. These projects rely on fundamental research programs in Seismic Hazards and Ground Motion, Geotechnical Engineering, Structures and Systems, Risk and Reliability, Intelligent and Protective Systems and Socioeconomic Issues. The interrelationships between fundamental and applied studies as shown in Figure 2, are described in the following paragraphs.

Earthquake damage to buildings and their nonstructural systems and components can lead to substantial loss of life and economic disruption due to property damage and business interruption. To address this problem the Center's **Buildings Project** is directed towards two classes of buildings that are common to the United States, unreinforced masonry and lightly reinforced concrete buildings. Both types are vulnerable to moderate and even low levels of shaking. Objectives are to develop rational methods for evaluating these structures and to devise cost-effective methods for retrofitting to provide adequate seismic resistance. To achieve this, studies have included the definition of the seismic and geotechnical hazard, refining conventional strengthening methods, applying smart materials and innovative protective systems, assessing fragility functions, improving loss estimation methodologies, examining relationships between building type and mortality and injury, and establishing reliable inventory methods for existing building stock. The **Nonstructural Components Project** focuses on minimizing damage to equipment and contents of buildings by improving methods of analysis and design and development of innovative support and restraint systems for specialized items such as computer mainframes, hospital equipment and museum artifacts.

Damage sustained by lifelines in recent earthquakes and the multimillion dollar losses that result from service disruption, have reinforced the need to perform comprehensive studies on these elements of the built environment. Motivated by the damaging potential for fire following earthquakes, NCEER's **Lifelines Project** began by performing detailed studies of the water supply systems in San Francisco and Memphis. These studies were later expanded to oil and gas transmission lines in the central United States and more recently to include electric power and telecommunications systems. Geotechnical studies have focused on liquefaction and other sources of large ground deformation which have serious consequences for buried lifelines. Systems modeling, fragility assessments, the economic impacts of service disruption and societal consequences of releases of hazardous materials have also been examined.

Transportation systems are major lifelines which form the focus of the **Highway Project.** All components of both new and existing highway systems are being studied, including short and long span bridges. Again, the research areas include defining the seismic and geotechnical hazards, developing both conventional and innovative retrofit methods by using smart materials and protective systems, applying risk assessment methodologies to highway networks, defining importance and performance criteria, improving prioritization algorithms, and estimating the highway stock. The development of manuals for the evaluation and retrofit of existing lifelines and the preparation of performance-based standards for all new lifelines are long term goals.

Thus, the NCEER research program uses a matrix approach to the development of earthquake mitigation advancements. Vertical systems link discipline-specific research findings with user needs, horizontal systems allow integration of multidisciplinary issues, and hybrid systems combine vertical and horizontal systems approaches to broaden the potential applicability of the research findings to a more extensive base of user communities.

Case studies and demonstration projects also play a valuable role in the NCEER research program, allowing opportunities to validate assumptions and models and to evaluate the practical application of technical solutions. NCEER places a strong emphasis on post-earthquake reconnaissance investigations and international cooperative research. Both have, in selected instances, provided opportunities for the types of case study research which are so valuable to the research process.

To effectively disseminate the findings of the Center's research program, NCEER conducts an active program in knowledge and technology transfer. The program consists of traditional activities such as publications, computational tools, workshops, conferences, graduate student education, and professional development courses for the practitioner. NCEER also supports less traditional mechanisms for the transfer of research findings, such as the NCEER Information Service, which provides relevant reference materials to interested users through a computerized, query-driven bibliographic database. Other methods which are considered nontraditional include the involvement of end users on research teams, participation of academic researchers in codes and standards development, and establishing active research partnerships with industry.

Because of the nature of the Center's approach to research, its accomplishments range from the discovery of basic knowledge to the increased public awareness of the earthquake hazard. In addition to its general achievements in the area of research management by interdisciplinary coordinated teams, optimized spending of available research funds and the raising of seismic awareness, the Center also reports some specific technical accomplishments. These achievements toward earthquake hazard mitigation include: enhanced understanding of potential seismic hazards under varying site conditions across the US, the development of design and guideline recommendations for buildings, bridges, lifelines and nonstructural components at State and Federal levels; research, development and implementation of innovative approaches to structural vibration control; and assessments of risk and impact of earthquakes on socioeconomic systems. A review of selected accomplishments which have resulted from NCEER research is provided in the following section.

LESSONS LEARNED FROM RECENT DISASTERS AND THEIR IMPACT ON NCEER MITIGATION RESEARCH

Santa Cruz, California - 1989

In 1989, a magnitude 7.1 earthquake occurred about 60 miles to the southeast of the Bay area cities of San Francisco and Oakland, California resulting in loss of life, and damage to the socioeconomic and civil infrastructure of the area. But for the civic attention of the community to baseball's World Series, the toll would have likely been greater. Because of the seismic hazard of the area, NCEER had already been involved in research projects focused on the area, but the 1989 event, provided an opportunity to validate some of these ongoing studies and to redirect some other relevant NCEER research to benefit from lessons learned from the Santa Cruz earthquake, known popularly as Loma Prieta. Among these studies are the collection of ambient ground motion measurements at critically damaged bridge sites for future site response studies, evaluations of bridge design and performance, application of an existing building stock inventory methodology for eastern US urban areas to Bay area damage estimations, execution of a case control study on morbidity and mortality, and a workshop to examine the effectiveness of preparedness, response, and recovery of the Bay area to the 1989 event.

The serviceability of water supply systems in San Francisco has been studied by NCEER since 1987, along with their impact on the control of fire following earthquakes. The objectives of the study

(O'Rourke, 1990) have been to promote advanced computer graphic techniques for lifeline systems and to validate these methods through application to a real system, while demonstrating potential improvements which can be made to existing systems to improve seismic performance. This cooperative study has involved NCEER researchers at Cornell and EQE, as well as end users in the Water and Fire Departments of the City of San Francisco. A computer code developed at Cornell for the hydraulic analysis and seismic evaluation of water delivery systems, GISALLE (**G**raphical **I**nteractive **S**erviceability **A**nalysis for **L**ife**l**ine **E**ngineering), has been structured to account for various flow patterns likely to occur in damaged systems. The code has been used to help improve the seismic capabilities of the auxiliary water supply system and the emergency plans of the San Francisco Fire Department. The strong working relationship established between NCEER researchers and officials of the City of San Francisco Water and Fire Departments as a result of this effort was instrumental in the passage of a $46.2 million bond issue to improve the seismic response of the city's water supply system prior to Loma Prieta.

The computer code was later validated against pressures and flows demonstrated by damaged systems in the Marina District after the Loma Prieta earthquake and was proven to be consistent with these measurements. The effort has emphasized the contribution of site response in damage to lifeline systems and that areas of high likelihood of water main breakage during earthquakes can be identified with existing modeling techniques. Emergency response can then be augmented by the availability of portable water supply systems such as those which were put in place by the City of San Francisco prior to the 1989 event.

Loma Prieta also caused disruption to utility services in areas closer to the epicenter, such as the community of Watsonville, which experienced damage and disruption to its electric power and water supply systems that strained the emergency response capabilities of the city. NCEER researchers conducted a followup study to determine how a community provides essential services to its residents in spite of lifeline damage and to examine the mechanisms in place to restore those services (Isenberg *et al.,* 1990). The most significant problem observed in Watsonville was the failure of electric power and the importance of availability of electric generators to enable continued operation of the local hospital, water well pumps and sanitary system lift stations. Loss of lifeline service, and, in particular, that of electrical power, was a primary source of damage and economic interruption to businesses in the area.

Earthquake injury epidemiology may be defined as the study of the distribution of death and injury in earthquakes and the causes of fatal and nonfatal injuries. Causal mechanisms may range from structural and nonstructural performance to emergency response and occupant behavior. An ongoing NCEER study (Jones *et al.*, 1994) on earthquake injury epidemiology redirected its attention to the Bay area after Loma Prieta to assess the relative risk for physical injury associated with different physical environments in disaster and post-disaster phases and to estimate the absolute risk of physical injury mortality and morbidity associated with the earthquake. The overall objective of the effort is to provide data for casualty and loss estimations in California and to establish methodologies which are transferrable to other areas of the US.

Nansei-Oki, Japan - 1993

NCEER researchers continued their investigation of the causes and effects of post-earthquake fire after the July 12, 1993 Hokkaido Nansei-Oki, Japan earthquake. The area impacted by the 7.8 event was affected by ground shaking, followed within five minutes by a damaging tsunami. Damage caused primarily by the tsunami resulted in widespread fire that was largely responsible for the destruction of the town of Aonae. Among the lessons learned from reconnaissance studies conducted after the event (Yanev and Scawthorn, 1993) are that the triple hazard presented by earthquake shaking followed by tsunami and fire can be devastating to modern cities. Tsunami walls constructed were found to be inadequate, as was the tsunami warning system in the area of the epicenter.

Northridge, California - 1994

The Northridge, California earthquake of January 17, 1994 was revealing in many ways. While those historically seismically vulnerable structures performed as expected, some structures designed to more exacting standards and some with better typical performance histories suffered unexpected damage. More specific vulnerabilities of lifelines were revealed, with additional attention to pipe materials and connection techniques. Lifeline networks have expanded as warranted by the rapid growth of Los Angeles, but lifelines vulnerabilities may place unexpected burdens on response and recovery mechanisms in place.

The Northridge earthquake has provided yet another opportunity to confirm findings from ongoing research studies and to redirect other studies to explore some of the developments which have been unique to this event. Some of these activities are summarized in the following paragraphs.

Correlation of analytical and observed seismic performance of selected building structures. The purpose of this project is to obtain a quantified picture of performance of building structures damaged or just shaken by the January 17 earthquake, correlated with the costs of repair or replacement and perceived levels of acceptable risk. Information generated by the study will be used for development of design guidelines based on acceptable performance. New computational tools developed by NCEER for evaluation of inelastic seismic response will be validated against observed damage data. It is expected that the results of this research will be immediately disseminated toward the development of performance-based design criteria by NEHRP and ATC-33. Other products which will be developed for the engineer in practice include reports correlating observed and calculated performance, and computational software which can be used for quantification of seismic performance.

Earthquake performance of water supply and electric power systems. Accurate evaluations of the seismic performance of lifeline systems are critical to the minimization of potential physical damage after an anticipated earthquake, as well as reduction of the attendant repair-restoration work and resultant system interruption. The objective of this study will be to validate, calibrate and improve selected tools, methods, and procedures for analyzing and evaluating the seismic performance of lifelines by comparing the predicted performance with the actual system performance after the Northridge earthquake. For this evaluation, performance is measured in terms of physical damage, system functional performance, extent of post-earthquake repair/restoration effort and attendant down-time.

NCEER Researchers will work closely with Los Angeles Department of Water & Power (LADWP) to develop improved procedures that can then be used to better predict lifeline performance in future earthquakes for the LADWP and other lifeline systems. The same tools, methods and procedures will be used to evaluate the seismic performance of the LADWP systems after the 1971 San Fernando earthquake, to examine the effectiveness of retrofit measures which were implemented after the 1971 event.

Performance of the nonstructural contents of selected buildings and their implications for the NEHRP Building Provisions. The Northridge earthquake is viewed as one of the most information rich events for addressing issues related to the observed seismic behavior of nonstructural components. The objectives of this NCEER study are to assess the effectiveness of recently recommended design criteria accounting for various dynamic effects on nonstructural components and to evaluate the performance of a selected group of seismic restraints used in the affected area. It is expected that these studies will lead to the development of safer and more rational design guidelines for operating these components in a seismic environment.

Influence of the spatial variation in the ground motion on the performance of the I-5/SR14 Separation and Overhead. Spatial variation in the ground motion at the site of the southbound separation and overhead at the I-5/SR14 highway interchange may have contributed to the collapse of this structure. In the first part of this NCEER project, ground motions at the site are estimated based on the aftershock records and evaluation of the seismology of the region. This is then followed by a series of structural analyses of increasing complexity to determine the consequences of spatial variation on structural performance. The results of this study will have application to the design of future long-span and multi-span structures as well as the retrofit of existing structures of this type.

On a more general note, much of NCEER's work will, to some degree, relate to or benefit from, the lessons learned from the Northridge earthquake. One particular example is an NCEER study which examines the hazard perception, mitigation and preparedness activities of lifeline organizations in Southern California. In addition, the study builds on work initiated in 1992 to survey business disruption and impact following the Northridge earthquake, to assess the effectiveness of disaster assistance programs and identify problems encountered by businesses during the recovery process.

Another ongoing effort involves NCEER researchers and colleagues at the Southern California Gas Company. the objective of the collaborative effort is to identify potential vulnerabilities of the Company's extensive gas system. The information will also be used by management to develop preparedness response and repair strategies.

Additional ongoing efforts may not only shed additional light on issues and problems commonly raised by many earthquakes, including Northridge, but will be able to provide tangible evidence from the January 17 event to validate existing hypotheses and problem-solving approaches.

Hurricane Andrew, Florida - 1992

As a result of NCEER's ongoing research in the area of lifelines, NCEER affiliates were selected by FEMA to conduct a study of the effects of Hurricane Andrew on critical lifelines facilities, including electric power, water and sewage, telecommunications, transportation, gas and oil. Hurricane Andrew hit South Florida on August 24, 1992, causing $15 - 20 billion in damage and leaving well in excess of 100,000 people homeless. Andrew ranks as the one of the worst US natural disasters since the 1906 San Francisco earthquake, with a major impact on critical lifeline facilities.

The study involves collection of information on direct physical damage to lifelines, actual post-hurricane performance of lifeline systems, interaction between lifelines and the recovery of lifeline systems. The information will contribute to an assessment of the overall economic impact associated with lifeline disruption. The perishable case study information will be permanently archived by NCEER for use by others in future related studies.

In a related effort, NCEER researchers have also explored the adaptability of social accounting matrices for the assessment of overall economic damage arising from primary and secondary effects of earthquakes. These matrices represent structures which serve as nodes or links in an economic system. The frameworks are then used to simulate potential vulnerabilities of the system when impacted by a disaster and to model potential recovery strategies. The transferral of this methodology to hurricanes is being evaluated. Work conducted in the Homestead, Florida area in the aftermath of Hurricane Andrew inventoried the extent of damage which will later be integrated with other data using statistical and indirect methods to infer distribution of risk. The objective will be to develop a method by which risk models can be validated and can be used to effect response and recovery programs.

The Homestead study has indicated that hard information on damage is still unavailable months after the event, confirming the need for more emphasis to be placed on the prioritization of data collection, after disasters.

MULTIPLE HAZARDS RESEARCH IN TAIWAN: ITS RELEVANCE AND IMPACT ON INTEGRATED EARTHQUAKE HAZARDS MITIGATION IN THE UNITED STATES

The impact of the seismic hazard which exists in Taiwan is further enhanced by frequent typhoons, heavy rainfall and a rugged topography which often leads to subsequent flooding or landslides. It has been estimated that, between 1961 and 1982, natural disasters in Taiwan were responsible for the deaths and injury of 456 people annually, with a further loss of nearly 12,000 homes each year. A major effort on disaster mitigation and prevention was initiated by the National Science Council in 1981 with the establishment of the Multiple Hazards Mitigation Research Program. This interdisciplinary effort began a five-year research plan in 1982 focusing on the following areas: meteorology, flood mitigation, seismology, earthquake engineering, and landslide prevention. In 1987, the second five-year plan expanded into the additional areas of geotechnical engineering (incorporating landslide prevention), industrial hazards mitigation, and social and economic studies for hazards mitigation.

Among the scientific objectives of the second five-year plan are to provide ground motion data for critical engineering design purposes, to use ground motion data to refine risk estimates to existing construction, to improve building codes to provide guidelines for reasonable earthquake resistant design and reinforcing methods, to use information on geotechnical conditions and behaviors to improve zoning and siting procedures, and to study the processes which contribute to the implementation of hazard reduction strategies as well as those which influence quick and efficient post-disaster recovery. These overall goals parallel many of those defined by disaster researchers in the United States and therefore provide a natural forum for exchange of valuable findings on disaster and hazards management between Taiwan and the US.

The following paragraphs describe two areas of the research which have been conducted under the umbrella of this research plan which address both fundamental and problem-focused issues and have particular relevance to interdisciplinary research agenda of NCEER.

Seismic Monitoring for Site Characterization

Recent California earthquakes have underscored the effect that stratigraphic heterogeneities and topographic irregularities can have on the propagation of ground motion. Lessons learned from

Mexico City have substantially highlighted the impact of site effects on the performance of structures and the loss of life and property that can result. Knowledge of subsurface strata structure are needed to more fully understand the complexities of observed strong ground motion.

Two damaging earthquakes occurred in Taiwan in 1986, a magnitude M_L 6.5 in the city of Hualien, followed some months later by a M_L 6.8 event offshore Hualien. While damage was experienced in Hualien, both events caused more damage in Taipei, located in an alluvial basin about 110 km. from the epicenter. Studies carried out in Taiwan have collected seismic data from an array of instruments placed in areas which are particularly vulnerable and have demonstrated site amplifications of seismic ground motion.

The SMA-1 accelerograph network of Taiwan was initiated in 1972 by the Institute of Sciences, Academia Sinica. Instruments are distributed throughout the island, installed in free-field sites, in buildings and building basements, at dam sites and other locations. A second network has also been maintained by the Taiwan Central Weather Bureau. Further regional detail is provided by the SMART1 array of accelerograms in northeast Taiwan, near the city of Lotung, and the Lotung Large Scale Seismic Test (LSST) three-dimensional array, installed by the Taiwan Power Company in cooperation with the Electric Power Research Institute (EPRI). An additional 31 strong motion stations have been installed in the Taipei basin and installation of a regional strong motion monitoring network is anticipated.

The Large Scale Seismic Test Program (Chen *et al.*, 1988) was established to study the effect of soil-structure interaction on the response of a nuclear power plant containment facility. Methodologies developed by the program have been viewed as effective and efficient in solving the three-dimensional dynamic response of a soil-structure system.

Additional studies have contributed to fundamental knowledge on the correlation between excess pore water pressure and seismic waves, dynamic characteristics of soil deposits, shear wave velocities of alluvial deposits, spatial variation and directionality of seismic waves, and modeling of scattering effects due to geological irregularities of subsurface structure. Data, models and approaches generated by this research may prove to be useful to US studies which seek to characterize seismic risk at various sites. Understanding of site response is essential to assess and improve the seismic performance of lifelines and to determine appropriate response spectra.

Innovative Structural Control

Structural control research has been identified as a research need by the National Center for Research in Earthquake Engineering in the Republic of China. Research studies to date on active and passive control have been largely analytical; expanded experimental facilities now offer the opportunity to validate analytical models through laboratory and field testing.

In passive control (base-isolation) studies, analytical studies will investigate the effects of long period motion, randomness of parameters and interaction between horizontal ground motion and vertical torsional and rocking modes, as well as design issues. Experimental investigations will include validation tests, large model testing and materials properties testing.

Analytical studies of active control methodologies examine optimal location of sensors, time delay in control system and control algorithms, randomness in system parameters, and reliability issues. Experimental emphases will be placed on testing of full-scale models, evaluating the effect of aging on structural performance, the reliability of active control methodologies, and development of efficient analytical methods.

A cooperative study between NCEER and the National Science Council of Taiwan (Chang *et al.*, 1993) has investigated some of the practical issues associated with the application of viscoelastic dampers in buildings. An issue of concern has been the effectiveness of the devices under high environmental temperatures. An objective of the study has been to develop a design procedure for viscoelastic dampers, taking into account ambient temperature, by following a modal strain approach. Studies suggest that this approach can be reliably used to predict the equivalent structural damping which the dampers provide to the structure even at very high temperatures. The experimental and analytical aspects of this study have combined to develop a rational procedure which could be used for design and retrofit of structures with added viscoelastic dampers.

Recent research (Tsai, 1993) has also investigated the effectiveness of welded steel triangular plates for seismic energy dissipation in structures. Experimental results have indicated that a properly-welded steel triangular-plate added damping and stiffness device (TADAS) can sustain a large number of inelastic cycles without evidence of stiffness or strength degradation.

Findings from studies such as these contribute further to the collective understanding of potential applications of structural control throughout the world as economic earthquake resistant methodologies.

The two extensive efforts described in the preceding sections are examples of studies which are compatible with those efforts of US researchers who also seek to better understand strong ground motion and seismic characteristics for various sites to influence the selection and regulation of design criteria, and to develop innovative technologies and approaches to reduce damage to structures. These studies reflect the need to identify and address fundamental knowledge gaps, not only to advance the state of knowledge but to contribute to problem-driven research investigations. Both the research findings of these Taiwan studies, as well as the coordinated approach used to conduct these programs should be useful to researchers in the US and throughout the world.

IMPLEMENTATION: FORGING BENEFICIAL SCIENTIFIC RELATIONSHIPS BETWEEN ACADEMIA AND THE PRACTICING COMMUNITY

It is well recognized that the implementation of research findings and application of potential seismic solutions is critical to the effectiveness of natural hazards mitigation efforts. Knowledge transfer is a complex and dynamic process which is influenced by many external variables which in turn affect the ultimate utilization of research findings. Experts in this area of communication agree that one of the most effective vehicles for knowledge utilization is through social interaction (Lagorio *et al.*), involving researchers and users from many different disciplines to maximize exchange of information on respective needs, potential applications and limitations. The principal assumptions of this model for social interaction are that the greater the systematic organization and coordination between user and researcher, the more likely that the research knowledge will be used; also, the greater the number and variety of end user communities, the more likely the user is to be innovative and to use new ideas. Such two-way exchange is particularly useful in the natural hazards field because it helps to delineate these end user communities, the potential relevance of research findings, and the timing of knowledge transfer needed to meet the needs of the users.

A number of obstacles exist, however, which may prevent the social interaction needed for knowledge transfer and utilization. The collaboration envisaged is not always one which is built of a natural partnership. The earthquake community continues to emphasize the need for earth scientists to interact with engineers, for engineers to work more closely with social scientists, and for researchers to more actively involve potential end users in the research process. Successful collaborative relationships must often be developed over time and demand a basis of mutual trust. In addition to awareness among collaborators of a recognized need and demand for research, and the development of practices and applications, users must also appreciate the credibility and cost benefits of the technological solutions.

Other obstacles include insufficient resources for research and development, such as limited funding, availability, expertise and motivation of staff, and lack of time. Implementation of engineering solutions is complicated by the need for a balanced consideration of technical, societal, and political issues and the existence of complex processes which must be negotiated to implement innovations. The presence of well-informed advocates within the process is viewed as essential to effective implementation.

It is clear that an outstanding factor in overcoming these obstacles is the education of many constituencies. Awareness of seismic risk and potential solutions is needed, not only among policy makers who regulate and fund, but among professionals who may face the need to implement these solutions in practice. In identifying fundamental knowledge gaps, a research consensus must be reached between earth scientists, engineers, social scientists and economists. To conduct useful problem-focused research, researchers must more fully appreciate the needs of the practicing professional, and the demands which control their ability to address these needs. Education of these varied groups can be accomplished through a number of different mechanisms such as: involvement of researchers and policy makers in problem-focused research projects; public awareness outreach activities; publications; workshops and conferences; continuing professional development courses to improve technical competencies of the professional; and involvement of professional associations.

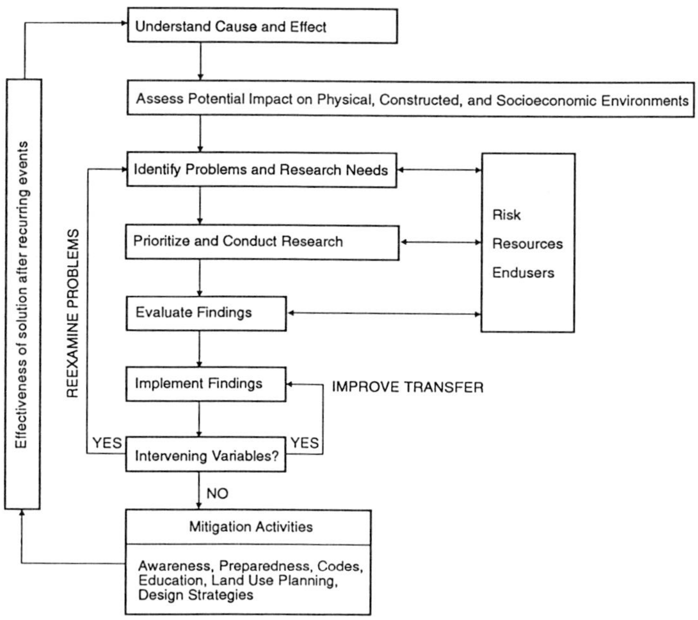

Fig. 3. From Natural Hazard to Mitigation - An Iterative Process

CONCLUSION: SYSTEMATICALLY INTEGRATED EARTHQUAKE HAZARD RESEARCH - AN EFFECTIVE MODEL FOR MULTIPLE HAZARDS MITIGATION?

Natural hazard mitigation can only be realized after a long process; from awareness and understanding of the problem, through problem-focused research, to the development of potential solutions, and to transfer this knowledge to implement the solution. Further, as shown in Figure 3, the process is not a linear one, but one which may be modified substantially at intervals within the process by changes in the needs and resources of users, the responsiveness of the research community, the regulatory process or the effectiveness of an application under earthquake conditions. A systematic and interdisciplinary, cross-professional research program is viewed by many as a mechanism which can potentially minimize the time between idea and concept and life-saving application. The approach links research with practice to advance the state-of-knowledge and the state of practice, while providing participants access to physical and intellectual resources. The systems approach demands that research spending be prioritized, thus insuring timeliness of effort for user needs. It also establishes mechanisms and networks to capitalize on windows of opportunity that can further hazard mitigation research. While much remains to be accomplished, it is hoped that NCEER's experiences in earthquake research and knowledge transfer will serve as useful examples for a possible approach to multiple hazards mitigation.

REFERENCES

Chang, K-C., T-T. Soong, M-L. Lai and E.J. Nielsen (1993). Viscoelastic Dampers as Energy Dissipation Devices for Seismic Applications. *Earthquake Spectra, Volume 9*, number 3, 371-387.

Chen, H-T., C-H. Loh and D-S. Juang (1990). Structural Control Research at National Center for Earthquake Engineering Research of the Republic of China. *Proceedings of the US National Workshop on Structural Control Research*, G.W. Housner and S.F. Masri, editors, 104-106.

Chen, C-H., W-Y. Jean, Y-J. Lee and J. Penzien (1989). Soil-structure Interaction Correlation of Field Test Results with Analytical Predictions. *Proceedings of the Ninth World Conference on Earthquake Engineering, Volume 3*, 703-708, Tokyo-Kyoto.

Isenberg, J., M.T. Phipps and C.R. Scawthorn (1990). Watsonville Regional Study: Interaction among Damaged Lifelines. *Proceedings: Putting the Pieces Together - The Loma Prieta Earthquake One Year Later*, pp. 112-121, Oakland, California.

Jones, N., G. Smith, and R. Wagner (1994). Morbidity and Mortality in the Loma Prieta Earthquake: A Review of Recent Findings. *Research Accomplishments 1986-1994*, pp. 95-106, National Center for Earthquake Engineering Research, Buffalo, NY.

Lagorio, J., R. A. Olson, S. Scott, and J. Shefner (1991). Knowledge Transfer in Earthquake Hazard Reduction: A Challenge for Practitioners and Researchers. Center for Environmental Design Research Publication Number #CEDR-01-91, University of California, Berkeley, California.

O'Rourke, T.D. and M.C. Palmer (1994). The Northridge, California Earthquake of January 17, 1994: Performance of Gas Transmission Pipelines. *Technical Report NCEER-94-0011*, National Center for Earthquake Engineering Research, Buffalo, New York.

O'Rourke, T.D., C.R. Scawthorn, F.T. Blackburn and T.S. Dickerman (1990). Response of the San Francisco Water Supply System During the 1989 Loma Prieta Earthquake. *Proceedings: Putting the Pieces Together - The Loma Prieta Earthquake One Year Later*, pp. 102-111, Oakland, California.

O'Rourke, T.D., H.E. Stewart, F.T. Blackburn and T.S. Dickerman (1990). Geotechnical and Lifeline Aspects of the October 17, 1989 Loma Prieta Earthquake in San Francisco. *Technical Report NCEER-90-0001*, National Center for Earthquake Engineering Research, Buffalo, New York.

The Northridge, California Earthquake of January 17, 1994: General Reconnaissance Report (1994). *Technical Report NCEER-94-0005*, (J.D. Goltz ed.), National Center for Earthquake Engineering Research, Buffalo, New York.

Tsai, K-C., H-W. Chen, C-P. Hong and Y-F. Su (1993). Design of Steel Triangular Plate Energy Absorbers for Seismic-Resistant Construction. *Earthquake Spectra, 9*(3), 505-528.

Tsay, C.-Y. (1988). Disaster Prevention Research in Taiwan. *Proceedings of the CCNAA-AIT Joint Seminar on Research and Application for Multiple Hazards Mitigation, Volume I*, Taipei, Taiwan.

Yanev, P.I. and C.R. Scawthorn (1993). Hokkaido Nansei-oki, Japan Earthquake of July 12, 1993. *Technical Report NCEER-93-0023*, National Center for Earthquake Engineering Research, Buffalo, New York.

ABSOLUTE ACCELERATION FEEDBACK CONTROL OF STRUCTURES

L. L. Chung,* C. S. Yeh,* L. Y. Wu** and T. G. Jin**

** National Center for Research on Earthquake Engineering, National Taiwan University, Taipei, Taiwan*
*** Department of Civil Engineering, National Taiwan University, Taipei, Taiwan*

ABSTRACT

Direct absolute acceleration feedback control is considered to be unfeasible even though accelerometers are favorable sensors to measure the dynamic structural responses from the viewpoint of measurement. In this paper, absolute acceleration feedback control becomes feasible when the feedback data is extended to cover the absolute accelerations of the previous time steps. Multi-step acceleration feedback control algorithm is derived through optimization process such that a certain prescribed quadratic performance index is minimized. Control forces are simply generated from absolute acceleration information of the current and previous time steps multiplied by the pre-calculated optimal feedback gain matrix. The control effectiveness of the proposed control algorithm is verified through numerical simulation to be close to that of state feedback control algorithm. Simple on-line calculation and availability of accelerometers make the proposed control algorithm favorable to real-time control implementation.

KEYWORDS

Active structural control; absolute acceleration feedback; earthquake engineering.

1. INTRODUCTION

Civil engineering structures are characterized by their large number of degrees of freedom. If the parameters of the structures are upgraded by state feedback control, all states of the structures including displacement and velocity of each degree of freedom have to be available (Chung *et al.* 1989). However, from the viewpoint of economy, data processing and on-line calculation, it is impossible and impractical to have all states measured (Chung and Soong 1987). Therefore, limited-state measurement is one of the practical problems in active structural control and direct output measurement feedback is one of the solutions to this problem (Balas 1979; Chung *et al.* 1993). Among all sensors measuring the dynamic responses of the structures, accelerometer is the most well-developed one. Absolute accelerations of the structures can be measured by mounting the sensors directly on the structures without using any reference frame (Nagarajaiah *et al.* 1993). However, direct absolute acceleration feedback control is considered to be unfeasible. In this paper, absolute acceleration feedback control algorithm is developed. Absolute acceleration feedback control becomes feasible by extending the feedback data to cover the absolute acceleration information of the previous time steps in addition to the current absolute acceleration information. Structural control system is described in discrete-time form because of the unavoidable trend to use digital computer for the on-line calculation of the control forces. Using absolute accelerations of the current time step and absolute accelerations of the previous time steps as feedback data, optimal direct multi-step absolute acceleration feedback control algorithm is derived such that the quadratic performance index is minimized. The control forces are simply generated from the absolute acceleration output measurements of the current and previous time-steps multiplied by the pre-calculated constant feedback gain matrix. The control effectiveness of the proposed algorithm is verified to be close to that of state feedback control. However, the restraint of the choice of sensor types is released. Simple on-line calculation and small number of sensors make the proposed control algorithm favorable to real implementation.

2. CONTROL ALGORITHM

When an n-DOF discrete-parameter structural system is attacked by ground acceleration $\ddot{x}_0(t)$, its equation of motion is

$$\mathbf{M}\ddot{\mathbf{x}}(t)+\mathbf{C}\dot{\mathbf{x}}(t)+\mathbf{K}\mathbf{x}(t)=-\mathbf{M}\mathbf{1}\ddot{x}_0(t) \tag{1}$$

where $\mathbf{x}(t)$ is the $n\times 1$ relative displacement vector, $\mathbf{1}$ is the n-vector with all elements equal to one, and $\mathbf{M}$, $\mathbf{C}$ and $\mathbf{K}$ are the $n\times n$ mass, damping and stiffness matrix, respectively. If output measurements are limited to absolute accelerations which are utilized for the on-line calculation of the control forces, the absolute accelerations are related with relative displacements and relative velocities by

$$\ddot{\mathbf{x}}_a(t)=\ddot{\mathbf{x}}(t)+\mathbf{1}\ddot{x}_0(t)=-\mathbf{M}^{-1}\mathbf{K}\mathbf{x}(t)-\mathbf{M}^{-1}\mathbf{C}\dot{\mathbf{x}}(t) \tag{2}$$

where $\ddot{\mathbf{x}}_a(t)$ is $n\times 1$ absolute acceleration vector.

After active control devices are implemented, the structure is counteracted by the control forces $\mathbf{u}(t)$. The governing equation of the structural system can be modeled as

$$\mathbf{M}\ddot{\mathbf{x}}(t)+\mathbf{C}\dot{\mathbf{x}}(t)+\mathbf{K}\mathbf{x}(t)=\mathbf{B}_1\mathbf{u}(t)-\mathbf{M}\mathbf{1}\ddot{x}_0(t) \tag{3}$$

where $\mathbf{B}_1$ is the $n\times p$ location matrix of control forces.

Represented in state-space form, the second-order differential equation (3) is changed to first-order differential equation as

$$\dot{\mathbf{z}}(t)=\mathbf{A}_c\mathbf{z}(t)+\mathbf{B}_c\mathbf{u}(t)+\mathbf{f}_c\ddot{x}_0(t) \tag{4}$$

where $\mathbf{z}(t)=\begin{bmatrix}\mathbf{x}(t)\\ \dot{\mathbf{x}}(t)\end{bmatrix}$, $\mathbf{A}_c=\begin{bmatrix}\mathbf{0} & \mathbf{I}\\ -\mathbf{M}^{-1}\mathbf{K} & -\mathbf{M}^{-1}\mathbf{C}\end{bmatrix}$, $\mathbf{B}_c=\begin{bmatrix}\mathbf{0}\\ \mathbf{M}^{-1}\mathbf{B}_1\end{bmatrix}$ and $\mathbf{f}_c=\begin{bmatrix}0\\ -\mathbf{1}\end{bmatrix}$.

In state space, the absolute accelerations $\ddot{\mathbf{x}}_a(t)$ are just a linear combination of the structural states $\mathbf{z}(t)$ by

$$\ddot{\mathbf{x}}_a(t)=\mathbf{D}_c\mathbf{z}(t) \tag{5}$$

where $\mathbf{D}_c=\begin{bmatrix}-\mathbf{M}^{-1}\mathbf{K} & -\mathbf{M}^{-1}\mathbf{C}\end{bmatrix}$ is the $n\times 2n$ absolute acceleration output matrix.

During real-time control, digital computer is used for the on-line calculation of the control forces which are calculated once every sampling period, Δt. The digital signals are converted into zero-order-hold analog signals which are piecewise step functions over sampling intervals, and are applied to the structure. Therefore, it is more logical and more realistic for the structural control system to be modeled in a discrete-time fashion as

$$\mathbf{z}[k+1]=\mathbf{A}\mathbf{z}[k]+\mathbf{B}\mathbf{u}[k]+\mathbf{f}\ddot{x}_0[k] \tag{6}$$

where $\mathbf{A} = e^{\mathbf{A}_c \Delta t}$, $\mathbf{B} = \mathbf{A}_c^{-1}(\mathbf{A} - \mathbf{I})\mathbf{B}_c$ and $\mathbf{f} = \mathbf{A}_c^{-1}(\mathbf{A} - \mathbf{I})\mathbf{f}_c$.
The corresponding discrete-time output equation is

$$\ddot{\mathbf{x}}_a[k] = \mathbf{D}\mathbf{z}[k] \tag{7}$$

where $\mathbf{D} = \mathbf{D}_c$ is the $n \times 2n$ discrete-time acceleration output matrix.

The optimal active control forces $\mathbf{u}[k]$ are found such that the classical quadratic performance criterion

$$J = \sum_{k=0}^{\infty} \{\mathbf{z}^T[k]\mathbf{Q}\mathbf{z}[k] + \mathbf{u}^T[k]\mathbf{R}\mathbf{u}[k]\} \tag{8}$$

is minimized. In the above equation, $\mathbf{Q}$ is the $2n \times 2n$ symmetric positive semi-definite weighting matrix for the structural responses, $\mathbf{R}$ is the $p \times p$ symmetric positive definite weighting matrix for the input control forces, and superscript T denotes transpose of a matrix.

Since the performance of direct acceleration feedback control is not satisfactory, aside from the absolute acceleration output measurements of the current time step $\ddot{\mathbf{x}}_a[k]$, the absolute acceleration output measurements of the previous time steps $\ddot{\mathbf{x}}_a[k-i]$ are also utilized for the on-line calculation of the control forces $\mathbf{u}[k]$. The control forces are further assigned to be linearly related to the absolute acceleration output measurements of the current time step and the previous time steps by

$$\mathbf{u}[k] = \sum_{i=0}^{m} \mathbf{G}_i \ddot{\mathbf{x}}_a[k-i] \tag{9}$$

where $\mathbf{G}_i$ is the $(p \times n)$ ith-step absolute acceleration output feedback gain matrix.

Since the absolute acceleration output measurements of the m previous time steps are stored and used for the on-line calculation of the control forces, the previous absolute acceleration output information can be lumped together with the current absolute acceleration output information as

$$\bar{\ddot{\mathbf{x}}}_a[k] = \begin{bmatrix} \ddot{\mathbf{x}}_a[k] \\ \ddot{\mathbf{x}}_a[k-1] \\ \ddot{\mathbf{x}}_a[k-2] \\ \vdots \\ \ddot{\mathbf{x}}_a[k-m+1] \\ \ddot{\mathbf{x}}_a[k-m] \end{bmatrix} \tag{10}$$

where $\bar{\ddot{\mathbf{x}}}_a[k]$ is the $n(m+1) \times 1$ augmented absolute acceleration vector.

With the above lumped absolute acceleration output information, the control forces $\mathbf{u}[k]$ as stated in equation (9) are linearly related with the augmented absolute acceleration output vector $\bar{\ddot{\mathbf{x}}}_a[k]$. Therefore, the control law can be rewritten as

$$\mathbf{u}[k] = \mathbf{G}\bar{\ddot{\mathbf{x}}}_a[k] \tag{11}$$

where $\mathbf{G} = [\mathbf{G}_0 \quad \mathbf{G}_1 \quad \mathbf{G}_2 \quad \cdots \quad \mathbf{G}_{m-1} \quad \mathbf{G}_m]$ is the $p \times n(m+1)$ multi-step absolute acceleration feedback gain matrix.

By introducing the augmented state vector $\bar{\mathbf{z}}[k]$, the augmented absolute acceleration output equation can be expressed as

$$\bar{\ddot{\mathbf{x}}}_a[k]=\bar{\mathbf{D}}\bar{\mathbf{z}}[k] \tag{12}$$

where $\bar{\mathbf{z}}[k]=\begin{bmatrix}\mathbf{z}[k]\\ \mathbf{z}[k-1]\\ \mathbf{z}[k-2]\\ \vdots\\ \mathbf{z}[k-m+1]\\ \mathbf{z}[k-m]\end{bmatrix}$ and $\bar{\mathbf{D}}=\begin{bmatrix}\mathbf{D}&\mathbf{0}&\mathbf{0}&\cdots&\mathbf{0}&\mathbf{0}\\ \mathbf{0}&\mathbf{D}&\mathbf{0}&\cdots&\mathbf{0}&\mathbf{0}\\ \mathbf{0}&\mathbf{0}&\mathbf{D}&\cdots&\mathbf{0}&\mathbf{0}\\ \vdots&\vdots&\vdots&\ddots&\vdots&\vdots\\ \mathbf{0}&\mathbf{0}&\mathbf{0}&\cdots&\mathbf{D}&\mathbf{0}\\ \mathbf{0}&\mathbf{0}&\mathbf{0}&\cdots&\mathbf{0}&\mathbf{D}\end{bmatrix}$.

Followed by the augmented state vector $\bar{\mathbf{z}}[k]$ mentioned above, the corresponding state equation (6) is also augmented as

$$\bar{\mathbf{z}}[k+1]=\bar{\mathbf{A}}\bar{\mathbf{z}}[k]+\bar{\mathbf{B}}\mathbf{u}[k]+\bar{\mathbf{f}}\ddot{x}_0[k] \tag{13}$$

where $\bar{\mathbf{A}}=\begin{bmatrix}\mathbf{A}&\mathbf{0}&\mathbf{0}&\cdots&\mathbf{0}&\mathbf{0}\\ \mathbf{I}&\mathbf{0}&\mathbf{0}&\cdots&\mathbf{0}&\mathbf{0}\\ \mathbf{0}&\mathbf{I}&\mathbf{0}&\cdots&\mathbf{0}&\mathbf{0}\\ \vdots&\vdots&\vdots&\ddots&\vdots&\vdots\\ \mathbf{0}&\mathbf{0}&\mathbf{0}&\cdots&\mathbf{0}&\mathbf{0}\\ \mathbf{0}&\mathbf{0}&\mathbf{0}&\cdots&\mathbf{I}&\mathbf{0}\end{bmatrix}$, $\bar{\mathbf{B}}=\begin{bmatrix}\mathbf{B}\\ \mathbf{0}\\ \mathbf{0}\\ \vdots\\ \mathbf{0}\\ \mathbf{0}\end{bmatrix}$ and $\bar{\mathbf{f}}=\begin{bmatrix}\mathbf{f}\\ \mathbf{0}\\ \mathbf{0}\\ \vdots\\ \mathbf{0}\\ \mathbf{0}\end{bmatrix}$.

In terms of the augmented state vector $\bar{\mathbf{z}}[k]$, the original performance index J in equation (8) becomes

$$J=\sum_{k=0}^{\infty}\left\{\bar{\mathbf{z}}^T[k]\bar{\mathbf{Q}}\bar{\mathbf{z}}[k]+\mathbf{u}^T[k]\mathbf{R}\mathbf{u}[k]\right\} \tag{14}$$

where $\bar{\mathbf{Q}}=\begin{bmatrix}\mathbf{Q}&\mathbf{0}&\mathbf{0}&\cdots&\mathbf{0}&\mathbf{0}\\ \mathbf{0}&\mathbf{0}&\mathbf{0}&\cdots&\mathbf{0}&\mathbf{0}\\ \mathbf{0}&\mathbf{0}&\mathbf{0}&\cdots&\mathbf{0}&\mathbf{0}\\ \vdots&\vdots&\vdots&\ddots&\vdots&\vdots\\ \mathbf{0}&\mathbf{0}&\mathbf{0}&\cdots&\mathbf{0}&\mathbf{0}\\ \mathbf{0}&\mathbf{0}&\mathbf{0}&\cdots&\mathbf{0}&\mathbf{0}\end{bmatrix}$ is the $2n(m+1)\times 2n(m+1)$ augmented response weighting matrix.

With multi-step absolute accelerations as feedback data, the optimization problem becomes: searching an optimal multi-step absolute acceleration feedback gain matrix **G** that minimizes the performance index J in equation (14) subject to the constraints of the augmented dynamic equation (13), the augmented absolute acceleration output equation (12) and the control law equation (11). The augmented absolute acceleration output equation (12) is substituted into the control law equation (11) and the result is further substituted into the augmented state equation (13). After rearrangement, the augmented state equation (13) is expressed as

$$\bar{\mathbf{z}}[k+1]=(\bar{\mathbf{A}}+\bar{\mathbf{B}}\mathbf{G}\bar{\mathbf{D}})\bar{\mathbf{z}}[k]+\bar{\mathbf{f}}\ddot{x}_0[k] \tag{15}$$

where $(\bar{\mathbf{A}}+\bar{\mathbf{B}}\mathbf{G}\bar{\mathbf{D}})$ is defined to be the effective augmented system matrix. It is observed that multi-step absolute acceleration feedback control can also upgrade the augmented system matrix from $\bar{\mathbf{A}}$ to $(\bar{\mathbf{A}}+\bar{\mathbf{B}}\mathbf{G}\bar{\mathbf{D}})$ and thus the system parameters to a favorable situation if the multi-step absolute acceleration feedback gain matrix **G** is properly chosen through optimization procedures.

It is seen that if the effect of the control forces make the structural response attenuate under arbitrary initial disturbance $\bar{\mathbf{z}}[0]$, it will work equally well for arbitrary seismic disturbance $\ddot{x}_0[k]$, and vice versa. Therefore, in order to simplify the derivation, only arbitrary initial disturbance is considered here. The arbitrary initial disturbance is further generalized to be random. Under random initial disturbance $\bar{\mathbf{z}}[0] = \bar{\mathbf{z}}_0$, the solution to the effective augmented state equation (15) is simplified as

$$\bar{\mathbf{z}}[k] = (\bar{\mathbf{A}} + \bar{\mathbf{B}}\mathbf{G}\bar{\mathbf{D}})^k \bar{\mathbf{z}}[0] = (\bar{\mathbf{A}} + \bar{\mathbf{B}}\mathbf{G}\bar{\mathbf{D}})^k \bar{\mathbf{z}}_0 \tag{16}$$

After equations (12), (11) and (16) are sequentially substituted into equation (14), it gives

$$J = \sum_{k=0}^{\infty} \bar{\mathbf{z}}_0^T [(\bar{\mathbf{A}} + \bar{\mathbf{B}}\mathbf{G}\bar{\mathbf{D}})^k]^T (\bar{\mathbf{Q}} + \bar{\mathbf{D}}^T\mathbf{G}^T\mathbf{R}\mathbf{G}\bar{\mathbf{D}})[(\bar{\mathbf{A}} + \bar{\mathbf{B}}\mathbf{G}\bar{\mathbf{D}})^k]\bar{\mathbf{z}}_0 = \bar{\mathbf{z}}_0^T \mathbf{H} \bar{\mathbf{z}}_0 \tag{17}$$

where the $2n(m+1) \times 2n(m+1)$ constant matrix $\mathbf{H}$ is defined as

$$\mathbf{H} = \sum_{k=0}^{\infty} [(\bar{\mathbf{A}} + \bar{\mathbf{B}}\mathbf{G}\bar{\mathbf{D}})^k]^T (\bar{\mathbf{Q}} + \bar{\mathbf{D}}^T\mathbf{G}^T\mathbf{R}\mathbf{G}\bar{\mathbf{D}})[(\bar{\mathbf{A}} + \bar{\mathbf{B}}\mathbf{G}\bar{\mathbf{D}})^k] \tag{18}$$

Since initial disturbance $\bar{\mathbf{z}}_0$ is random, the quadratic performance index J is also random. Therefore, instead of the performance index J itself, its expected performance index

$$\bar{J} = \mathrm{E}\{J\} = \mathrm{E}\{\bar{\mathbf{z}}_0^T \mathbf{H} \bar{\mathbf{z}}_0\} = tr(\mathbf{H}\bar{\mathbf{Z}}_0) \tag{19}$$

is minimized, where $\mathrm{E}\{\bullet\}$ denotes expectation, $\mathrm{tr}(\bullet)$ denotes trace of a square matrix, and $\bar{\mathbf{Z}}_0 = \mathrm{E}\{\bar{\mathbf{z}}_0\bar{\mathbf{z}}_0^T\}$ is the second statistical moment of the initial disturbance $\bar{\mathbf{z}}_0$. It is found that the constant matrix $\mathbf{H}$ satisfies the following equation

$$(\bar{\mathbf{A}} + \bar{\mathbf{B}}\mathbf{G}\bar{\mathbf{D}})^T \mathbf{H}(\bar{\mathbf{A}} + \bar{\mathbf{B}}\mathbf{G}\bar{\mathbf{D}}) - \mathbf{H} + (\bar{\mathbf{Q}} + \bar{\mathbf{D}}^T\mathbf{G}^T\mathbf{R}\mathbf{G}\bar{\mathbf{D}}) = \mathbf{0} \tag{20}$$

The optimization problem is now converted to one that minimizes the expected performance index $\bar{J}$ in equation (19) subject to the constraint of equation (20). Incorporated with the constraint equation (20), the Lagrangian $\bar{J}'$ can be introduced as

$$\bar{J}' = \mathrm{tr}(\mathbf{H}\bar{\mathbf{Z}}_0) + \mathrm{tr}\left\{\mathbf{L}[(\bar{\mathbf{A}} + \bar{\mathbf{B}}\mathbf{G}\bar{\mathbf{D}})^T \mathbf{H}(\bar{\mathbf{A}} + \bar{\mathbf{B}}\mathbf{G}\bar{\mathbf{D}}) - \mathbf{H} + (\bar{\mathbf{Q}} + \bar{\mathbf{D}}^T\mathbf{G}^T\mathbf{R}\mathbf{G}\bar{\mathbf{D}})]\right\} = \mathbf{0} \tag{21}$$

where $\mathbf{L}$ is the $2n(m+1) \times 2n(m+1)$ Lagrangian multiplier matrix. Because the performance index is quadratic, and $\bar{\mathbf{Q}}$ and $\mathbf{R}$ are positive semi-definite and positive definite matrices, respectively, the necessary and sufficient conditions for minimization of the Lagrangian $\bar{J}'$ are

$$\frac{\partial \bar{J}'}{\partial \mathbf{L}} = (\bar{\mathbf{A}} + \bar{\mathbf{B}}\mathbf{G}\bar{\mathbf{D}})^T \mathbf{H}(\bar{\mathbf{A}} + \bar{\mathbf{B}}\mathbf{G}\bar{\mathbf{D}}) - \mathbf{H} + (\bar{\mathbf{Q}} + \bar{\mathbf{D}}^T\mathbf{G}^T\mathbf{R}\mathbf{G}\bar{\mathbf{D}}) = \mathbf{0} \tag{22}$$

$$\frac{\partial \bar{J}'}{\partial \mathbf{H}} = (\bar{\mathbf{A}} + \bar{\mathbf{B}}\mathbf{G}\bar{\mathbf{D}})\mathbf{L}(\bar{\mathbf{A}} + \bar{\mathbf{B}}\mathbf{G}\bar{\mathbf{D}})^T - \mathbf{L} + \bar{\mathbf{Z}}_0 = \mathbf{0} \tag{23}$$

$$\frac{\partial \bar{J}'}{\partial \mathbf{G}} = 2\bar{\mathbf{B}}^T \mathbf{H}(\bar{\mathbf{A}} + \bar{\mathbf{B}}\mathbf{G}\bar{\mathbf{D}})\mathbf{L}\bar{\mathbf{D}}^T + 2\mathbf{R}\mathbf{G}\bar{\mathbf{D}}\mathbf{L}\bar{\mathbf{D}}^T = \mathbf{0} \tag{24}$$

With initial guess of the feedback gain matrix $\mathbf{G}$, say $\mathbf{G} = \mathbf{0}$, matrices $\mathbf{H}$ and $\mathbf{L}$ can be solved from equations (22) and (23). Then $\mathbf{G}$ is updated from equation (24). The acceleration feedback gain matrix $\mathbf{G}$ is obtained iteratively until relative error is acceptable.

3. NUMERICAL VERIFICATIONS

An idealized single-degree-of-freedom structure with tendon control system in place is illustrated to verify the feasibility of the absolute acceleration control algorithm numerically (Chung *et al.* 1988). The structure is subjected to seismic excitation and it is counteracted by the tendon control device implemented on the first floor. The control force produced by adjusting the motion of the servo-controlled electro-hydraulic actuator is transmitted to the structure through a set of four pre-tensioned tendons. Relevant parameters of the control system are listed in Table 1.

The results from eigen-analysis are listed in Table 2. In eigen-analysis, the enhancement of the control system is indicated by the system parameters such as natural frequencies and damping factors extracted from the eigen-values of the system matrix $\overline{\mathbf{A}}$ and the effective system matrix $(\overline{\mathbf{A}}+\overline{\mathbf{B}}\mathbf{G}\overline{\mathbf{D}})$ of the control system. With state feedback SFB where displacement and velocity of the structure are available for feedback information, the natural frequency is increased a little bit from 3.47 Hz to 3.69 Hz but the damping factor is upgraded from 1.24% to 24.09%. With direct absolute acceleration feedback AFB, the damping factor is increased to a small extent of 3.02% but the natural frequency is degraded to 2.89 Hz. This matches the conclusion that direct acceleration feedback control is unfeasible. If the absolute acceleration of the one time step previous to the current time step and the absolute acceleration of the current time step in addition are used as feedback data AM1FB, it becomes feasible. The damping factor is increased to a large extent of 10.80% and the degradation in natural frequency is alleviated. If the feedback information is extended to include the absolute accelerations of the two time steps previous to the current time step in addition to the absolute acceleration of the current time step AM2FB, the damping factor is further increased to 14.33%.

The results from frequency domain analysis are consistent with those from eigen-analysis as shown in Figures 1-2. In frequency domain analysis, the degree of vibration suppression is shown by the amplitude of the frequency response function of the control system. In the control case of SFB, the control effectiveness is realized as the peak of the displacement frequency response function is flattened. In the control case of AFB, the peak of the displacement frequency response function is lowered a little bit but the peak is shifted leftward. The control effectiveness is not guaranteed but depends on the frequency content of external disturbance. In the control case of AM1FB, the peak of the displacement frequency response function is flattened. In the control case of AM2FB, the peak of the displacement frequency response function is further flattened.

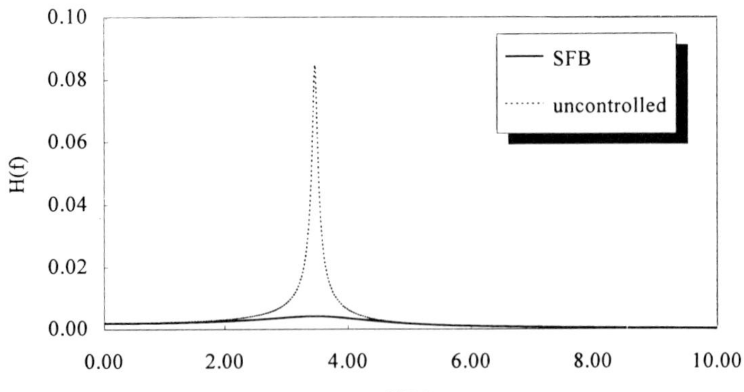

Fig. 1. Displacement Frequency Response Function of Structure with State Feedback

Table 1. Relevant Parameters of SDOF Control System

System parameter (1)	Parameter values (2)
Mass, m (kg)	2924
Stiffness, k (N/cm)	13900
Damping coefficient, c (N-sec/cm)	15.81
Tendon stiffness, k_c (N/cm)	3721
Tendon inclination, $\alpha(^\circ)$	36
Response weighting matrix, Q	$\begin{bmatrix} k & 0 \\ 0 & 0 \end{bmatrix}$
Control weighting matrix, R	$\beta / (16 k_c \cos^2 \alpha)$
Control weighting factor, β	10
Second moment of initial condition $\overline{\mathbf{Z}}_0 = \mathrm{E}\{\overline{\mathbf{z}}_0 \overline{\mathbf{z}}_0{}^T\}$	$\mathbf{I}$

Table 2. Control Effect for SDOF Structure

Control Case (1)	Natural frequency f(Hz) (2)	Damping ratio $\zeta(\%)$ (3)	Poles (4)	$\bar{J}(\times 10^7)$ (5)
Uncontrolled	3.47	1.24	$0.9737 \pm 0.2157i$	13.0
SFB	3.69	24.09	$0.9218 \pm 0.2111i$	1.3
AFB	2.89	3.02	$0.9781 \pm 0.1798i$	8.6
AM1FB	3.31	10.80	$0.9570 \pm 0.2007i$	2.5
AM2FB	3.29	14.33	$0.9506 \pm 0.1971i$	2.1

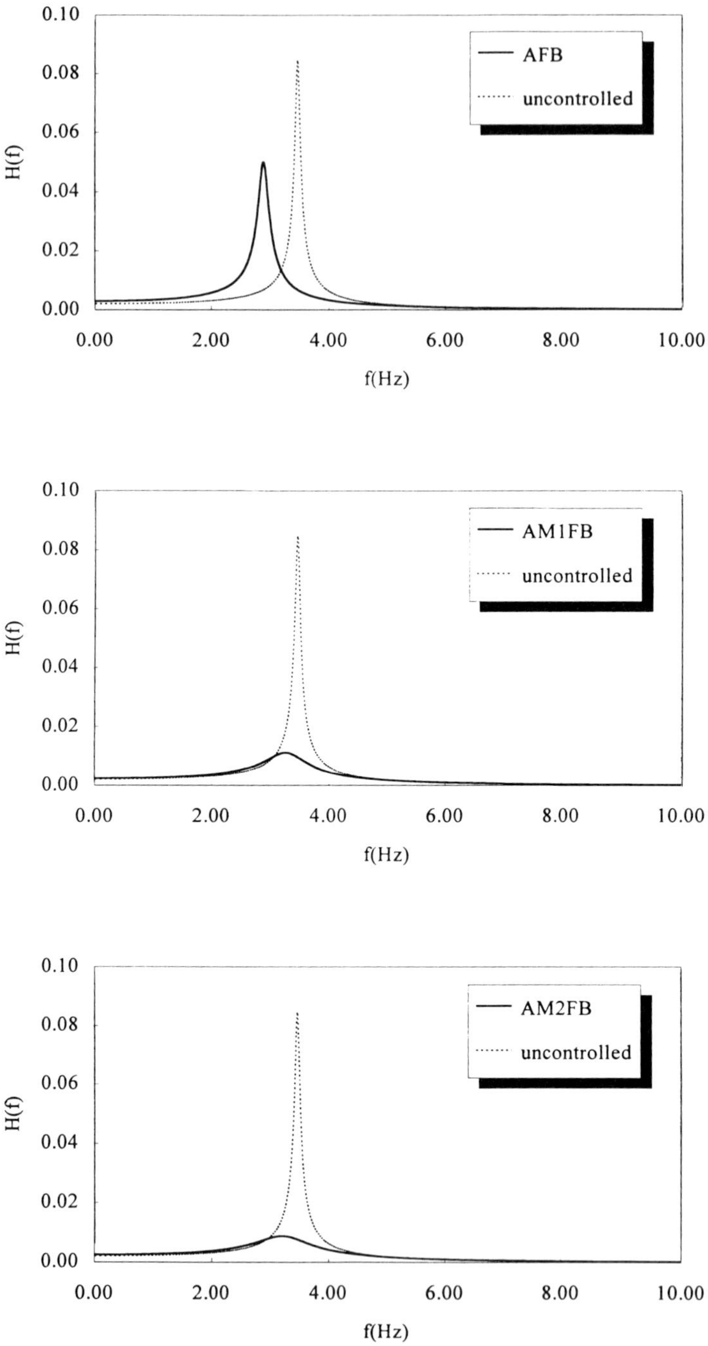

Fig. 2. Displacement Frequency Response Function of Structure with Acceleration Feedback

4. CONCLUSIONS

The velocity signal co-located with the controller has been considered to be the only essential data for direct output feedback control of the structures and direct absolute acceleration feedback control is found to be unfeasible even though accelerometers are favorable from the view point of measurement. However, when the feedback data is extended by storing previous absolute acceleration output measurements without using extra sensors, absolute acceleration feedback control becomes feasible. Multi-step absolute acceleration feedback gain matrix is obtained systematically through optimation procedure such that a certain quadratic performance index is minimized. The control effectiveness of the proposed control algorithm is verified numerically to be close to that of state feedback control algorithm. According to the proposed control algorithm, the control forces are simply generated from the absolute acceleration information of the current and previous time steps multiplied by the pre-calculated optimal feedback gain. Simple on-line calculation and availability of accelerometers make the proposed control algorithm favorable to the implementation of real-time control. Moreover, since the full-order mathematical model of the structure is considered throughout the whole derivation and no further model reduction is assumed, the spillover effect in control and observation due to modal control is eliminated. System stability is guaranteed as long as the structure is well described by the mathematical model.

ACKNOWLEDGMENTS

This research was supported in part by National Science Council of the Republic of China under grant No. NSC 82-0410-E-002-356. This support is greatly appreciated.

REFERENCES

Balas, M. J. (1979). Direct output feedback control of large space structures. *J. Astronautical Sci.*, **27**, 157-180.

Chung, L. L. and T. T. Soong (1987). Practical considerations in discrete-time structural control. *Vibration Control and Active Vibration Suppression,* D. J. Inman and J. C. Simonis, eds., ASME, New York, NY, 201-217.

Chung, L. L., A. M. Reinhorn and T. T. Soong (1988). Experiments on active control of seismic structures. *J. engrg. mech. ASCE*, **114,** 241-256.

Chung, L. L., R. C. Lin, T. T. Soong, and A. M. Reinhorn (1989). Experimental study of active control for MDOF seismic structures. *J. engrg. mech. ASCE*, **115**, 1609-1627.

Chung, L. L., C. C. Lin and S. Y. Chu (1993). Optimal direct output feedback of structural control. *J. eng. mech. ASCE*, **119**, 2157-2173.

Nagarajaiah, S., M. A. Riley and A. Reinhorn (1993). Control of sliding-isolated bridge with absolute acceleration feedback. *J. eng. mech. ASCE*, **119**, 2317-2332.

BRIDGE DAMAGE DIAGNOSIS BY VIBRATIONAL SIGNATURE ANALYSIS

G. C. Yao* and K. C. Chang**

** Department of Architecture, National Cheng Kung University, Tainan, Taiwan*
*** Department of Civil Engineering, National Taiwan University, Taipei, Taiwan*

ABSTRACT

An investigation has been carried out on the possibility of using MMS for damage location identification of bridges. The basic theory of MMS is reviewed first. MMS in an analytical model of a scaled down bridge is found to have a very useful feature of of large variation around the damaged locations. Experiments were carried out to prove that the analytical results are physically measurable and correct.

KEYWORDS

Moment mode shapes; vibration signal; system identification; bridge damage.

INTRODUCTION

Since the advance of steel manufacturing technology from the last century, steel structures have been widely adopted in all aspects of human lives such as steel bridges in public transportation. These steel bridges have several structural problems that gradually surfaced in the last decades and pose a major challenge to structural engineers. Fatigue cracks resulted from large repeated loading and member corrosion from aging are two common examples. Recently, with the earthquake experience increases, post-earthquake safety evaluation is another test to bridge engineers for a quick and accurate structural safety evaluation.

Small cracks in hidden locations on bridges caused a lot of trouble for engineers to observe them by eyesight as in the current practice. Therefore, automatic damage detection becomes an important research topic through out the world. Many studies (Fisher, 1981, Tarricone, 1990, Biswas et al., 1990) concentrated on the change of vibration signatures of bridges after damage took place. The advantage of this approach is in its non-destructive nature and the simplicity of its theory. However, one major obstacle is the low sensitivity of the state variables chosen for comparison (Mazurek, 1988, Alampalli et al., 1993).

Yao (Yao, et al., 1992) proposed a state variable called Strain Mode Shape (SMS) and has shown high sensitivity to minor damages in steel buildings. Theoretical and experimental investigation have provided evidence for its strength. In this paper, a new variable, Moment Mode Shape (MMS), a derivation from SMS, will be used to detect minor damages in a two-span model bridges. The numerical and experimental work will be presented to illustrate the effectiveness of MMS.

THEORY OF MMS

The concept and application of Strain Frequency Response Function (SFRF) have been widely discussed in recent years (McVerry, 1980). Its theoretical basis can be briefly summarized as follows. Let $\{\varepsilon\}$ and $\{x\}$ represent the strain and displacement vectors in a typical finite element formulation at certain locations of a structure. Then

$$\{\varepsilon\} = [\theta]\,\{x\} \tag{1}$$

defines a relationship between the strain and the displacement vectors. It states that strain vector can be derived from nodal displacement vectors by a transformation matrix, $[\theta]$. It is called strain-displacement transformation matrix in finite element analysis and is dependent upon the derivatives of shape functions. Strain in a beam element can be related to the internal force, such as bending strain can be calculated from bending moment. In general, small nodal displacements which is hard to measure experimentally will accompany larger internal force and therefore strain. It is the authors' intention to measure differences of $\{\varepsilon\}$ to represent internal force change as a means of damage diagnosis.

For a MDOF system, the equation of motion can be written as

$$[m]\,\{\ddot{x}\} + [c]\,\{\dot{x}\} + [k]\,\{x\} = \{f\} \tag{2}$$

where $[m], [c], [k]$ are the mass, damping, and stiffness matrices, respectively, and $\{f\}$ is the forcing function.

Solving the above equation in the frequency domain and using $j=\sqrt{-1}$, we obtain

$$\{X(\omega)\} = [\Phi]\left[\frac{1}{K_i\left(1-\frac{\omega^2}{\omega_i^2}+j2\zeta_i\frac{\omega}{\omega_i}\right)}\right]_{diag}[\Phi]^T\{F(\omega)\} \tag{3}$$

where $[M], [C], [K]$ are the corresponding modal matrices for $[m], [c], [k]$ while $[\Phi]$ is the mode shape matrix. $\{X\}$ and $\{F\}$ is the Fourier transform of $\{x\}$ and $\{f\}$, and ζ_i is the damping ratio of the ith mode.

Let

$$[e_i] = \left[\frac{1}{K_i\left(1 - \frac{\omega^2}{\omega_i^2} + j2\zeta_i \frac{\omega}{\omega_i}\right)} \right]_{diag}$$

then

$$\{X(\omega)\} = [\Phi][e_i][\Phi]^T\{F(\omega)\} = [H^D]\{F(\omega)\} \tag{4}$$

where $[H^D]$ is known as the Displacement Frequency Response Function(DFRF).

From (1) and (4) we can obtain

$$\{\epsilon(\omega)\} = [\theta]\ [\Phi]\ [e_i]\ [\Phi]^T\{F(\omega)\} = [H^s(\omega)]\ \{F(\omega)\} \tag{5}$$

where ϵ is the Fourier transform of ε and $[H^s]$ is the Strain Frequency Response Function (SFRF).

Strain mode shape (SMS) can be derived from SFRF easily. In the physical sense, SMS represents the relative strain magnitude at different measurement locations of a certain mode. If measured strain is caused by bending moment, the distribution of SMS can be transformed to the moment distribution and be called Moment Mode Shape (MMS). Mapping relations between DMS and MMS will be different from structure to structure depending on the boundary conditions and the dynamic properties.

From the discussions above, it is understood that any damage will create a change in a structure's displacement mode shape will display a change in its MMS as well. A structural crack before propagating to a major damage will only produce small change in DMS but will generally accompanied by a large MMS variation. Therefore the direct measurement of MMS will serve as a sensitive variable to detect small changes in the structural integrity if MMS before and after damage have been recorded. This has been found feasible both theoretically and experimentally in building structures (Yao et al., 1992).

BRIDGE MODEL

A two-span continuous bridge model whose total length is 6.4 meter with rollers on both sides and a hinge at the center, as shown in Fig. 1, is designed to study the damage diagnosis technique by the change of its MMS signature. Three steel I girders(H194 X 150 X 9 X 6) are the main structures and is topped with a concrete slab.

Two types of simulated crack whose length initiated from the bottom flange into the 1/4 web depth are produced at different locations. One is at the center of the right span the other is at 1/8 span length from the center support of the left span. In the mathematical

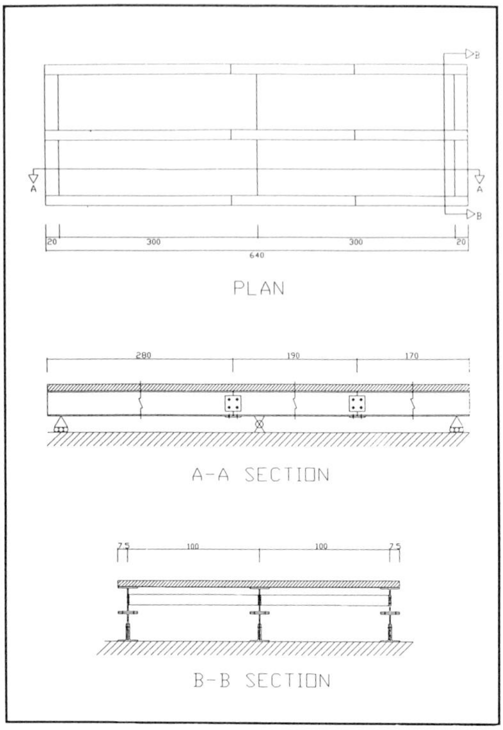

Fig. 1. Bridge model for testing.

model, discontinuous finite elements were used to simulate the cracks. In the experimental model, cracks were cut and assembled by reenforcement gusset plates at the center of the web and the bottom flanges. Cracks were simulated by the release of the bottom flange gusset plates, as shown in Fig. 1.

ANALYTICAL SIMULATION

The structural analysis software SAP90 is used to simulate the behavior of the steel bridge. Both flanges were modelled as beam elements while the web was simulated with SHELL elements with both bending and membrane DOF. The decks were modelled as the plate bending elements. The whole model and the web section is shown in Fig. 2.

MMS is calculated first by imposing a particular displacement mode shape on the structure to create a forced displacement. The internal force components produced in the structure can then be decomposed to obtain the MMS [12]. MMS from the undamaged structure is used as the baseline value to compare with the damaged MMS. Only the first two vertical modes were analyzed, their displacement mode shapes are shown in Fig.3 and Fig.4.

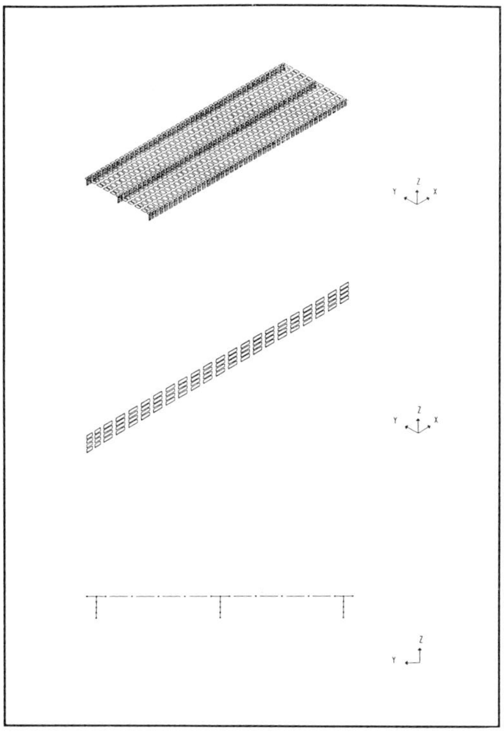

Fig. 2. Analytical Bridge Model.

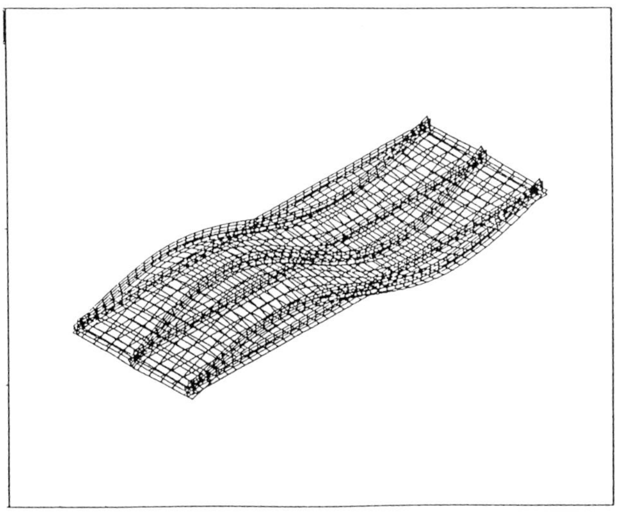

Fig. 3. Displacement mode shape of the first vertical mode.

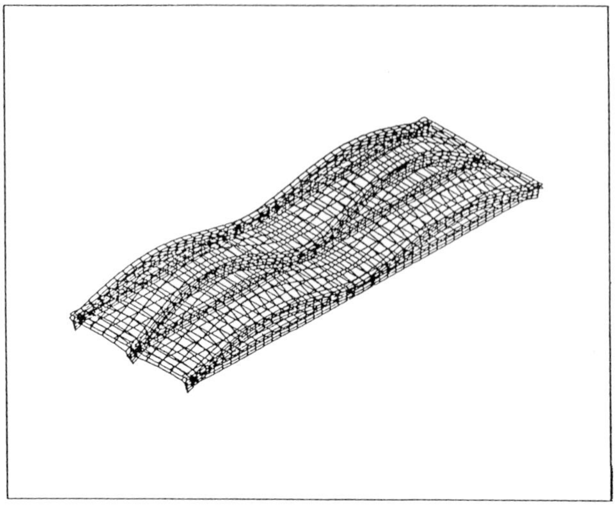

Fig. 4. Displacement mode shape of the second vertical mode.

EXPERIMENTAL VERIFICATION

Cracks are introduced into the beam sequentially one at a time by removing gusset plates on the pre-cut area. Their locations are shown in Fig.5. 1A and 2A are at the center of their span while 1B and 2B are at the 1/8 span length. The model bridge is excited by a quick release approach as shown in Fig.6. An initial displacement is imposed on the structure and vibration is taken when the cable is quickly set free.

The changed MMS of damage type 2A and 2B from the baseline line value are recorded and plotted in Fig. 7 and 8. The solid lines are the analytical results and the dotted lines are experimental. It can be found that when damage location is produced at the center of the span (2A), an obvious MMS variation from the baseline value can be observed. However, if damage took place at the location where MMS is small in the baseline value(2B), the results are not as obvious. The same conclusion can be observed on the results from damage type 1A and 1B as shown in Fig. 9 and 10.

It is found that damage location could be indicated by the change of its MMS value if its baseline value is sufficiently large. Therefore, 1A and 2A cracks are easily identifiable from its first mode MMS variation, and 1B and 2B are less sensitive. According to Fig.4, the second mode MMS of the bridge model has a large value around the center support where the first mode is small. It is a reasonable conclusion that a second mode MMS monitoring should be able to identify damage zones around 1B and 2B area. Owing to the limitation of experimental facilities, the second mode of the bridge can not be excited and the experimental verification will have to be tested in the future.

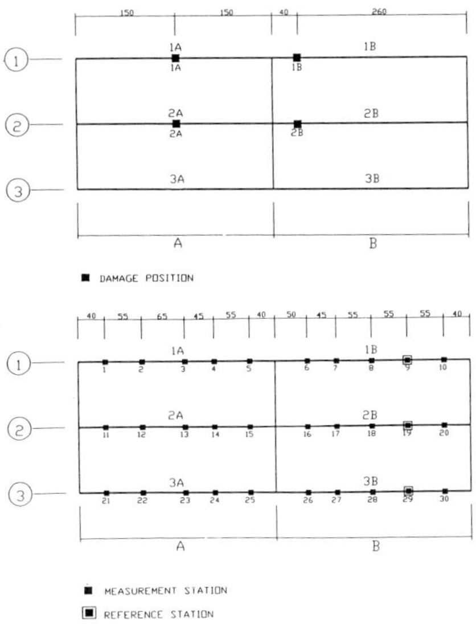

Fig. 5. Damage location and strain measurement location.

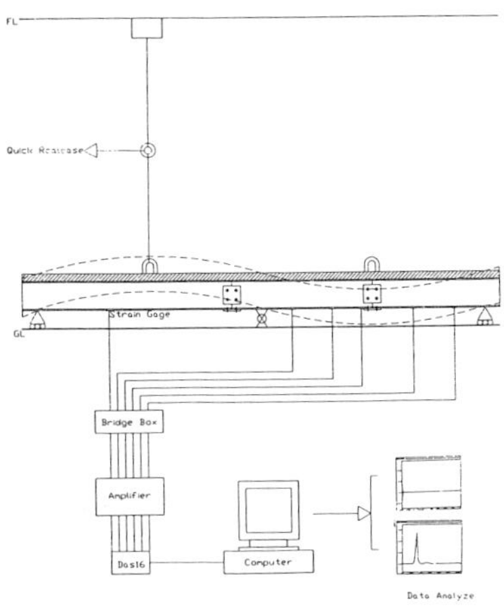

Fig. 6. Schematic drawing of the experimental setup.

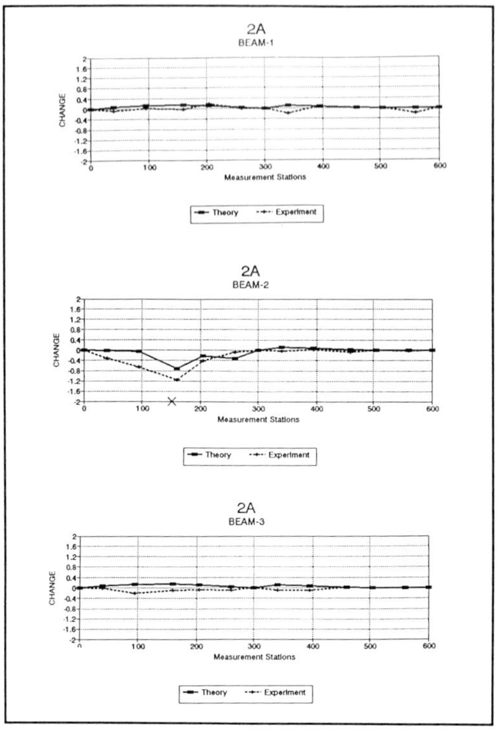

Fig. 7. MMS change comparison of damage at 2A.

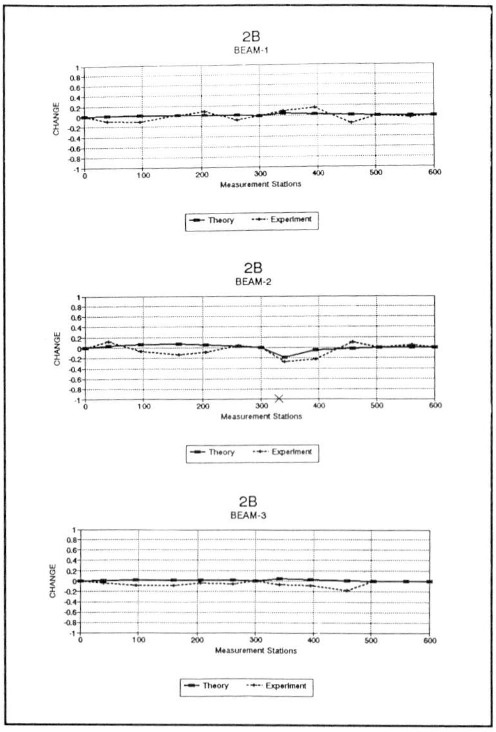

Fig. 8. MMS change comparison of damage at 2B.

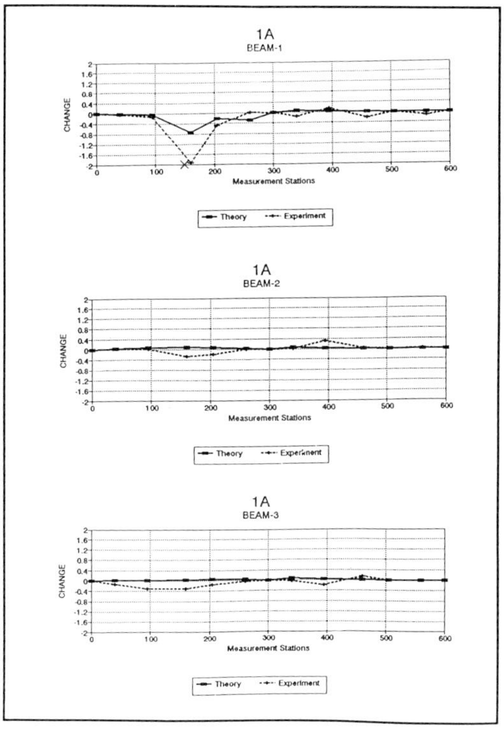

Fig. 9. MMS change comparison of damage at 1A.

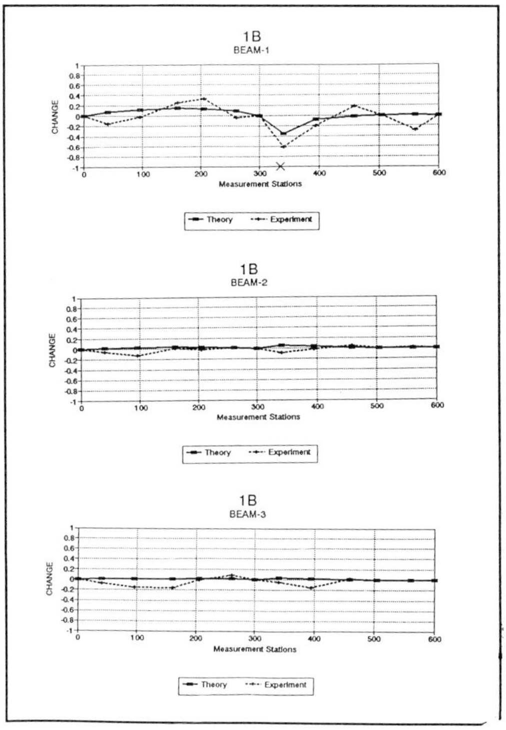

Fig. 10. MMS change comparison of damage at 1B.

CONCLUSIONS

According to the analytical and experimental study of a bridge model damaged at different locations, its vibrational characteristics of MMS is found to have the following features:

(1) MMS change as the damage signature shows that MMS change is greater in the vicinity of the damaged area and decrease farther away from the inflicted area.

(2) The first mode MMS serves as a good indicator for damage locations in the center of the span. The experimental results agrees well with the analytical predictions.

(3) Higher mode MMS could be used to detect possible damage locations where the first mode MMS has a small baseline value.

ACKNOWLEDGEMENT

Gratitude is given to the financial support from NSC in Taiwan by the NSC 83-0414-006-006. The assistance of the former graduate students Y.M. Su is the integral part of the research. His work is valuable to this research.

REFERENCES

Alampalli, S. and G.K. Fu(1993), Full scale dynamic monitoring of highway bridges, Structural Engineering in natural Hazards Mitigation Vol 1, pp 1602-1607.

Biswas, M., Pandey, A.K., and Samman, M.M.(1990),"Diagnostic Experimental Spectral/Modal Analysis of a Highway Bridge," The International Journal of Analytical and Experimental Modal Analysis, Jan.

Fisher, J.W.(1981), Inspecting Steel Bridges for Fatigue Damage, Fritz Engineering Laboratory Report No. 386-15, Lehigh University.

Mazurek, D.F.(1988), Monitoring Structural Integrity of Girder Bridges Through Vibration Measurement, Ph.D. dissertation, The University of Connecticut,

McVerry,G.H.(1980),"Structural Identification in the Frequency Domain from Earthquake Records", Earthquake Engineering and Structural Dynamics, Vol. 8,

Tarricone, P.(1990),"Bridges Under Surveillance", Civil Engineering, ASCE, May, pp. 48-51.

Yao, G.C., K.C. Chang and G.C. Lee(1992). Damage diagnosis of steel frames using vibrational signature analysis. Journal of Engineering Mechanics, Vol 118, No.9, 1949-1961.

Yao, G.C. and N. Lien (1994). Modal moment analysis for damage diagnosis of steel structures. Proc. of the Fifth US National Conf. on Earthquake Engineering. pp 497-506.

SEISMIC PERFORMANCE OF BRIDGES IN CALIFORNIA DURING THE JANUARY 17, 1994, NORTHRIDGE EARTHQUAKE

L. -H. Sheng

Office of Earthquake Engineering,
California Department of Transportation,
Sacramento, CA, U.S.A.

ABSTRACT

The California Department of Transportation (CALTRANS) started its bridge retrofit program after the 1971 San Fernando earthquake. After the 1989 Loma Prieta earthquake, a bridge retrofit program was established for all publicly owned bridges in the State of California. In this paper, seismic performance of bridges during January 17, 1994, Northridge earthquake is examined. Damages from collapsed bridges are discussed as well as other bridge structures.

KEYWORDS

Bridge damage; Northridge earthquake; Bridge performance; Lifelines performance;

Introduction

On January 17, 1994, at 4:31 a.m. Pacific Time a strong earthquake struck the San Fernando Valley at the Northridge area. The epicenter was at a latitude of 34°12.53'N and a longitude of 118°32.44'W, located approximately 19 miles northwest of the City of Los Angeles, registered 6.8 Ms (6.7 Mw) on the Richter scale. Although no direct ground rupture has been associated with the earthquake, extensive ground breakage was observed to the north of the epicenter in the northern section of the Santa Susana Mountains and in the San Fernando Valley. The quake was felt strongly as far as Las Vegas to the East, San Diego to the South, and San Luis Obispo to the North. Estimates of the damage caused by the quake have risen as high as $30 billion dollars. Three PEQITs (Post EarthQuake Investigation Teams) were dispatched the morning the earthquake occurred by the Division of Structures, Caltrans (California Department of Transportation), along with 15 maintenance structure evaluation teams to assess bridge damage and safety.

Strong Ground Motion

The Northridge earthquake was recorded extensively over the Los Angeles area on strong motion instruments. These instruments are operated by the California Strong Motion Instrumentation Program (CSMIP) (Figure 1), USGS, CalTech, USC, LADWP, and Southern Edison. The largest peak horizontal ground acceleration was recorded as 1.82g at Tarzana approximately 7 kilometers south of the epicenter.

Bridges in the Los Angeles County

There are 2292 state bridges and 1523 county bridges in the Los Angeles county based on the 1991 bridge data base. Most of these bridges are concrete bridges and were built before 1971. The following tables show the breakdown of bridges by type and year constructed.

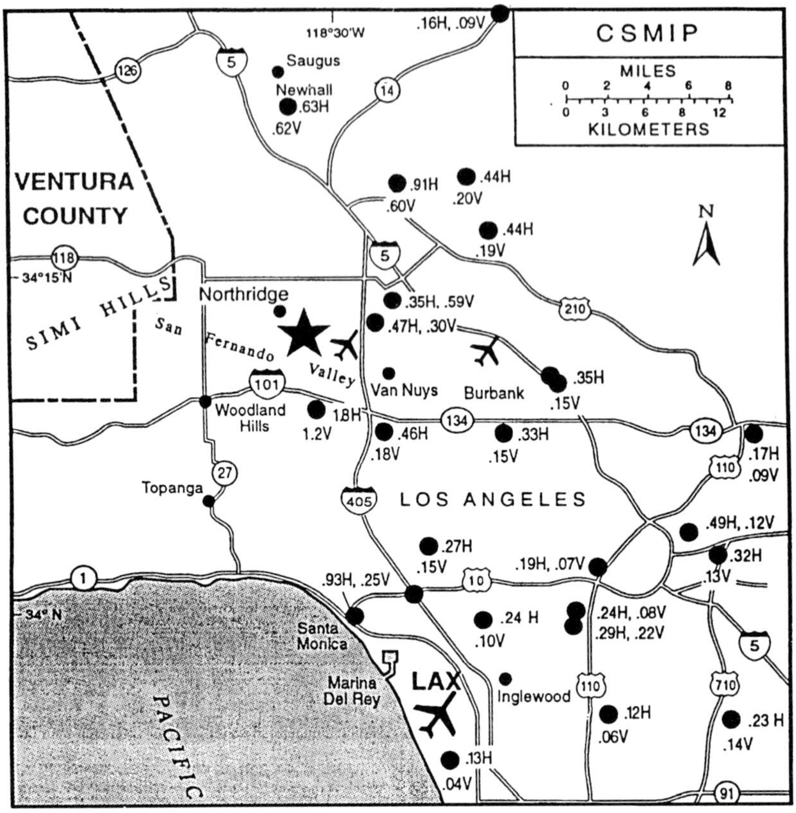

Fig. 1 Map of peak ground acceleration recorded at selected CSMIP stations

Bridge Types	State Bridges	County Bridges
Concrete Bridge	**1879**	**802**
Reinforced Concrete	**1247**	**574**
Box Girder	923	96
Girder	158	183
Slab	163	293
Precast	3	2
Prestressed Concrete	**545**	**175**
Box Girder	443	31
Girder	4	3
Slab	10	13
Precast	88	128
Steel Bridge	**144**	**262**
Girder	135	239
Truss	6	20
Arch	2	3
Suspension	1	0
Timber	**0**	**43**

Year of Constructed	State Bridges	County Bridges
before 1971	1825	1260
1971-1980	393	154
after 1980	74	109

Performance of Bridges

There were 173 bridges which had various degrees of damage. Six bridges at five locations collapsed, others suffered damage from column cracks to approach settlements (Figure 2).

Performance of Concrete Bridges

Several concrete bridges suffered major damage and six of them collapsed. The most significant damage is described in this paper.

Gavin Canyon UC (53-1797L/R)

Gavin Canyon Undercrossing (Figure 3) was a five span 3 frame structure located approximately 2 miles north of the intersection of Route 5 and 14. These were two separate bridges built in 1955, were retrofitted with hinge restrainers in 1974. Both bridges were built with a 66° skew at all abutments and bents with hinges in span 2 and span 4. Frame 1 and 3 of both bridges were CIP RC box girders. Frame 2 was CIP P/S box girder from the hinge in span 2 to the hinge in span 4. The bridges were supported on end diaphragm monolithic abutments with spread footings and columns 6'-3" x 10' with octagonal sections with a one way flare over a transition length of 20 feet. Transverse reinforcement consisted of #4@12" ties. Column footings had only a bottom mat of steel and were supported by 100 ton driven piles.

The bridges rotated and partially unseating the hinge spans. Four spans collapsed and the remainder was partially seated at the hinges. The large skew at this bridge reduced the 8" support length by half when the motion is perpendicular to the hinge. An early maintenance report indicated that this bridge had cracks at the hinge spans after the 1971 San Fernando earthquake.

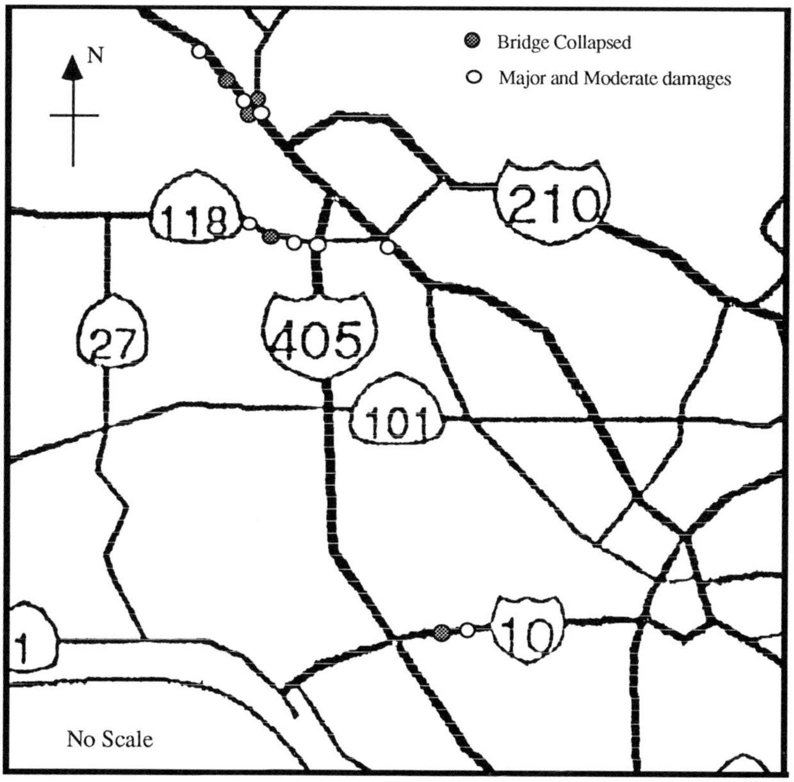

Fig . 2 Locations of major bridge damages

Fig. 3

Route 14/5 Separation (53-1960F)

The route 14/5 Separation connects southbound Route 14 to southbound I-5. This bridge had ten spans composed of five frames with a total length 1528 feet. The bridge was under construction in 1971 and was damaged in the San Fernando Earthquake. At that time, all of the columns had been constructed. The bottom slab and web stem had concrete placed from abutment 1 to the hinge in span 3. All concrete had been placed from the hinge in span 9 to abutment 11 without prestress. The superstructure consisted of alternating CIP P/S and RC box girder frames. Hinge seat width was 14". The bents consisted of single 12'x4' or 12'x6' octagonal columns with one way flare to a 26' width at the superstructure soffit. Main longitudinal column reinforcement was #18 bars with lateral reinforcement #5@12". This bridge was completed in 1974.

The first frame of this structure collapsed (Figure 4,5). Abutment 1 was unseated and laid 5 feet away from the abutment face. The major causes of failure were the damaged superstructure elements from the 1971 earthquake in conjunction with large stiffness changes between frames.

Fig. 4

Fig. 5

Bull Creek Canyon Channel Bridge (53-2206)

This bridge was designed in 1973 using the post 1971 updated seismic design guidelines and details. It was opened to public traffic in 1976. This bridge was approximately 256 feet in length and 200 feet minimum in width. The three span structure had a various skew over the lengths of the bridge:36°, 42°, and 47°. The middle span crossed over Bull Creek Canyon Channel which is a concrete lined channel.

The superstructure was a dual multi-celled CIP P/S box girder with a longitudinal joint near the center line of Route 118. The structure had a monolithic end-diaphragm type abutment with height of 8' to 9' tall supported by 16" diameter CIDH concrete piles. The multi-column bents constructed of 4' octagonal prismatic columns and supported by 16" diameter CIDH concrete pile footings. The main column reinforcement was #9 continuous from the bottom of footing into the bent cap. The transverse reinforcement was #5 spiral @3" at both ends of column and #5@12" spacing throughout the remain portion of the column. At bent 3, the column section changed shape at the top of the channel wall.

This structure suffered major damage. At bent 3, all columns failed at the top of the lined channel wall at ground line (Figure 6). At bent 2, the right bridge had 2 of the 5 columns fail near the top of column just below under the close spaced spiral where the transverse reinforcement was in the transition from a 3" pitch to a 12" pitch. The left bridge did not suffer as much damage as did the right bridge at bent 2. It is obvious that the channel lining dramatically reduced the effective column length and acted as a grade beam providing fixity at the intersection. In addition, the column reinforcing detail did not consider the presence of the concrete channel. It is believed that the damage at bent 2 was a secondary effect which occurred after the column failure at bent 3.

Mission-Gothic UC (53-2205)

This bridge was also designed in 1973 using the post 1971 updated seismic design guidelines and details. It was opened to public traffic in 1976. The structure consisted of two bridges, left and right. The 3 span left bridge was 506 feet in length and 98 feet in width. The 4 span right bridge was 566 feet in length and 98 feet in width. The structures had 44° right skew at the west abutments and 46° left skew at the east abutments

The superstructures were multi-celled CIP P/S box girders with a longitudinal joint near the center line of Route 118. Both left and right bridges were supported on elastomeric bearing pads at the abutments with an available seat width of 2 feet. The abutment footings were founded on 16" diameter 45 ton CIDH concrete piles. The multi-column bents consisted of two 6' octagonal columns with one way flare to 14' over a 12 foot transition length. The footings had both a top and bottom mat of steel and were supported by 16" diameter 70 ton CIDH concrete pile footings. All columns were pinned at bottom. The main column reinforcement was #11 continuous from the bottom of column into the bent cap. The transverse reinforcement was #5 spiral @3.5" spacing throughout the height of column.

Fig. 6

Because the wedge shape geometry in plan view, the bridge had limited abutment restraint in the transverse direction. The #11 reinforcement at the flare section created a stiffer section at the upper part of the column and increased the plastic shear demand below the flare. The large deflection at abutment permitted the end span to become unseated and the column to yield and consequently to fail in shear (Figure 7).

Le Cienega-Venice UC (53-1609)

This bridge is located on I-10, the Santa Monica freeway. The bridge was two structures separated by a longitudinal joint at the median. There were nine spans with total length of 872 feet. These bridges were designed in 1962 and completed in 1964.

The superstructure was a RC box girder. Skew angles varied from 5° at abutment 1 to 41° at abutment 10. Hinge seat width was 6". These bridges were retrofitted in 1978 with 1 1/4" diameter H.S. rods.

Fig. 7

Fig. 8

Many of the columns lost their confinement steel and their concrete cores disintegrated (Figure 8). Superstructures from Bent 3 to Bent 7 were resting on masonry buildings (mini storage units). Because these bridges are 16 miles away from the epicenter, it is believed that the soft site condition in the area enhanced the motions at the bridge. Further investigation has revealed that early river or lake formations at this location.

Damage Summary of Concrete Bridges

Damage to the concrete bridges can be classified into the following reasons;

Loss support seats
Large hinge or abutment skew angles
Non-ductile columns
Change of effective column length
Site condition effect

Performance of Steel bridges

Although less than 10% of the bridges in Los Angeles County are steel, three steel bridges sustained some damage. All three bridges are located on a 3 mile segment of I-5 near the City of Newhall.

Pico Lyon Overcrossing (53-1783)

This bridge was designed in 1964 and completed in 1967. The bridge is 318 feet in length and 108 feet in width. The superstructure of the bridge consists of a reinforced concrete deck which acts compositely with the supporting steel plate girders. The two span steel girders are continuous and supported on 4 flared single column reinforced concrete piers. Each pier supports two steel girders. The overcrossing has a 39.5° skew. The southern half of bridge is supported by steel rocker bearings at abutments and fixed bearing at bents whereas the northern half of bridge is supported by elastomeric pads. In the 1980's, seismic keys were installed to restrain girder movements of the southern half of bridge. The keys are thick steel plates bent into an L-shape.

During the Northridge earthquake the steel girders apparently moved in both longitudinal and transverse directions causing the bolts between the girder flange and keys to travel to the end of the slot. Additional movement caused concrete to spall and bend the L-shape keys. There was no sign of damage to the intermediate diaphragms in the bridge. However, all three end diaphragms at the east abutment had damage in the gusset plate connections (Figure 9), stiffener connections, and in the connection of the stiffener to the girder web. This damage most likely had been caused by a large horizontal force at the deck level. This unique superstructure behavior we first observed after the Cape Mendocino/Petrolia earthquake in 1992.

Fig. 9

Valencia Boulevard Overcrossing (53-1815)

The Valencia Boulevard Overcrossing was designed in 1965. The bridge has a total length of 246 feet and is 41 feet in width. The overcrossing is not skewed and has no horizontal curvature. Steel plate girders are continuous over two equal spans and are supported by a single flared reinforced concrete pier. Two water pipes are suspended under the deck.

Overall damage to the structure was minimal. Most damage occurred because of the restraint of the water pipe by the abutment. When the superstructure moved at the west abutment, the vertical member of the end diaphragm hit the 27 in diameter steel casing of the water pipe located under the deck (Figure 10). At the east abutment the movement of the superstructure caused the vertical member to barely touch the pipe. The impact of the diaphragm and pipe indicated that the relative movement of the superstructure and pipe during earthquake had been at least 10" for the east abutment and 7" for the west abutment.

Fig. 10

Santa Clara River Bridge

The Santa Clara River Bridge was designed in 1962 and completed in 1964. The bridge has seven simply supported spans and a total length of 741 feet. The left and right bridges have widths of 68 feet and 72 feet and are separated by a distance of 17 feet. The bridges have 42.5° skew angles. Each right and left bridge has six girders per span. There are 12 concrete wall piers providing the interior supports for the bridge. The reinforced concrete wall piers are 2 feet thick and have horizontal and vertical reinforcement on both faces. The top of the all wall piers is widened to 40" to form a wider beam to support the masonry plate of the elastomeric bearings of the steel plate girders. End and intermediate diaphragms connect the plate girders transversely. Because of the 42° skew of the bridge and the layout of the diaphragms, each diaphragm connects to the plate girders at different point along the span (Figure 11). The vertical stiffeners were not physically connected to the bottom flange as require by current design standards. It is important to note that the original drawings specified that the vertical intermediate stiffeners should be tight fit to the bottom flange. However, the fabricated girders at site had the vertical stiffeners terminated about 1" above the bottom flange. In 1983 cable restrainers were added to the four interior girders of each span.

The pier caps had severe seismic damage in the form of spalling of the concrete bearing surface supporting the elastomeric pads under the facia girders. The most severe damage occurred at the supports of the acute end of the pier cap (Figure 12). Hairline cracks were found in the web of the girders next to the vertical stiffeners. No crack had propagated into the flange or into the fillet weld connecting the web to the flange. Because a gap existed between the bottom of stiffeners and the bottom flange, out-of-plane bending of the web of the girders was allowed. It is possible that such bending over the past 30 years may have caused this area to develop fatigue cracks or steel embrittlement. The behavior of this bridge indicates the disadvantages of staggering the diaphragms in skew bridges. There was no visible damage in the members of the intermediate diaphragms. Horizontal and diagonal members buckled in the end diaphragms above pier 1 of the right bridge.

Damage Summary of Steel Bridges

Damages to the steel bridges can be classified into the following reasons:

Insufficient anchor bolt detail
Un-compatible movements
Staggered intermediate diaphragm
Buckling of end disphragm member

Performance of Retrofitted bridges

There are 269 bridges in Los Angeles County had been retrofitted and 378 additional bridges are either under retrofit construction or design. None of the retrofitted bridges exhibited any structural damage. The most significant example is the Cadillac Ramp Separation (53-1485F) and Bollona Creek (53-1671K). These bridges are located right between the Le Cienega-Venice UC and Fairfax-Washington UC. The most damage observed at these two bridges were cracks on slope paving and approach settlement.

Fig. 11

Fig. 12

Conclusion

Northridge earthquake caused the most damage to the highway bridges in California in recent history.

There are several factors involved in this. Strong shaking occurred in an urban area. The retrofit program has not yet been completed. Most bridges damaged by this earthquake were designed by old design codes. It also demonstrated that the retrofit design is effective.

Acknowledgment

The author wishes to thank Professor Hassen Astaneh-Asl for his assistant for providing damage information for steel bridges.

REFERENCES

A. Astaneh-Asl, B. Bolt, K. M. McMullin, D. Modjtahedi, D. Roupen and S. Cho. (1994) Seismic Performance of Steel Bridges During the 1994 Northridge Earthquake. UCB/CE-Steel-94/01. Department of Civil Engineering. University of California at Berkeley, Berkeley.

California Department of Transportation - Division of Structures. (1994) The Northridge Earthquake, Post Earthquake Investigation Report. California Department of Transportation, Sacramento.

California Department of Conservation, Division of Mines and Geology, Strong Motion Instrumentation Program. Fifth Quick Report on CSMIP Strong-Motion Data from the Northridge/San Fernando Vally Earthquake of January 17, 1994. California Department of Conservation, Sacramento.

STEEL CONNECTION FAILURES DURING THE NORTHRIDGE EARTHQUAKE

W. F. Chen

School of Civil Engineering, Purdue University,
West Lafayette, IN 47907, U.S.A.

ABSTRACT

The January 1994 Northridge Earthquake caused unprecedented damage at beam-to-column connections in modern steel moment resisting frames. Almost all of the moment connection failures observed in this Earthquake involved the standard welded flange-bolted web detail. This paper attempts to explain the possible causes of the observed damage by examining the basic deficiencies in the current approach to design for such connections and by proposing a possible strategy for repair and rehabilitation of these damaged frames.

KEYWORDS

Beam-to-column connections; Bolts; Buildings; Failures, Frames; Steels; Structural Engineering; Welds.

INTRODUCTION

The most widely used moment connection detail in U.S. practice for seismic-resistant framing is the welded flange-bolted web detail as shown in Fig. 1. The beam flanges are welded to the column flange using complete joint penetration single bevel groove welds. The beam web is field bolted to a single plate shear tab which is shop welded to the column flange. As shown in the inset of Fig. 1, in some cases, supplemental welds are provided between the shear tab and the beam web. The good performance of this connection detail has been confirmed by the full scale tests of isolated specimens in several US university laboratories. Thus, all major US seismic codes recommend this detail for ductile moment frames.

The intent of this design is for ductile yielding to occur away from the connection detail but the actual damage observed in the Northridge Earthquake indicates that there was a high stress concentration at the connections causing cracking to occur primarily in the beam tension flange fusion zone of the weld root with propagation out into the column flange as shown in Fig. 2. It appears that the bottom flange connection experienced much more stress than it was designed for. It was generally observed that the higher stress concentration combined with the possible notch conditions due to possible incomplete fusion or in-place backing bars at the joint leads to crack initiation and propagation. In some instances, this cracking propagated into the web of the column and into the panel zone. This suggests that improved performance of steel moment connections cannot be achieved by solely improving weld quality. Several factors that will affect the local stress computation at the connecting joint must be properly accounted for in design analysis, so that the detailing of the connection may be changed to reduce this high stress levels in the beam flanges near the connection in the design process. This is described in what follows.

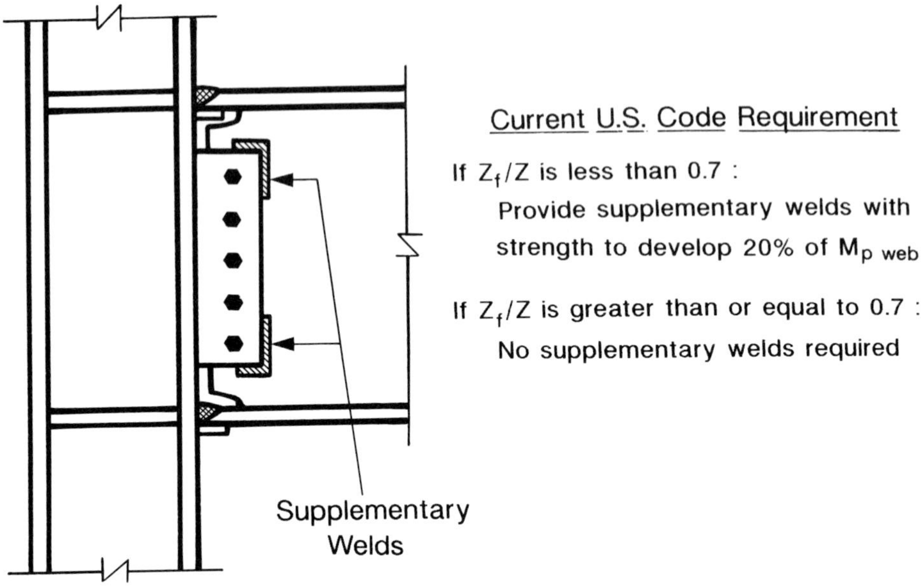

Figure 1. Standard welded flange-bolted web detail (AISC, 1994).

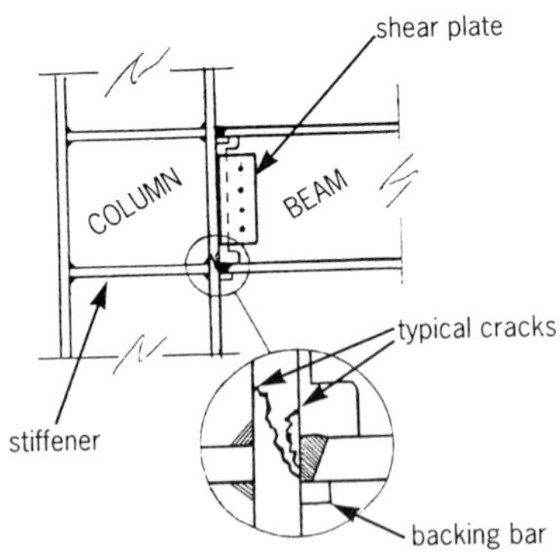

Figure 2. Connection where damage occured (AISC, 1994).

BASIC DEFICIENCIES IN THE DESIGN

The Effect of Composite Floor Slab

The presence of the concrete slab can result in substantially higher beam strength and stiffness than the usual bare frame design, and it also shifts the neutral axis of the composite beam toward the beam's top flange. Both effects will result in a much larger tensile force near the bottom flange of the beam. This added force is rarely accounted for in the current design of steel moment frames.

The Effect of Vertical Acceleration

The Northridge Earthquake subjected buildings to exceptionally large vertical and horizontal ground motion. The axial force in some columns of these buildings may be in tension due to a combination of overturning moment and vertical excitation. This will induce a triaxial tensile stress state near the center of the bottom flange of the beam in the column flange. This must play an important role in the failure process observed.

The Effect of Flexibility in Panel Zone

In modern seismic steel design codes, the panel zone is permitted to yield during a major earthquake in order to dissipate energy as long as the effect of panel zone deformation on frame stability is properly considered in the seismic design. Such yielding in panel zone shear deformation is often associated with the development of large strains at the connecting joint resulting in large local strains near the corners of the shear panel. This will result in a much larger local ductility demand in the weld and in the column flange. The large local ductility demand coupled with a substantial overstrength at the top of the beam due to the concrete slab and the development of a triaxial tensile stress state in the column flange due to the vertical acceleration is the likely cause of the failures observed.

AISC REPAIRS-INTERIM GUIDELINES

Two possible repair details of the welded flange-bolted web connection are shown in Figs. 3 and 4. Figure 3 is a all-welded *unreinforced* moment connection, whereas Fig. 4 is a connection *reinforced* with cover plates. For the unreinforced case, the plastic hinge in the beam will develop at the face of column, while for the reinforced case, the plastic hinge in the beam is forced away from the face of the column and away from the flange groove welds. Limited testing of both types of connections indicates *unreliable* performance for the unreinforced case [Fig. 5, Specimen 1] and good performance for the *reinforced* case [Fig. 6, Specimen 3]. Subsequent tests for the connections reinforced with ribs have also shown good performance [Fig. 7, Specimen 6]. AISC therefore recommends the repair procedures that consist of replacing damaged welds and/or material with other reinforcement or other enhancement to provide a reliable lateral load resisting system. Details of this recommendation with supporting test results are summarized in a recent AISC Northridge Steel Update I report (October, 1994).

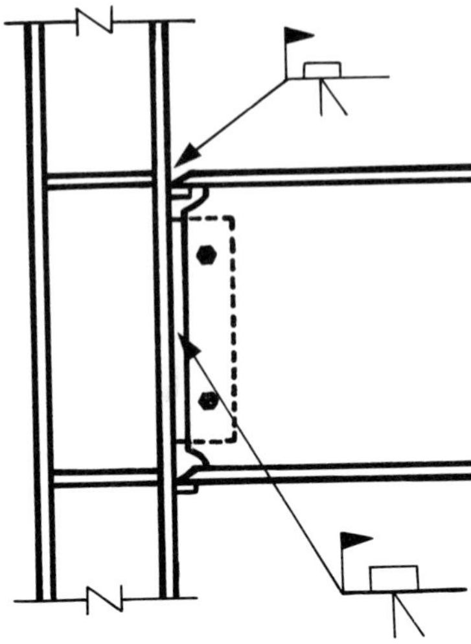

Proposed Solution 1 :
All-Welded Moment Connection

Figure 3. An all-welded unreinforced moment connection (AISC, 1994).

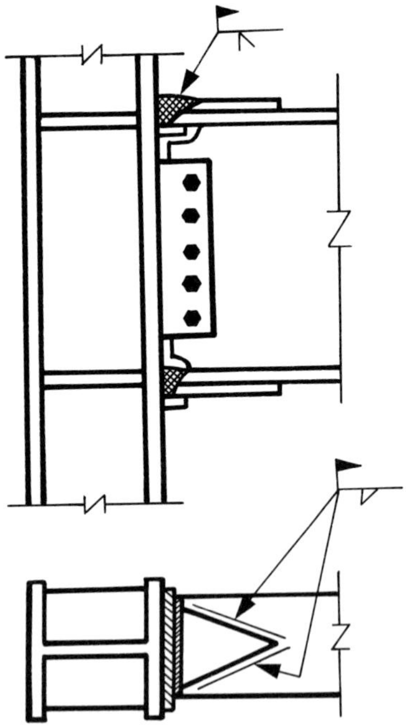

Figure 4. A connection reinforced with cover plates (AISC, 1994).

Proposed Solution 2 :
Reinforced Moment Connection

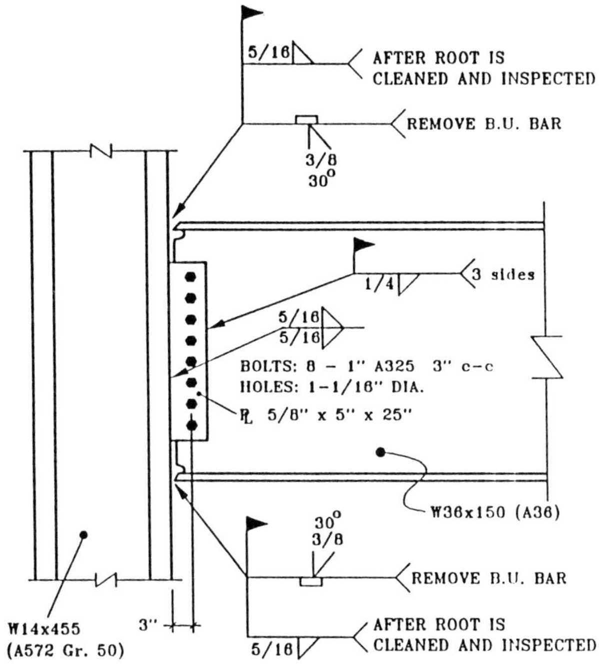

Figure 5. Unreliable performance for the unreinforced case (AISC, 1994).

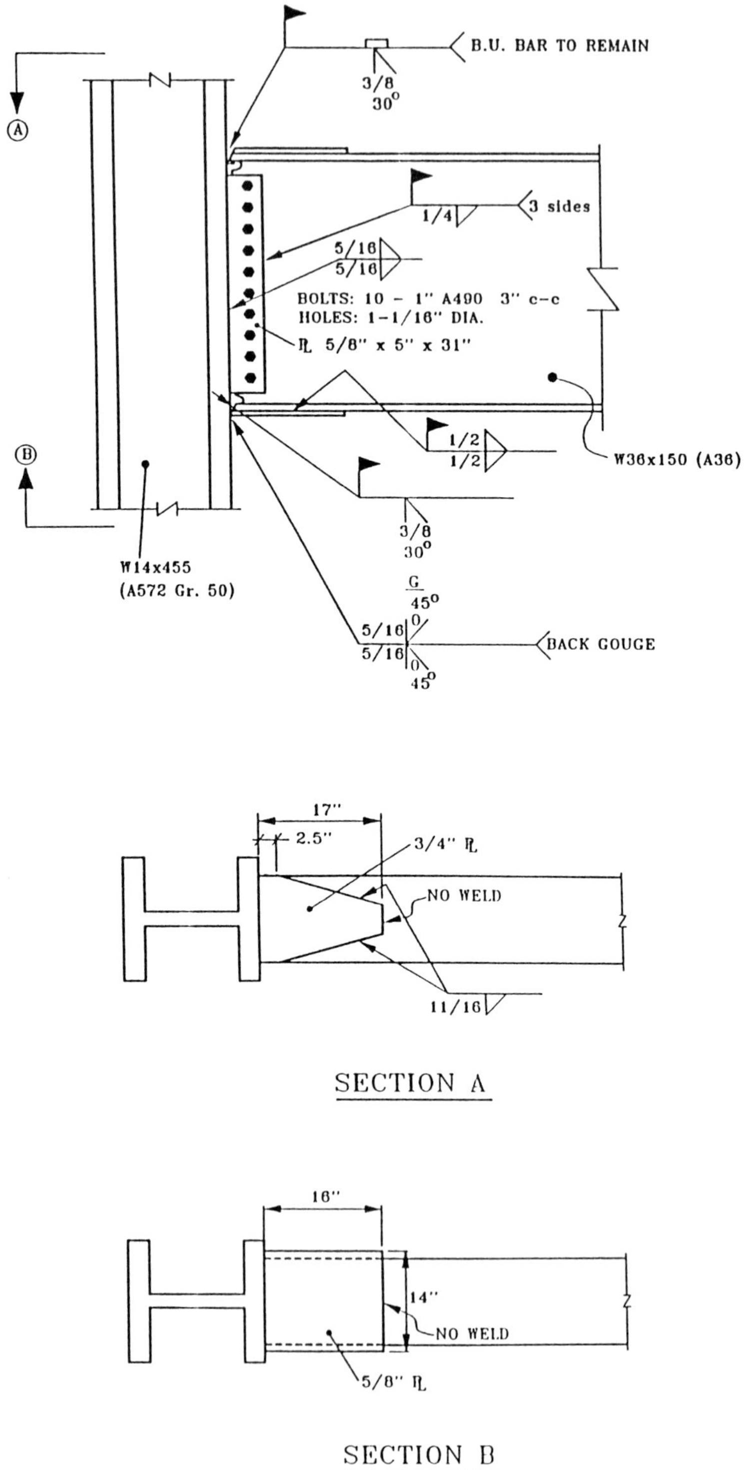

Figure 6. Good performance for the reinforced case with cover plates (AISC, 1994).

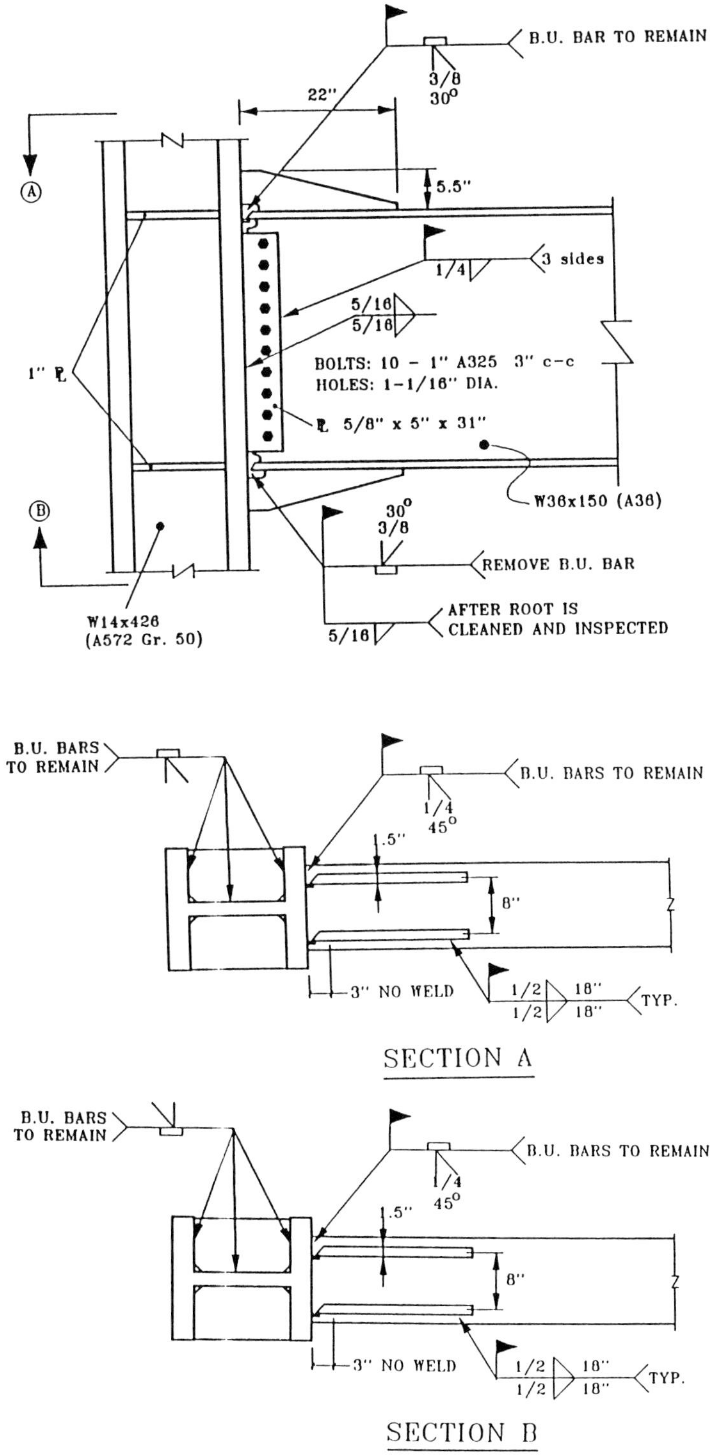

Figure 7. Good performance for the reinforced case with ribs (AISC, 1994).

ALTERNATIVE DESIGN

Mixed Use of Moment/Semi-Rigid Connections

Most cracked steel frame buildings as a result of the Northridge Earthquake are all found in relatively new structures, and there are no cracks reported in steel buildings more than 10 years old. This is in part due to less redundancy employed in modern steel moment frames, because most of these modern moment connections are concentrated in perimeter frames in as few as one or two bays of the perimeter. The reduced redundancy leads to the use of heavy column section (W14x455) and large beam section (W36x210) which make welds and thick plates more susceptible to heat affected damage, laminate tearing and quality control issues.

In the mixed use of moment and semi-rigid structural system, the plastic hinges are intentionally moved away from a few critical locations and spreaded over a well-controlled semi-rigid connection area. This will certainly reduce the brittle failure modes of either welds or heated affected zones near the joints. Furthermore, the semi-rigid connections can be practically inserted into the existing frame and converted it into a semi-rigid system. It can therefore be used to repair or retrofit an existing or damaged steel structure.

Generally, flexibly jointed frames are preferred for low-rise multibay frames. For high-rise tall buildings, the mixed use of rigid and semi-rigid connections can be used to reduce the number of moment connections. This has been demonstrated in a recent study by Kishi et al. (1994) using four combinations of connections for a four-bay eight-story frame as shown in Fig. 8. Two types of connections are used: top- and seat-angle connections with/without double web-angle. Building drifts under service load and floor moments in columns under factored load are analyzed. It is found that the semi-rigid connections combining with rigid connections can be used economically for high-rise building design. The Case 4 combination of semi-rigid and moment connections results in the largest stiffness gain of the structural system among the four cases shown in Fig. 8.

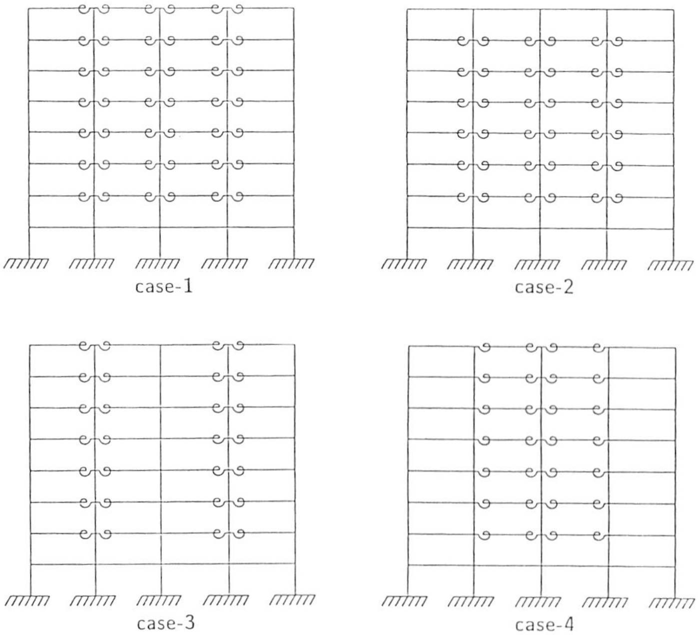

Figure 8. Combinations of rigid and semi-rigid connections (Kishi et al. 1994).

End-Plate Moment Connections

Since welded moment connections are sensitive to welding, moment end-plate connections as shown in Fig. 9 may provide a possible alternative. End-plate connections eliminate expensive field welding and may be used to repair the damaged joints or retrofit the existing structures.

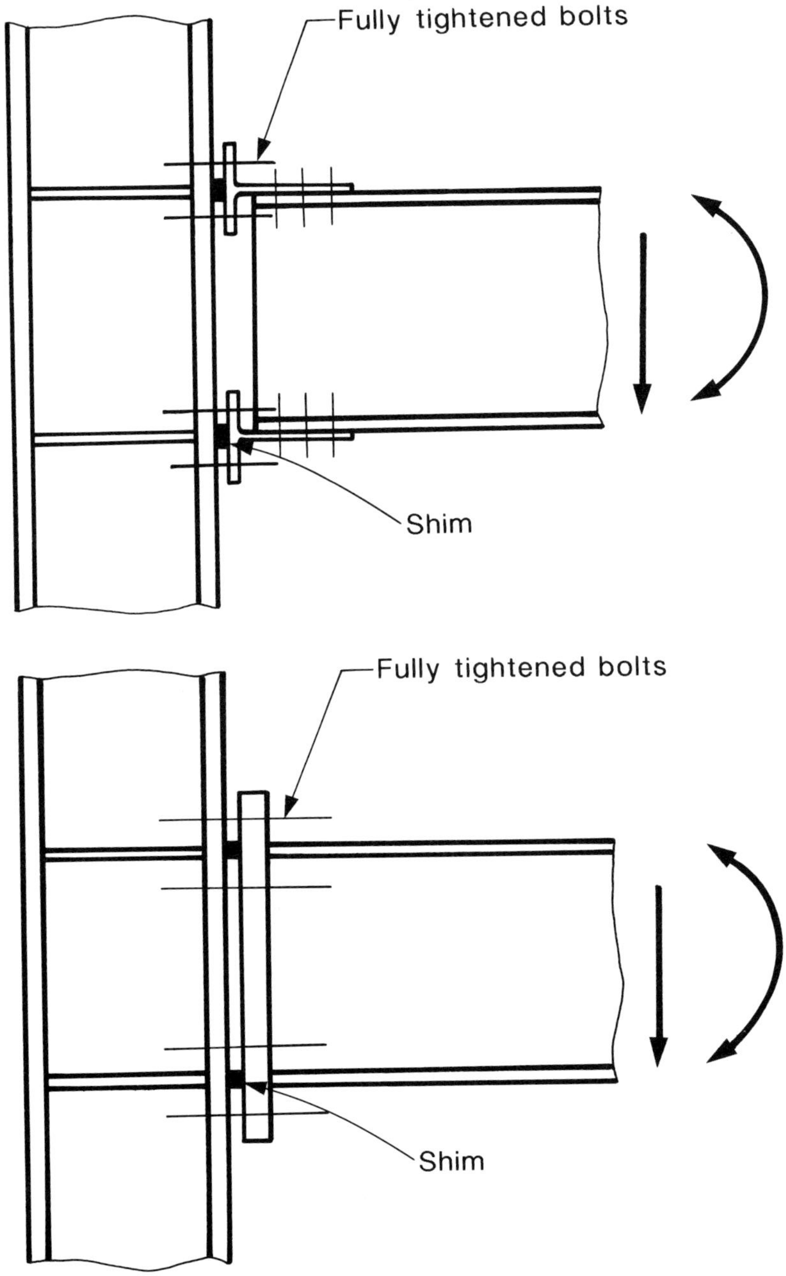

Figure 9. Moment end-plate connection.

Limiting Forces Transmitted to Connections

To force the plastic hinge to develop away from the face of column, the connections at critical locations must be protected from possible brittle failure modes. To this end, connections cannot be simply treated as an isolated element and must be designed as an element in the whole structural system. This is illustrated in Fig. 10. By controlling the forces that can be transmitted to critical connections in a building frame during a major earthquake, the ductility demand on these connections will be limited while the required ductility will be provided by the adjoining girders at the joint. To achieve this, advanced analysis based on second-order inelastic methods must be utilized. These advanced analysis methods have been well developed in recent years (Chen and Toma, 1994), but much more remains to be done.

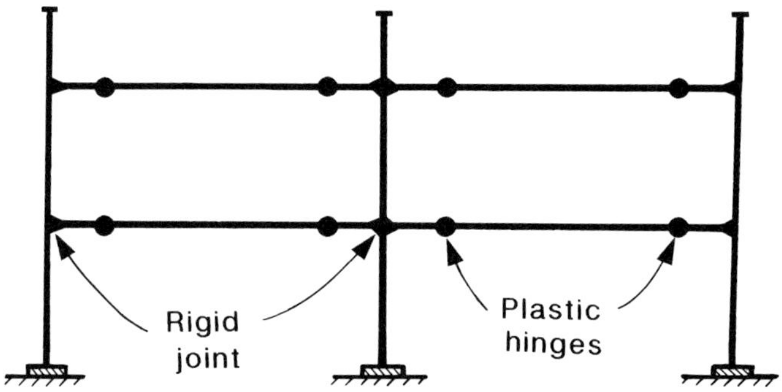

Figure 10. Limiting forces transmitted to connections.

REFERENCES

AISC (1994). Northridge Steel Update I, Chicago, IL, October. See also, *Modern Steel Construction*, December, 1994, pp. 22-31.

Chen, W.F. and S. Toma (1994). *Advanced Analysis of Steel Frames: Theory, Software, and Applications,* CRC Press, Boca Raton, FL.

Kishi, N., W. F. Chen, Y. Goto and R. Hasan (1994). Design of Tall Buildings with Mixed Use of Rigid and Semi-Rigid Connections, Structural Engineering CE-STR-94-11, School of Civil Engineering, Purdue University, West Lafayette, IN.

ANALYSIS OF BUILDING SEISMIC RECORDS USING ARTIFICIAL NEURAL NETWORKS

W. R. Chang and S. T. Mau

*Department of Civil and Environmental Engineering,
University of Houston, Houston, TX 77204-4791, U.S.A.*

ABSTRACT

An artificial neural network was used to analyze building acceleration records obtained during earthquakes. The neural network contained a buffer layer with six delay data points. The hidden layer contained either a nonlinear activating function or a simple linear activation function. The neural network was first tested on a model ten-story building frame using real earthquake ground acceleration and simulated response at each floor based on a second order linear mathematical model. The training of the network was successful and the fundamental vibration frequency and damping ratio were identified by exciting the network with harmonic acceleration. The neural network was then applied to a set of real data during the 1989 Loma Prieta Earthquake from a four-story building. The data exhibited pronounced time-variation in the building characteristics. The neural network was not able to simulate this time-variation behavior over a long duration of thirty seconds but was accurate in characterizing behavior within short duration windows of two to four seconds.

KEYWORDS

System identification; neural networks; time-variation; building records; earthquake; nonlinear behavior.

INTRODUCTION

Building seismic records are typically used either for the verification of mathematical models or for the characterization of dynamic properties. In the latter, usually some form of system identification is utilized. Studies on recent earthquake records using system identification methods reveal that building responses often exhibit a kind of nonlinearity that is the result of the time-varying nature of the building characteristics (Li and Mau, 1991, Li and Mau, 1992). The time-varying nature may come from the change of building stiffness due to disengagement of nonstructural elements, non-damaging cracking of reinforced concrete members, or the partial damage of structural elements. In applying system identification methods to these building records, it is difficult to get close match to the recorded motions throughout the duration of the record unless the records are divided into short duration windows, allowing changes of parameters from window to window. The purpose of this study is to explore an alternative approach using artificial neural networks (ANN).

Using neural networks for system identification is certainly not new (Doo, 1989, Narendra and Parthasarathy, 1990, Billings *et al.*, 1992). There are even some attempts in damage detection using neural network (Wu *et al.*, 1992, Elkordy *et al.*, 1993). In most of these previous works, however, simulated data were used for training. By using simulated data, the system is well defined mathematically and the neural networks training is often successful. In this study, a set of real earthquake data is used. The building characteristics are not known. The neural networks training is expected to be more difficult. Whether the neural network is able to model the building behavior is a key question to be answered.

In the following sections the selection of a neural networks paradigm is presented first. The results of applying the neural network to simulated earthquake data are described next. Finally, the neural network is applied to a set of real earthquake data and the results are presented.

NEURAL NETWORK PARADIGM

There are more than twenty neural networks paradigms in use today (Simpson, 1990). The ability of neural networks to approximate any function to a high degree of accuracy with a sufficient number of hidden nodes is well established (Hornik *et al.*, 1989). In adopting a neural networks paradigm for the current study, one factor to be considered is that we will use only acceleration time histories. Time

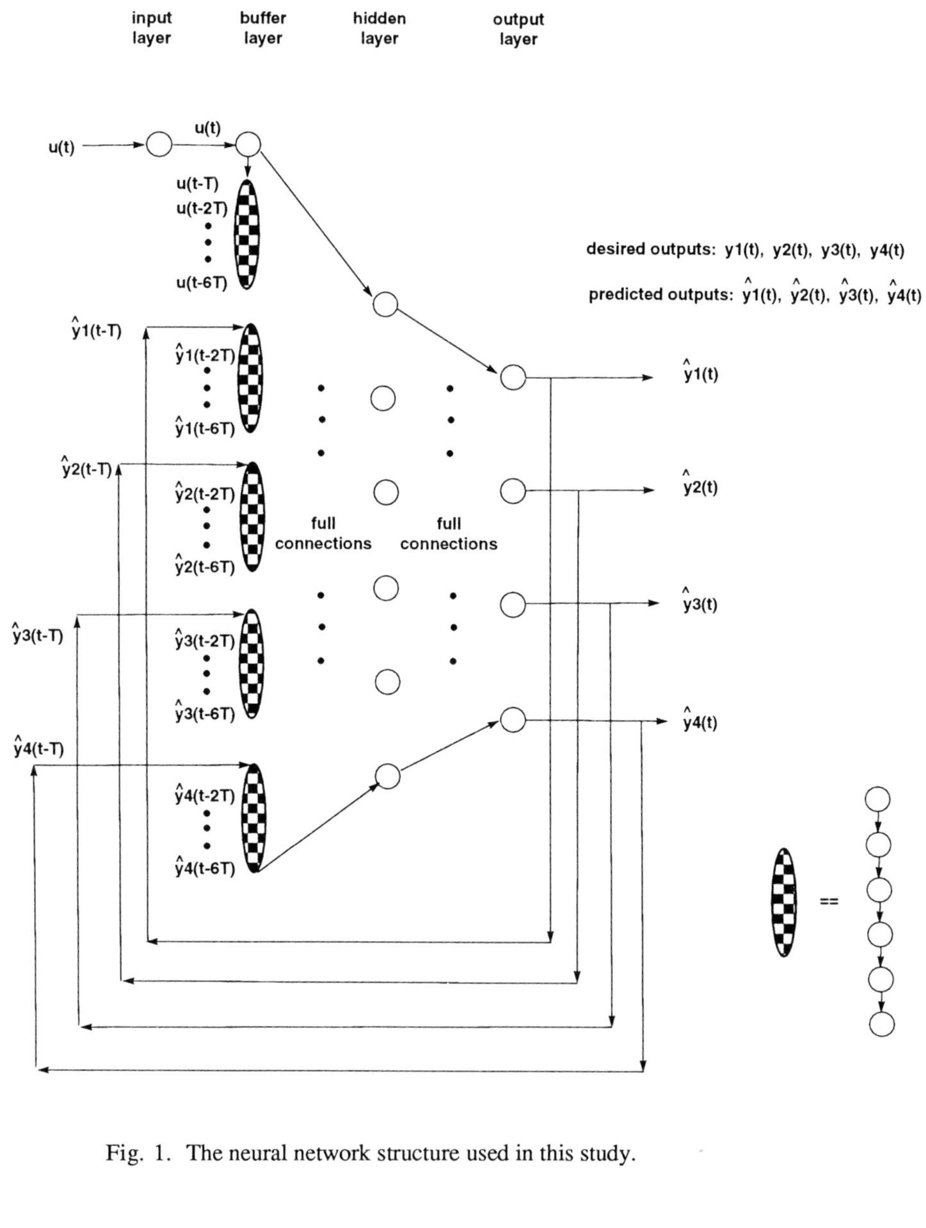

Fig. 1. The neural network structure used in this study.

histories of velocity and displacement are derived quantities from the acceleration records. Depending on how they are derived, noises of low frequency nature may be introduced (Hudson, 1979). By using acceleration time histories alone, the initial velocity and initial displacement at the beginning of a time window become implicit unknowns. Thus, the neural networks paradigm selected should be able to overcome this problem.

In using neural networks as a means of system identification for building seismic records, we treat the excitation records as the input to the network and the response record as the output of the network. The neural network paradigm selected is a modified model based on one proposed by Werbos (Werbos et al., 1992). As shown in Fig. 1, the input layer consists of one or more input nodes. Each node receives a time history u(t) which represents the acceleration time history, typically recorded at the ground level of a building. The buffer layer contains six nodes for each incoming time history. In the hidden layer, six

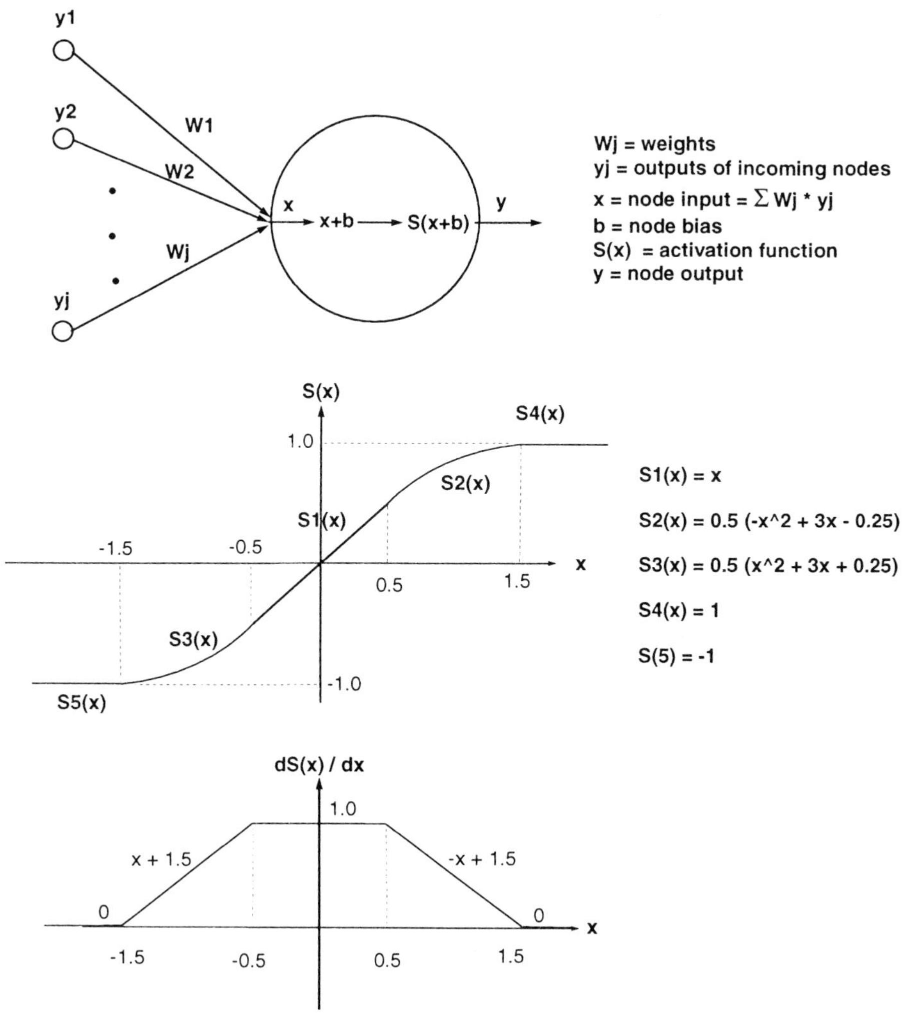

Fig. 2. A hidden node and its activation function.

nodes were shown in Fig. 1. In both the buffer layer and the hidden layer, other choices of the number of nodes were explored. The chosen number of six was found to be adequate. The output layer consists of the recorded time histories of building responses, typically at the roof level and other levels above the ground level. The predicted output time histories also become the feedback signal at the buffer layer as shown in Fig. 1.

In the hidden layer, the activation function is nonlinear in general. For purpose of comparison, we also include an option of a linear hidden layer. For a nonlinear hidden layer, a simple nonlinear activation function is shown in Fig. 2. Other commonly used activation functions such as hyperbolic functions were also tried. Results of the application of this particular neural networks paradigm and its variations are presented below.

RESULTS USING SIMULATED DATA

The actually recorded ground acceleration time history from a ten-story building in California is used as the excitation to a model building frame consisted of ten stories and a single bay. This model building frame was used in a separate study and found to be adequate in simulating the response of real building frames (Mau and Aruna, 1994). The properties of the model building frame were chosen in such a way that the fundamental vibration frequency is close to one Hz. The response accelerations at all levels were then calculated using a modal superposition method that included all ten modes of vibration. The mathematical model of the frame was that of a typical second order viscously damped system with a uniform modal damping ratio of 2%. The calculated data were 30-second long with a sampling time interval of 0.02 seconds. The calculated responses at the roof and the second floor are shown in Fig. 2 in solid lines. This set of dated is used for the purpose of testing the neural networks paradigm for its ability to be trained to simulate the behavior of a linear second order dynamic system of constant parameters. The activation function of the hidden layer is kept linear for this purpose.

Several different training strategies were tried. All use the ground record as input and four upper level records as output. As a first trial, the whole 30-second records were used for training. The resulting fit to the records is illustrated in Fig. 3 by showing the computed response at the roof and the second level in dashed lines. The best fit was at the roof where an almost perfect fit was achieved. The worst fit was at the second level where a still very close fit was achieved. The slight discrepancy at the second level may be attributed to the higher mode contributions. To investigate the ability of the network in simulating higher mode behavior, the characteristics of the trained network was revealed by exciting it with steady harmonic input and recording the steady-state response of the network. By changing the frequency of the excitation systematically, an amplitude frequency response curve was obtained. This is show in Fig. 4 in the form of acceleration amplification ratio of the roof response to the ground excitation. As can be seen, the exact curve displays three peaks of decreasing amplitude as expected. The neural network was able to reproduce the first two peaks very closely but erred on the third peak. Since higher modes usually have diminished contribution to the overall response, the error in the higher mode is not very significant in the simulation of overall behavior. The closeness of frequency response curves near the first peak is illustrated in Fig. 5 in an enlarged horizontal scale.

Table 1. Fundamental frequency and corresponding damping ratio determined from the trained neural network using different number of 2-second time windows

No. of windows	Frequency (Hz)	Damping ratio
1	0.9880	0.0175
2	0.9940	0.0180
3	0.9940	0.0198
4	0.9950	0.0211
15	1.0015	0.0208
Exact	1.0016	0.0200

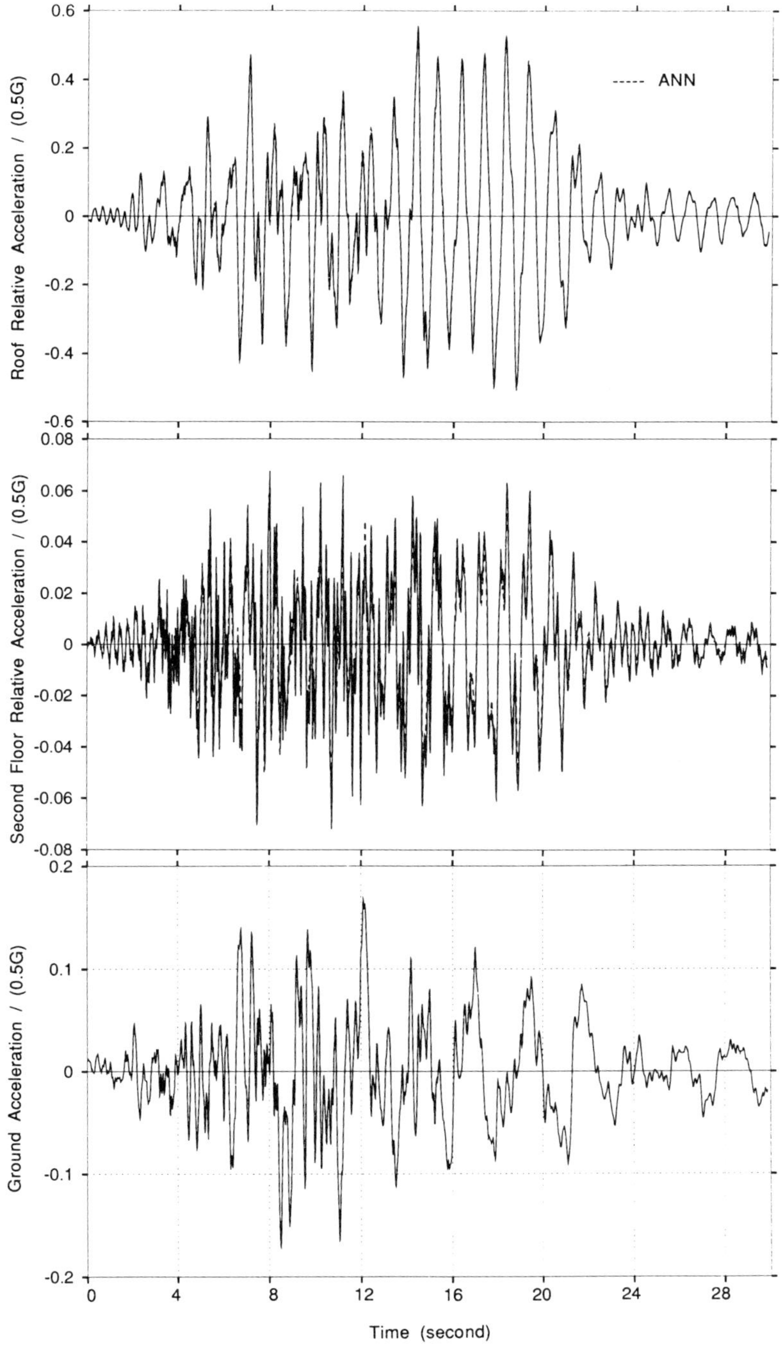

Fig. 3. Excitation ground acceleration and responses at the second level and the roof.

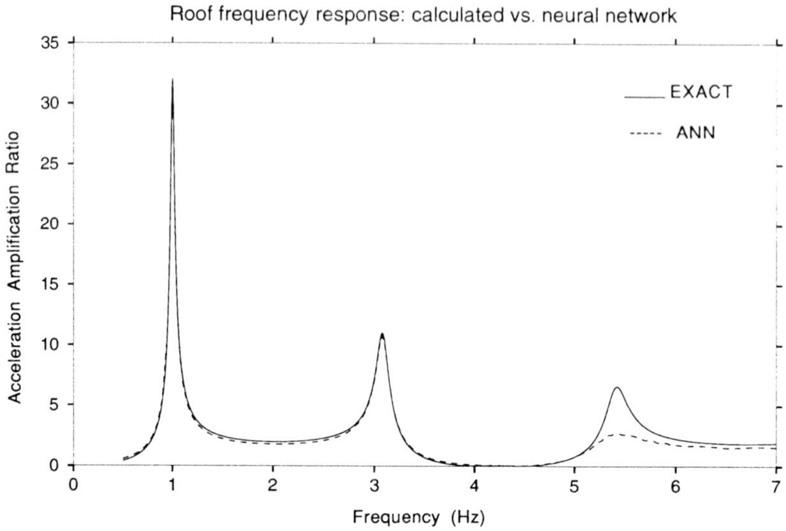

Fig. 4. Frequency response function of the system.

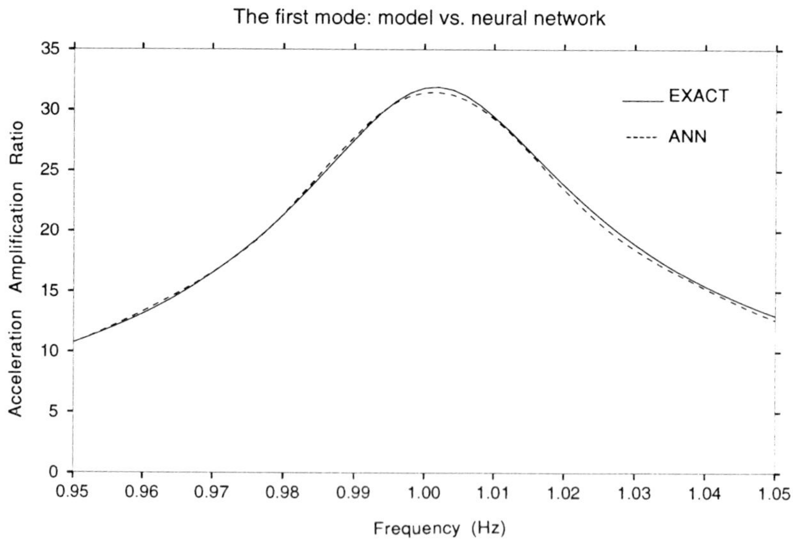

Fig. 5. Frequency response function near the first peak.

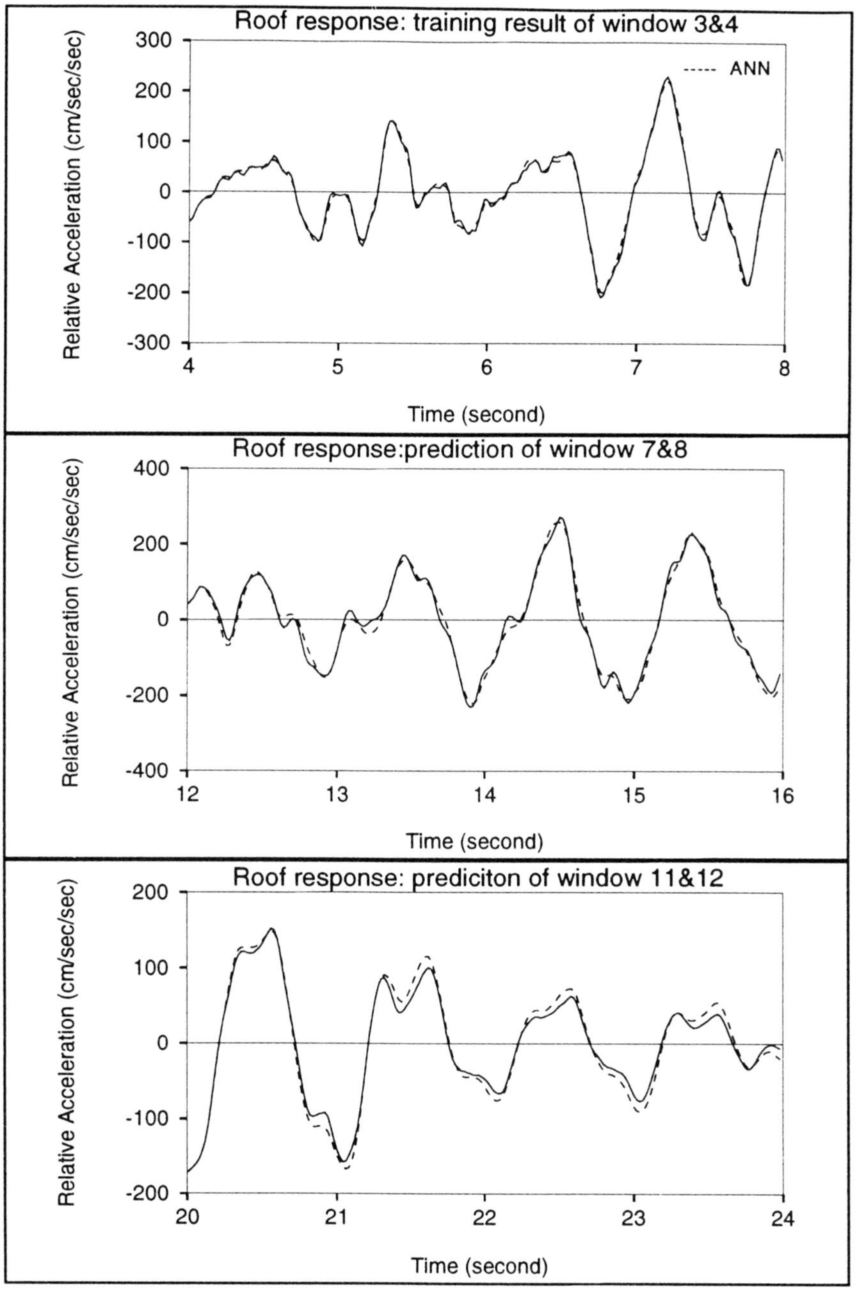

Fig. 6. Results of training using windows 3 and 4 and prediction at other times using the simulated data.

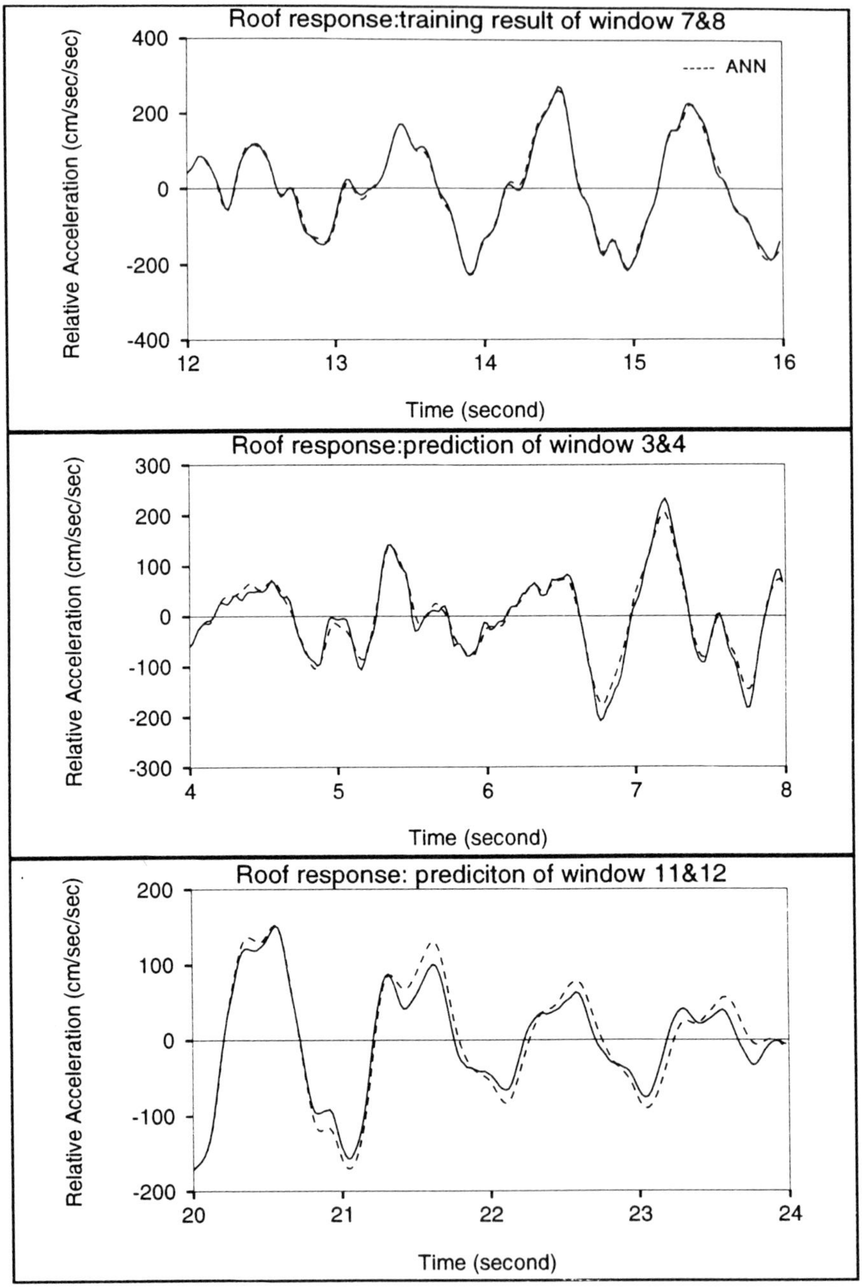

Fig. 7. Results of training using windows 7 and 8 and prediction at other times using the simulated data.

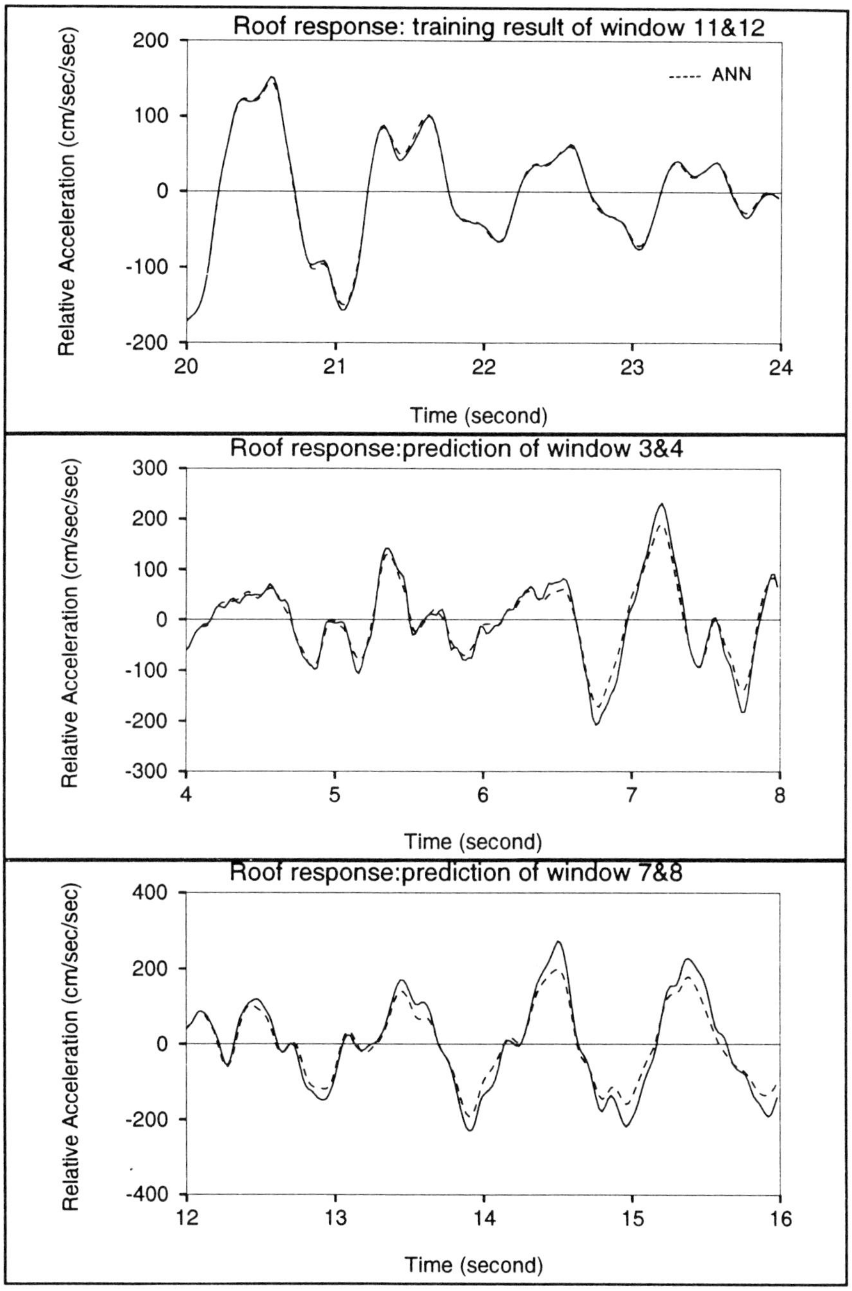

Fig. 8. Results of training using windows 11 and 12 and prediction at other times using the simulated data.

The frequency response curve produced from the trained network is used to extract modal frequency and damping values using the well known half-power bandwidth method. The fundamental frequency and the corresponding damping ratio were found to be 1.0015 Hz and 0.0208, respectively. Comparing to the exact values of 1.0016 Hz and 0.0200, the neural network results are very good indeed.

To see whether shorter duration of record may be used to train the neural network, the 30-second records were divided into fifteen 2-second windows. Training of the network was conducted using one, two, three, or four consecutive windows. The results on the fundamental frequency and the corresponding damping ratio are summarized in Table 1. It is seen that the fundamental frequency is more accurately predicted than the corresponding damping ratio, a conclusion similar to that in conventional system identification. It seems that for a fundamental frequency close to 1 Hz, two windows of a total duration of four seconds and sampling interval of 0.02 seconds are needed for the training of the neural network.

To see whether neural networks trained with two windows capture the main features of overall response of the system, we used trained networks to predict the responses at other windows. The results are shown in Figs. 6, 7, and 8. In each of these figures, the neural network trained with two consecutive windows was used to predict the response of a two-window duration at other times. The closeness of these predictions in all three figures indicates that the neural networks trained with four-second duration of records do capture the main features of the second order dynamic system. This conclusion is meaningful only because the system does not change its characteristics in time. When real earthquake data is used, however, the situation may be different as shown in the following section.

RESULTS USING EARTHQUAKE DATA

The recorded east-west direction accelerations at the roof and the second level of a 4-story steel hospital building were used as output for neural networks training. The ground level accelerations in the east-west direction was used either as the sole input or in conjunction with the acceleration in the north-south direction and the torsional rotation to form a set of two to three input motions to the neural network. A previous study of this set of records using conventional system identification revealed a significant change of fundamental frequency during the course of a 32-second duration (Li and Mau, 1992). Thus, the purpose of the current study is to see whether the selected neural network is able to capture the changing nature of the behavior of the building.

The nonlinear activation function shown in Fig. 2 was used for the hidden layer. Several training strategies were tried, all using the whole 32-second records. The different trials include increasing the number of nodes in the buffer layer from six to ten, increasing the number of nodes in the hidden layer from six to ten, using different activation functions such as hyperbolic functions, and using more than the east-west ground motion to include north-south ground motion and torsional ground motion as input. In all these trials, the results were similar--the fit in the strong motion portion (between 9 second and 26 second) was good with small mismatches in amplitude and the fit outside the strong motion portion was not good. Figure 9 illustrates the results of the artificial neural network training using one, two, and three ground motions as input. It is also found that the results are similar to that obtained with a linear activation function. In other words, the nonlinear activation function in the hidden layer does not improve the ability of the network in simulating the behavior of this building.

The frequency response curves of the two trained networks, one with the nonlinear activation function and the other with a linear activation function are shown in Fig. 10. The lack of a clearly defined second peak is understandable. The previous study by Li and Mau (1992) established that the predominant motion was in the first mode. Thus, the lack of clear definition of the second mode is not a source of error. The fundamental frequency and the corresponding damping ratio obtained from the two curves related to the east-west ground excitation are 1.55 Hz and 0.085 for the linear network and 1.52 Hz and 0.078 Hz for the nonlinear network. The differences are insignificant.

The source of error is in the network's inability to follow the change in the building behavior. This is demonstrated by using two windows to train a network and using the resulting network to predict motions at other times as shown in Figs. 11, 12, and 13. In all cases, the prediction was way off, signaling that the building behavior has changed from one time duration to another.

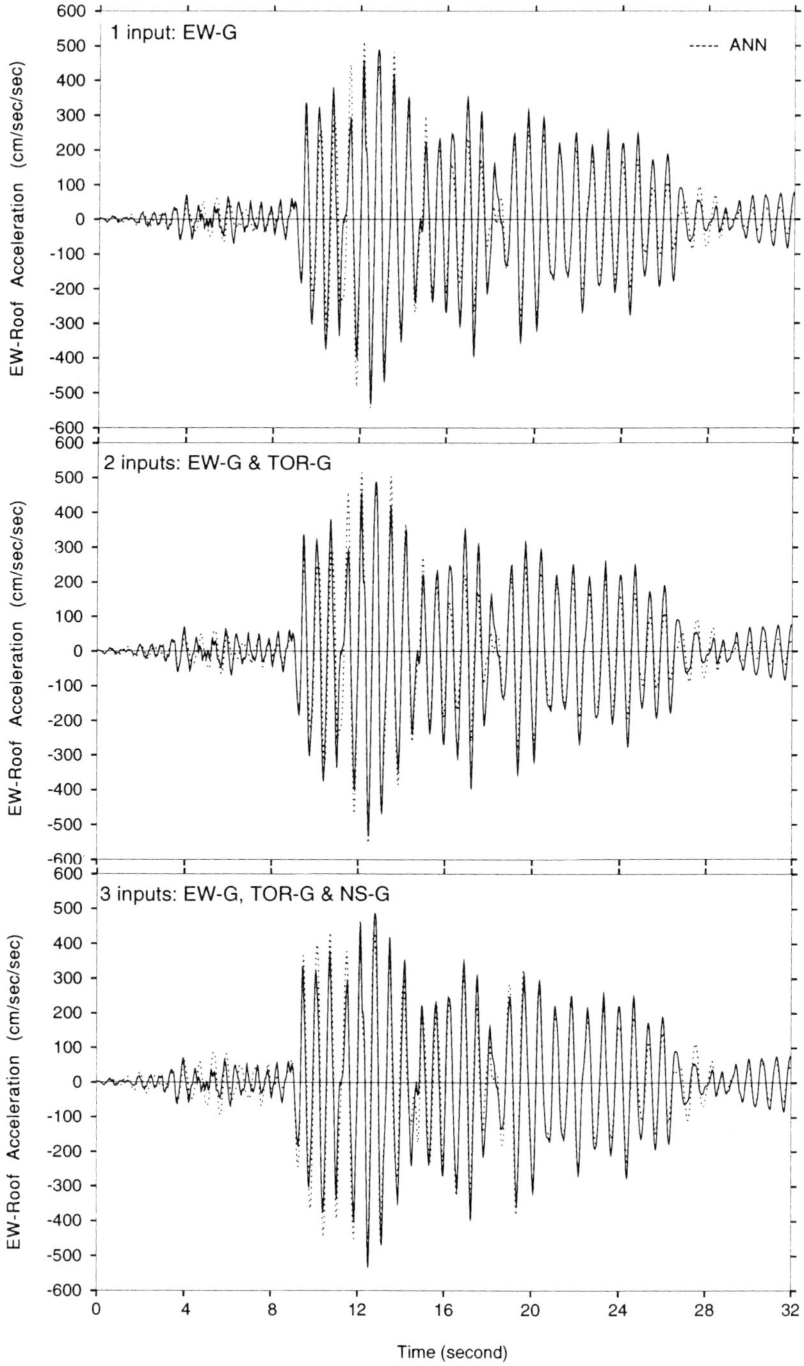

Fig. 9. Results of training using earthquake data with three different input schemes.

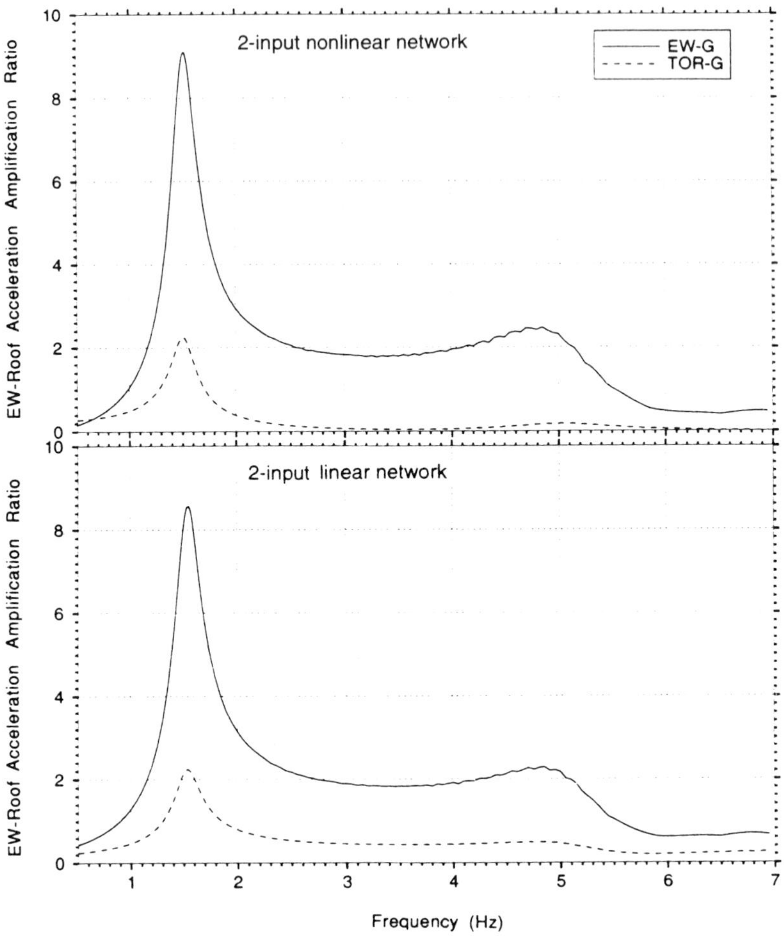

Fig. 10. Frequency response functions of two trained networks using the earthquake data.

CONCLUSIONS

An artificial neural network has been studied for its ability in simulating the behavior of a second order linear dynamic system and a real building system. In the case of a constant parameter second order system, the neural network was able to capture the main features of the system. The fundamental modal frequency and damping ratio were accurately estimated by exciting the network with harmonic motions and analyzing the steady-state responses. The network trained by a four-second window can be used to predict the response at other times with good accuracy.

In the case of real earthquake data from a building known for its changing behavior during an earthquake, the neural network was not able to follow the changing nature of the building response. Such a changing behavior of a building is usually attributed to the change of stiffness due to disengagement of nonstructural elements and may not be modeled by any simple continuous function in time. Further study is needed to determine whether artificial neural networks are able to simulate this kind of nonlinear behavior.

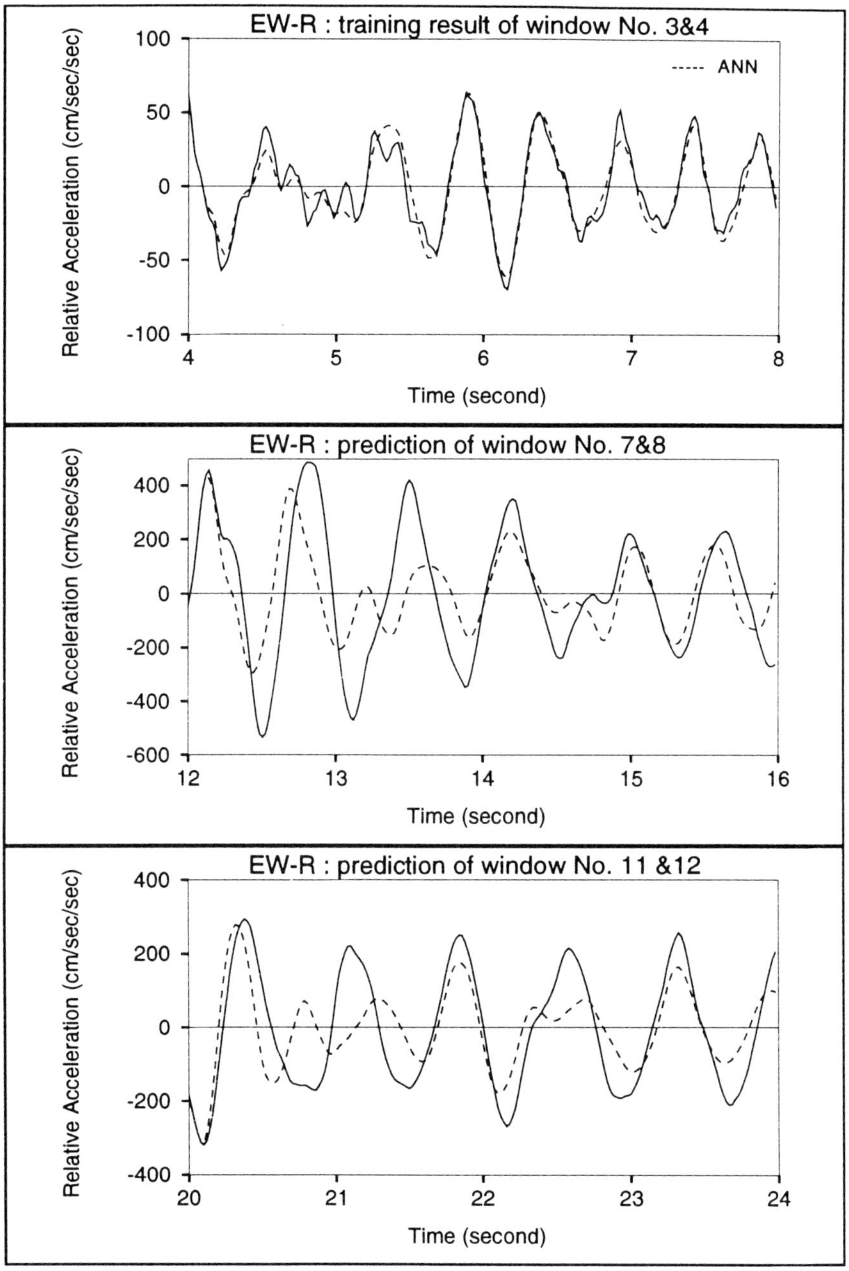

Fig. 11. Results of training using the earthquake data at windows 3 and 4 and predictions at other times.

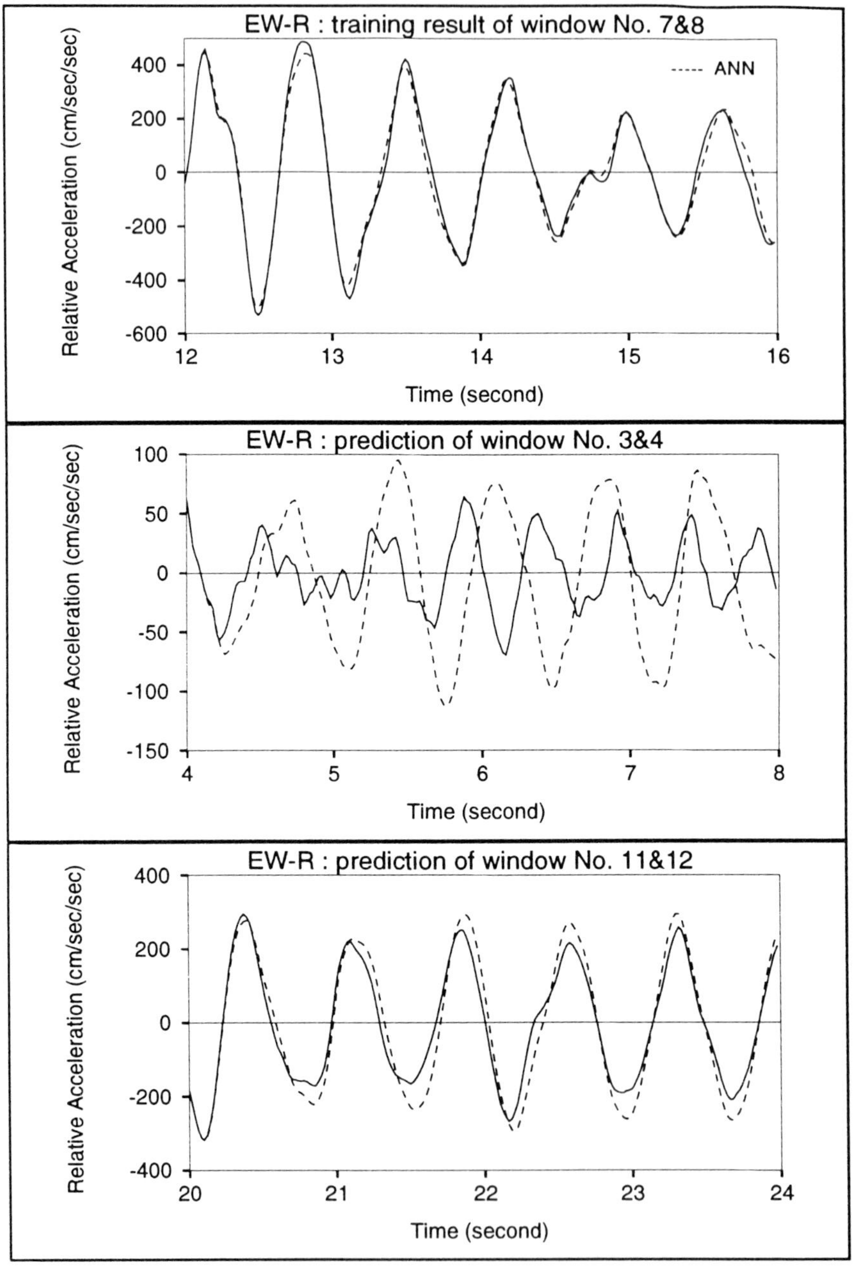

Fig. 12. Results of training using the earthquake data at windows 7 and 8 and predictions at other times.

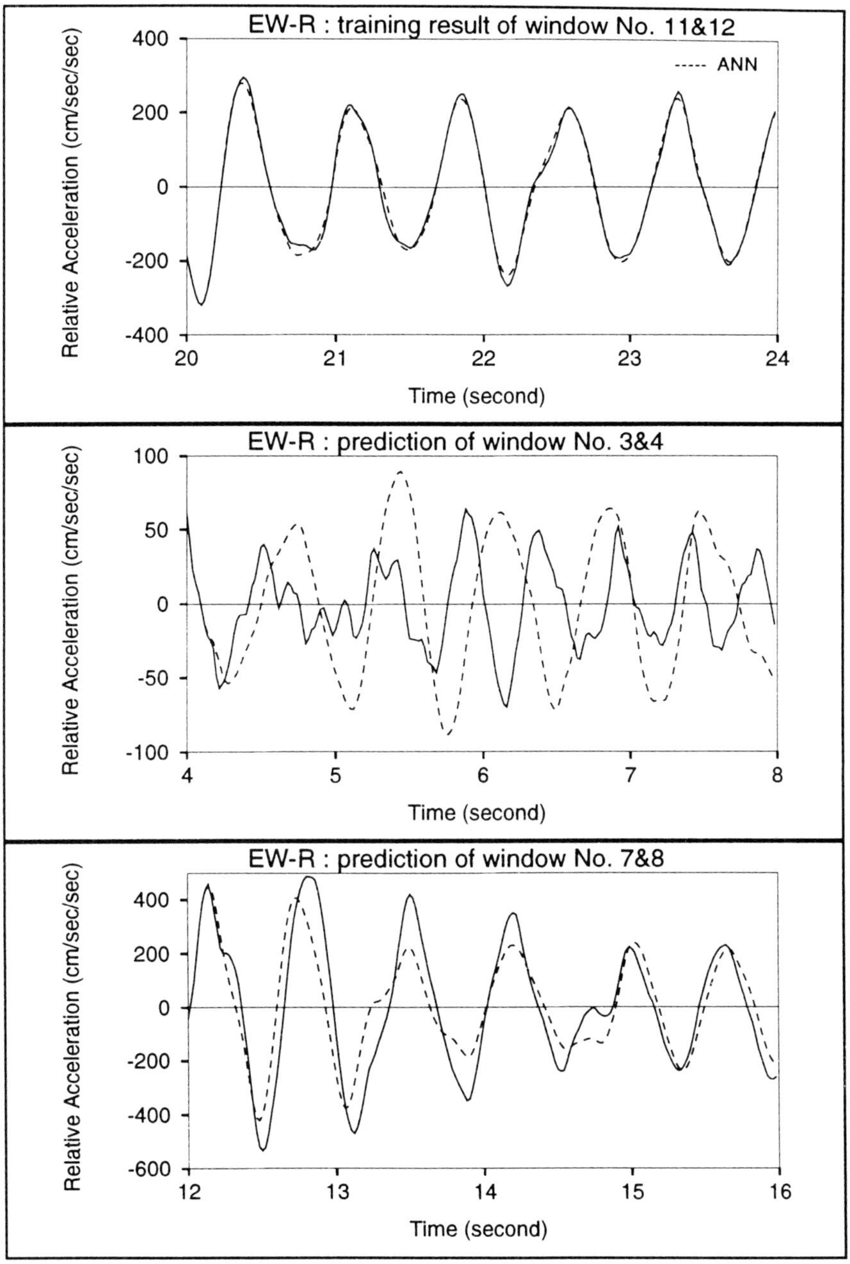

Fig. 13. Results of training using the earthquake data at windows 11 and 12 and predictions at other times.

REFERENCES

Billings, S. A., H. B. Jamaluddin, and S. Chen (1992). Properties of neural networks with applications to modelling nonlinear dynamic systems, *International Journal of Control*, 55, 193-224.

Doo, L. B. (1989). Neural networks for system identification and pulse control of linear systems, Master's Thesis, Carnegie Mellon University.

Elkordy, M. F., K. C. Chang, and G. C. Lee (1993). Neural networks trained by analytically simulated damage states, *Journal of Computing in Civil Engineering, ASCE*, 7, 130-145.

Hornik, K., M. Stinchcombe, and H. White (1989). Multilayer feedforward networks are universal approximations, *Neural Networks*, 2, 359-366.

Hudson, D. E. (1979). *Reading and Interpreting Strong Motion Acclerograms*, Earthquake Engineering Research Institute, Berkeley, California.

Li, Y. and S. T. Mau (1991). A case study of mimo system identification applied to building seismic records, *Earthquake Engineering and Structural Dynamics*, 20, 1045-1064.

Li, Y. and S. T. Mau (1992). Anatomy of the Loma Prieta Earthquake records of two steel buildings using mimo system identification, *Proc. Ninth Conf. Engrg. Mechanics, ASCE*, College Station,Texas, pp. 689-692.

Mau, S. T. and V. Aruna (1994). Story-drift, shear, and otm estimation from building seismc records, *Joural of Structural Engineering, ASCE*, 120, 3366-3385.

Narendra, K. S. and K. Parthasarathy (1990). Identification and control of dynamic systems using neural networks, *IEEE Transactions on Neural Networks*, 1, 4-27.

Werbos, P., T. McAvoy, and T. Su (1992). Neural networks, system identification, and control in the chemical process industries, In: Handbook of Intelligent Control (D. A. White and D. A. Sofge, Ed.),Chap. 10, Van Nostrand Reinhold,New York.

Wu, X., J. Ghaboussi, and J. H. Garrett, Jr. (1992). Use of neural networks in detection of structural damage, *Computers & Structures*, 42, 649-659.

Simpsom, P. K. (1990). *Artificial Neural Systems: Foundations, Paradigms, Applications, and Implementations*, Pergamon Press.

Landslides and Rockslides

SEISMIC STABILITY OF COHESIVE SOIL SLOPES

T. D. Stark,* I. A. Contreras* and I. M. Idriss**

** Department of Civil Engineering, University of Illinois at Urbana-Champaign, IL, U.S.A.*
*** Department of Civil Engineering, University of California at Davis, CA, U.S.A.*

Abstract

The paper presents a re-evaluation of the Fourth Avenue landslide in Anchorage during the 1964 Alaskan earthquake. The re-evaluation reveals that the slide was caused by a large undrained strength loss and development of an undrained residual strength condition in the Bootlegger Cove clay. This mechanism differs from the liquefaction mechanism proposed by Shannon and Wilson (1964), Seed and Wilson (1967), and Seed (1968). A constant volume ring shear apparatus was developed to measure the undrained peak and residual shear strength of cohesive soils. The results of constant volume ring shear tests on Bootlegger Cove clay are presented and compared to back-calculated shear strength values. The comparison shows that slide blocks, which moved less than one-half foot mobilized the undrained peak shear strength. Slide blocks that moved more than one-half foot mobilized the undrained residual strength. The results of this study suggest that the peak undrained shear strength should be used to evaluate the seismic stability of cohesive soil slopes. If sliding will be triggered due to earthquake shaking, an undrained residual strength will be mobilized and should be used to estimate the permanent lateral displacements.

Keywords

soft clays, seismic slope stability, undrained shear strength, ring shear tests

Introduction

A significant number of landslides involving cohesive soils have occurred during earthquakes. Some of the earthquakes in which these slides have occurred are the New Madrid earthquake of 1811, the Chilean earthquake of 1960, the Alaskan earthquake of 1964, and most recently the Saguenay earthquake of 1988. The most notable landslides are the Fourth Avenue, L-Street, Government Hill, First Avenue, Romig Hill, and Turnagain Heights in Anchorage that were caused by the 1964 Alaska earthquake (Seed, 1968) and the Sainte-Thècle and Saint-Adelphe landslides that took place during the

1988 Saguenay earthquake (Lefebvre et al., 1992). These landslides have lead to an interest in the undrained peak and residual shear strength of cohesive soils. It is anticipated that these slides were caused by an undrained failure and post-peak strength loss in the cohesive soil involved in the slides.

Following the 1964 Alaska earthquake, many of the landslides that occurred in Anchorage were intensively studied. In particular, the Fourth Avenue landslide, prompted considerable investigation because of the large destruction resulting from the slide. The main failure mechanism proposed in the resultant literature was liquefaction of sand lenses (Shannon and Wilson, 1964; Seed and Wilson, 1967; Seed, 1967,1968; Wilson, 1967). A re-evaluation of the Fourth Avenue slide by Woodward Clyde Consultants (1982), revealed that the slide was not caused by liquefaction but by a large undrained strength loss in the Bootlegger Cove clay. However, an apparatus that could measure the entire undrained post-peak strength loss and the collapse of the soil structure was not available during the 1982 study. Stark and Contreras (1994) developed a constant volume torsional ring shear apparatus that allows the measurement of the magnitude and rate of undrained post-peak strength loss, and the undrained residual strength. The amount of lateral movement in the Fourth Avenue slide was either several feet or less than a few inches. It is postulated that in areas where the lateral movements were greater than one-half foot, the shear strength of the Bootlegger Cove clay was reduced to the undrained residual value. This paper presents a re-evaluation of the Fourth Avenue landslide based on tests results from the new constant volume ring shear apparatus.

Fourth Avenue Landslide

The Fourth Avenue slide occurred during the great 1964 Alaska earthquake. This earthquake had an epicenter approximately 130 km east of Anchorage. The earthquake magnitude was estimated as 8.5 (Woodward-Clyde Consultants, 1982) and the surface ground motion levels at Anchorage were estimated to be 0.15 to 0.2 g (Newmark, 1965; Housner and Jennings, 1964; and Shannon and Wilson, 1964). However, no accelerograms of the earthquake shaking were obtained. The duration of the ground motions in Anchorage was reported to range from 4 to 7 minutes, with strong, potentially damaging shaking lasting approximately 2 to 3 minutes (Steinbrugge, 1970; Housner and Jennings, 1964).

The slide mechanism was primarily horizontal translatory which is characterized by lateral spreading and graben development. Two grabens were created east of C street. The survey of ground movements suggested that by the end of the earthquake, another graben had begun to form between Fourth and Fifth Avenues at D and E streets (Wilson, 1967). The greatest damage to structures developed within and adjacent to the grabens, and along the pressure ridge that developed at the toe of the slide between First and Second Avenues. In contrast, very little damage was suffered by buildings and streets that were located on the sliding mass.

Based on analyses of the slide conducted shortly after the earthquake, the zone of shearing was found to be at a depth of approximately 60 to 70 ft below the ground surface. This depth is at or near the interface between a fine sandy layer and the underlying soft Bootlegger Cove clay. At that time it was not clear whether the slide occurred through the fine sandy layer because of liquefaction or through the soft Bootlegger Cove clay layer because of undrained failure (Shannon and Wilson, 1964). This uncertainty regarding the failure mechanism was in part due to the relatively limited data, particularly on the sands, available at that time.

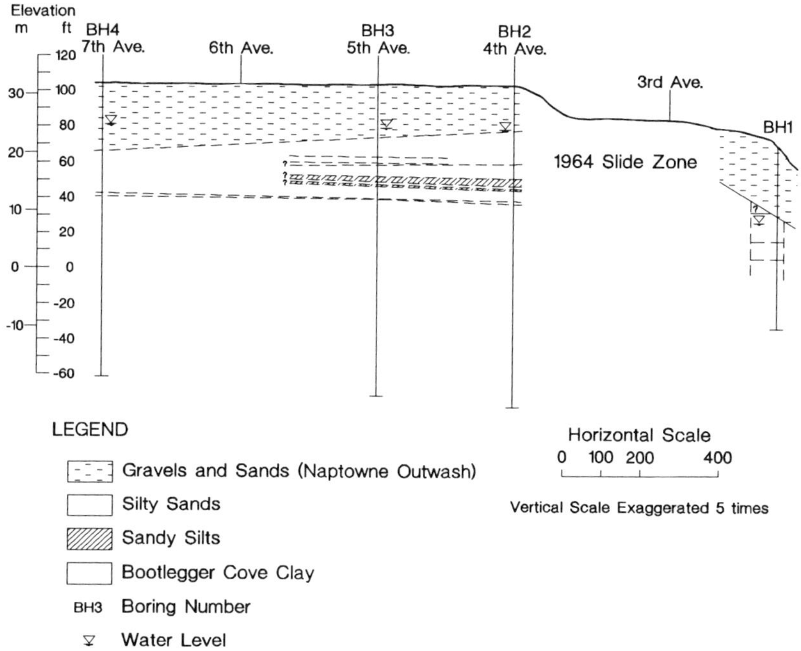

Fig. 1. Subsurface Conditions Along D Street (After Idriss, 1985)

The subsurface conditions in the slide area are illustrated in Fig. 1, which is a cross section along D Street. From the subsurface investigations conducted after the earthquake (Shannon and Wilson, 1964) and the 1982 Woodward Clyde Consultants investigation, the following subsurface conditions were obtained:

1.) The upper most zone consists of very dense gravels and sands (Naptowne Outwash). The thickness of the Naptowne Outwash generally ranges from 25 ft to 40 ft. The SPT blow counts in the outwash are typically greater then 60 blows/ft.

2.) An interbedded zone underlies the Naptowne Outwash and consists of alternating layers of clay, silty sands, and sandy silts to a maximum depth of 62 ft below the ground surface (elevation 40 ft). This interbedded zone corresponds to the upper portion of the Bootlegger Cove Formation. The silty sands and sandy silts in this zone are not continuous and are dense to very dense. The SPT N_1 blow counts range from 21 to 70 blows/ft. The clays in this zone are stiff to very stiff probably due to desiccation (Updike, 1988). Overconsolidation ratios (OCR) of about 3 to 4 were measured for the clays in the interbedded zone (Woodward Clyde Consultants, 1982).

3.) Below the interbedded zone, a soft clay is found to the maximum depths explored (about 150 ft below the ground surface). This soft clay belongs to the intermediate zone of the Bootlegger Cove Formation. This soft clay displays an OCR of 1.2 to 1.5. The intermediate zone of the Bootlegger Cove Formation has been identified with zones of sensitive clays with values of sensitivity ranging from 3 to 11.

As noted earlier, the evaluation of the Fourth Avenue slide after the earthquake indicated that the zone of shearing was located near the lower boundary of the interbedded zone. The liquefaction potential of the silty sand and sandy silt layers in the interbedded zone was evaluated (Woodward-Clyde Consultants, 1982) using the empirical procedure based on field observations (Seed and Idriss, 1981). The modified blow counts for these layers (Fig. 2) correspond to a minimum factor of safety against liquefaction of approximately 1.6. As a result, it was concluded that the sandy silt or silty sand layers in the interbedded zone did not liquefy, and thus was not the cause of the slide. Therefore, the cause of the slide was thought to be an undrained failure of the soft Bootlegger Cove clay.

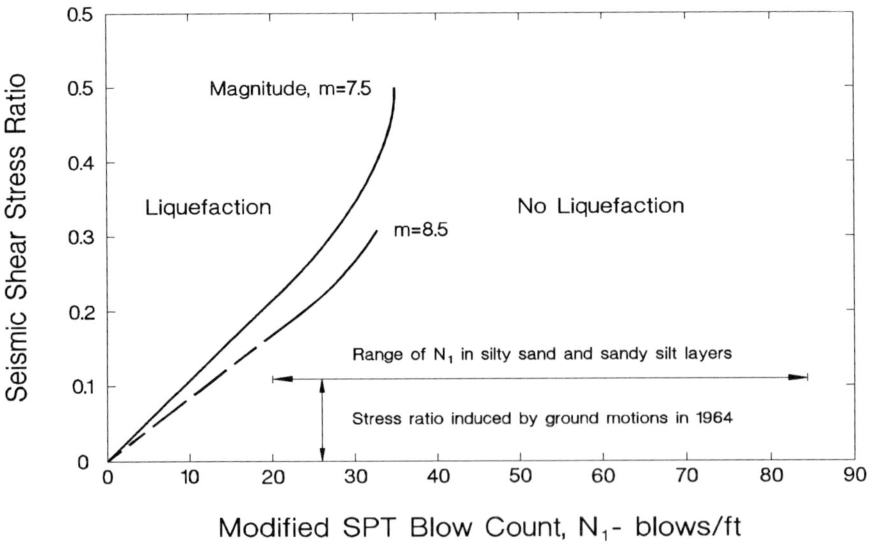

Fig. 2 Liquefaction Potential in the Sandy Strata in the Interbedded Zone During the 1964 Alaska Earthquake (After Idriss, 1985)

Shear Strength of the Soft Bootlegger Cove Clay

A summary of undrained peak shear strength data available for the Bootlegger Cove clay from the investigations conducted after the earthquake (Shannon and Wilson, 1964) and the 1982 Woodward-Clyde Consultants study is shown in Fig. 3. Figure 3 includes relationships for triaxial compression (TC), triaxial extension (TE), direct simple shear (DSS), and laboratory vane shear (LV) modes of shear. In general, the relative positions of the TC, DSS, TE, and LV data points agree well with the average lines proposed by Jamiolkowski et al. (1985), Chandler (1987), and Mesri (1989).

As indicated earlier the slide mechanism was primarily horizontal translatory. As a result, the direct simple shear apparatus allows the closest laboratory simulation of the stresses and deformations imposed by the earthquake shaking and translational sliding on soil elements in the field. The undrained shear strength ratio from the DSS test, $S_{u(DSS)}/\sigma'_{vc}$ in Fig. 3 varies from 0.18 to 0.24 with an average of 0.21. Changes to this strength can result due to cyclic shear strains induced by the earthquake loading and/or due to large displacements during slide movements (Idriss, 1985).

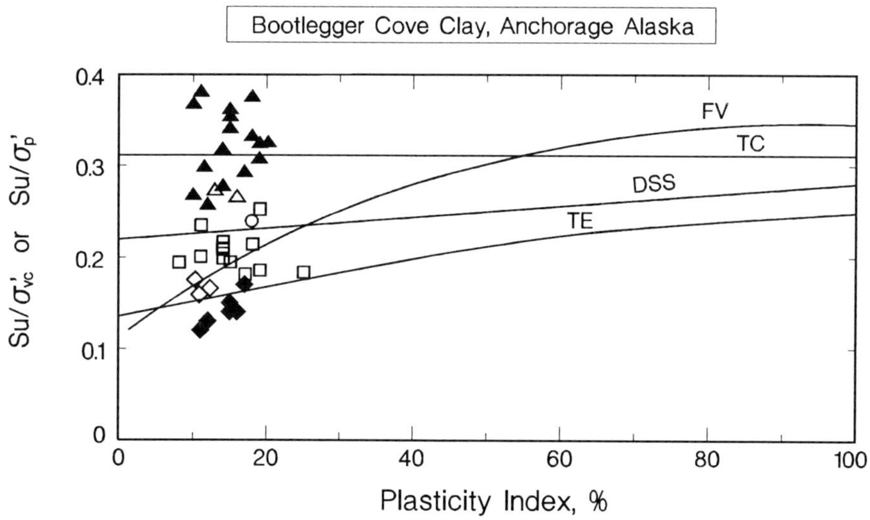

Test	Shannon & Wilson (1964)	Woodward-Clyde (1982)
Triaxial Compression	▲	△
Direct Simple Shear	N/A	□
Triaxial Extension	N/A	○
Laboratory Vane	◆	◇

Fig. 3. Summary of Undrained Peak Strength Ratios of Bootlegger Cove Clay

The potential for strength changes due to cyclic loading on Bootlegger Cove clay was investigated by Woodward-Clyde Consultants (1982) using cyclic direct simple shear tests followed by post-cyclic static tests. It was found that reduction of the strength due to cyclic loading is rather small and generally does not exceed more than 25 to 30 percent of the original static strength. Lefebvre and LeBoeuf (1989), concluded that even extrasensitive clays are not believed to suffer significant loss of shear strength during the cyclic loading representing a moderate earthquake.

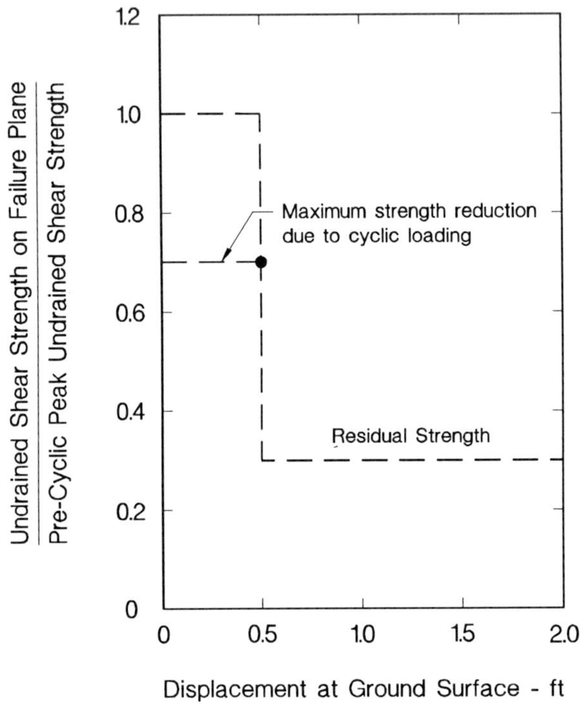

Fig. 4. Variation of Undrained Shear Strength on Potential Failure Plane as a Function of Displacement at the Ground Surface (After Idriss, 1985)

Reduction in shear strength can also occur in cohesive soils due to large deformations that result in clay particles rearranging into a preferred orientation, such that only the frictional resistance of parallel soil particles is available. This minimum strength is termed the residual strength. If sufficient deformation takes place during an earthquake, a residual strength may be achieved under undrained conditions. However, at the time of the 1982 Woodward-Clyde Consultants investigation a laboratory apparatus and test procedure for measuring the undrained residual strength of clays was not available. As a result, during the 1982 investigation the results of miniature laboratory vane shear tests, cone penetration tests, and engineering judgment were used to estimate the undrained residual strength.

Examination of the laboratory vane shear tests and the cyclic DSS tests lead to the relationship between undrained shear strength and ground surface displacement shown in Fig. 4.

It can be seen from Fig. 4 that for displacements less than one-half foot the undrained shear strength is reduced a maximum of 30 percent of the static peak strength due to the effects of cyclic loading. Idriss (1985) concluded that in the Fourth Avenue landslide the slide blocks that moved less than one-half foot underwent a shear strength reduction up to 30 percent. If sufficient deformations take place during an earthquake (i.e., more than one-half foot) the shear strength is further reduced to the residual value. Idriss (1985) concluded that this strength reduction corresponds to 70 percent of the peak value.

Constant Volume Ring Shear Apparatus

To investigate the peak and residual undrained shear strengths, the original Bromhead (1979) ring shear apparatus was modified by Stark and Contreras (1994) to conduct constant volume tests. The modifications include a mechanism for decreasing the normal stress during shear, such that the volume change is negligible during shear, and fabricating a new specimen container to allow undisturbed specimens to be trimmed directly into the container.

The ring shear specimen is annular with an inside diameter of 70 mm and an outside diameter of 100 mm. Drainage is provided from the bottom by a 6.5 mm thick annular bronze porous stone that is secured by four screws. The specimen is confined radially by the specimen container, which is 10 mm deep. Drainage is provided from the top by an annular bronze porous stone, which is attached to the loading platen. The porous stone is secured to the loading platen using four screws. It has a thickness of 5.7 mm and aids the transfer of shear stress to the top boundary of the soil specimen. The porous stone is serrated to develop a strong interlock with top boundary of the soil specimen. Two intersecting groups of serration lines are grooved in the porous stone. Spacing between lines in each group is 2 mm. The indentation angle is 45 degrees with a depth of 0.7 mm. These serrations prevent slippage at the loading platen/soil interface during shear. The annular bronze porous stone also has an inner and outer groove machined at 1 mm from the bottom edge of the ring to accommodate two O-rings. These O-rings close the gap between the top annular ring and the walls of the specimen container to minimize the amount of soil extrusion that takes place during shear.

The unload mechanism developed to maintain a constant specimen volume during shear, and thus conduct an undrained test, is illustrated in detail in Fig. 5. It consists of a steel bracket that is connected to the end of the loading arm. The loading arm applies the dead weight on the hanger to the top of the specimen. Attached to the top of the bracket, is a sensitive load cell. Connected to the top of the load cell, is a threaded steel rod 9 mm in diameter. This threaded steel rod goes through a 14 mm diameter hole in an aluminum beam that is connected to the frame of the original apparatus by two screws. A nut is threaded to the top of the steel rod. An electronic readout device is connected to the load cell, such that the load carried by the load cell can be determined. The nut is threaded and adjusted during shear to reduce the amount of normal load transferred to the loading platen. The nut is adjusted such that a negligible amount of vertical displacement of the soil specimen occurs during shear.

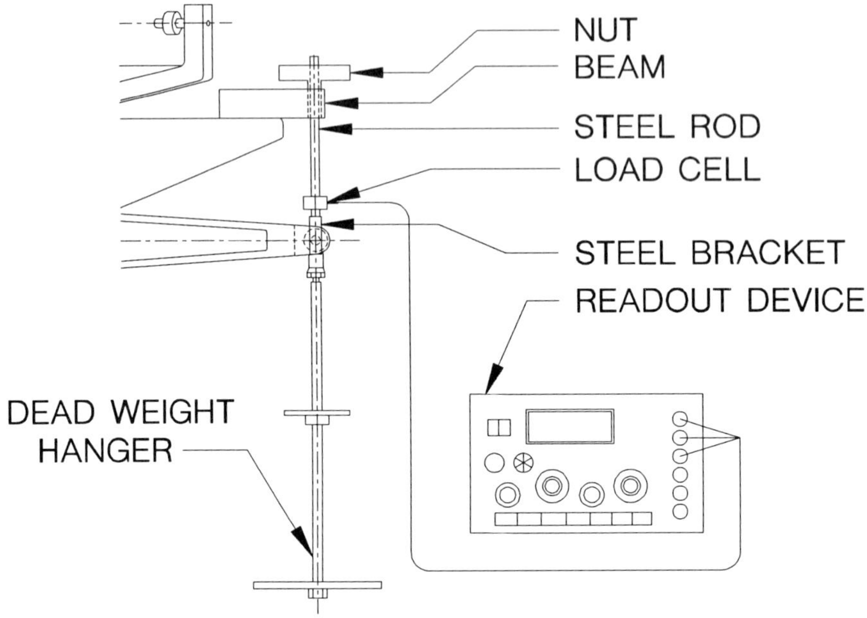

Fig. 5. Unload Mechanism for Ring Shear Apparatus

Constant Volume Ring Shear Test Procedure

In the constant volume ring shear apparatus, the specimen is sheared by rotating the specimen container past the stationary loading platen at a constant rate of 0.018 mm/minute. The procedure described by Gibson and Henkel (1954) is used to determine the shear displacement rate. This displacement rate is selected such that the pore-water pressures in the specimen are zero throughout the test. The normal stress is reduced during shear to maintain a constant specimen height or volume. It is assumed that the decrease in applied vertical stress during shear is equivalent to the increase in shear-induced pore-water pressure that would occur in an undrained test with constant vertical stress. The validity of the pore-water pressure assumption for constant volume tests was verified by Dyvik et al. (1987) for the direct simple shear apparatus and by Berre (1981) for the triaxial apparatus.

During a constant volume test the height of the specimen is monitored using the dial gauge that measures vertical displacement of the loading platen. The nut connected to the steel rod in the unload mechanism (Fig. 5) is used in combination with the vertical dial gauge to adjust the normal stress, such that the specimen thickness remains constant during shearing. Adjustments are made by rotating the nut an amount required to maintain zero vertical displacement of the soil specimen. When the nut is rotated, a portion of the dead weight is transmitted to the rod and the load cell indicates the magnitude of this load. Details of the constant volume ring shear apparatus can be found in Stark and Contreras (1994).

Constant Volume Ring Shear Test Results

Undisturbed samples of Bootlegger Cove clay were obtained using a 100 mm diameter sampler at the Fourth Avenue Slide. This Bootlegger Cove clay has the following index properties: (1) natural water content of 34 percent; (2) liquid limit of 36 percent; (3) plasticity index of 15 percent; and (4) clay-size fraction of 49 percent.

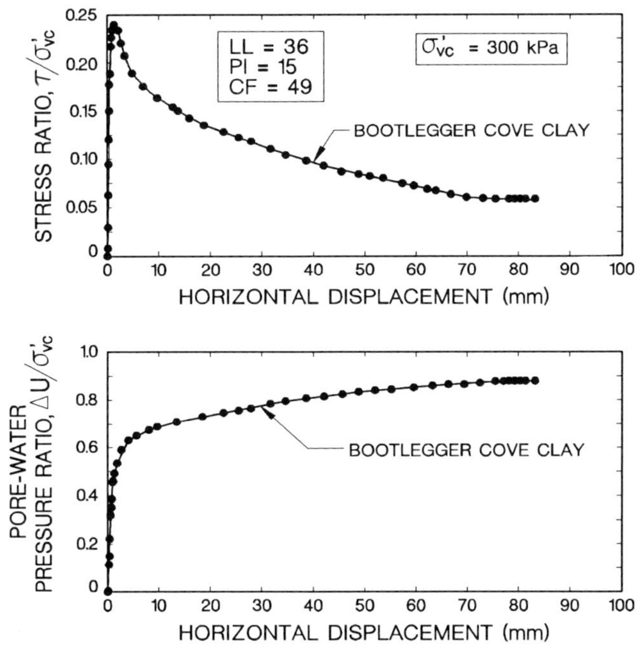

Fig. 6. Constant Volume Ring Shear Test on Bootlegger Cove Clay

Figure 6 shows a typical set of results of a constant volume test on a specimen of Bootlegger Cove clay. Figure 6 shows the stress ratio-horizontal displacement relationship. The stress ratio is defined as the shear stress, τ , divided by the consolidation stress, σ'_{vc}. It can be seen that the undrained peak stress ratio is reached after approximately 1.5 mm of horizontal displacement and the undrained residual stress ratio is reached at a displacement of about 80 mm. The undrained peak stress ratio, $S_{u(RS)}/\sigma'_{vc}$, is approximately 0.23. These $S_{u(RS)}/\sigma'_{vc}$ ratio is within the range of values of $S_{u(DSS)}/\sigma'_{vc}$ obtained by Woodward Clyde Consultants (1982), and represents a typical ratio for soft clays of this plasticity (Fig. 3).

After the undrained peak stress ratio is mobilized in the ring shear apparatus, shear displacement along the failure surface continues and the stress ratio starts to decrease with increasing displacement. Shear displacement along the failure surface causes orientation of soil particles parallel to the direction of shear. The movement and arrangement of particles along the failure surface is accompanied by a change in pore-water pressure. This continues until the undrained residual strength is mobilized. It can be seen from Fig. 6 that the undrained residual strength ratio is mobilized at a shear displacement of approximately 80 mm. The undrained residual stress ratio is measured to be approximately 0.06. It can be seen that this undrained residual stress ratio is significantly lower than the peak undrained stress ratio. It is postulated that this large post-peak strength loss, approximately 70 percent, led to the large lateral displacements observed during the earthquake. This undrained residual shear stress ratio is in agreement with the strength ratio reported by Idriss (1985) for blocks that move more than one-half foot.

Figure 6 also presents the pore-water pressure ratio-horizontal displacement relationship measured for Bootlegger Cove clay in the constant volume ring shear apparatus. The pore-water pressure is assumed to be equal to the decrease in normal stress. It can be seen that the pore-water pressure ratio increases and then becomes essentially constant at the residual condition.

At the undrained residual strength condition, it is possible to compute the drained residual friction angle, ϕ'_r, of a clay from the results of a constant volume ring shear test using the following equation:

$$\tan \phi'_r = \frac{\tau_r}{\sigma'_r} \qquad (1)$$

where τ_r is the residual shear stress and σ'_r is the effective stress at the residual condition. The value of ϕ'_r computed using Equation (1) is approximately 25 degrees. The measured ϕ'_r from a series of drained ring shear tests on the same Bootlegger Cove clay, following the procedure described by Stark and Eid (1994), is also 25 degrees. Therefore, this agreement reinforces the validity of the results and the constant volume test procedure.

Conclusions

Re-evaluation of the Fourth Avenue landslide that occurred in Anchorage, Alaska during the 1964 earthquake reveals that the slide was caused by a large undrained strength loss and development of a residual strength condition in the Bootlegger Cove clay. This mechanism differs from the liquefaction mechanism postulated immediately after the earthquake. A constant volume ring shear apparatus was developed to measure the undrained peak and residual shear strength of cohesive soils. The results of constant volume ring shear tests on Bootlegger Cove clay are presented and compared to the undrained shear strengths back-calculated for the slide blocks. The comparison shows that slide blocks, which moved less than one-half foot mobilized the undrained peak shear strength. Slide blocks that moved more than one-half foot mobilized the undrained residual strength. The results of this study suggest that the peak undrained shear strength should be used to evaluate the seismic stability of cohesive soil slopes. If sliding will be triggered due to earthquake shaking, an undrained residual strength will be mobilized and should be used to estimate the permanent lateral displacements.

Acknowledgments

This study was performed as part of the United States Geological Survey, Grant No. INT1434-92-G-2178. The support of this agency is gratefully acknowledged. The writers also acknowledge G. Mesri for his many valuable suggestions.

References

Berre, T. (1981). "Comparison of Undrained and Constant Volume Triaxial Tests of Plastic Clay from Drammen." Internal Report 56103-23, Norwegian Geotechnical Institute, Oslo.

Bromhead, E.N. (1979). "A Simple Ring Shear Apparatus." Ground Engineering, Vol. 12, No. 5, pp. 40-44.

Chandler, R.J. (1987). "The in situ measurement of the undrained shear strength of clays using the field vane." International Symposium on Laboratory and Field Vane Strength Testing, American Society for Testing and Materials, Tampa, FL.

Dyvik, R., Lacasse, S., and Raadim B. (1987). "Comparison of Truly Undrained and Constant Volume Direct Simple Shear Tests." Geotechnique, Vol. 37, No. 1, pp. 3-10.

Gibson, R.E. and Henkel, D.J. (1954). "Influence of Duration of Tests at Constant Rate of Strain on Measured 'Drained' Strength." Geotechnique, Vol. 4, No. 1, pp. 6-15.

Housner, G.W., and Jennings, P.C., (1964). "Generation of Artificial Earthquakes", J. of Soil Mechanics and Foundation Engineering. Div., ASCE, Vol. 90, pp. 113-150, Feb.

Idriss, I.M., (1985). "Evaluating seismic risk in engineering practice", Proceedings of the 11th International Conference on Soil Mechanics and Foundation Engineering, San Francisco, Vol. 1, pp. 57-153.

Jamiolkowski, M. Ladd, C.C., Germaine, J.T., and Lancellotta, R. (1985). "New developments in field and laboratory testing of soils." Proceedings of the 11th International Conference on Soil Mechanics and Foundation Engineering, San Francisco, Theme paper, 66 p.

Lefebvre, G., and LeBoeuf, D. (1989). "Stability threshold for cyclic loading of saturated clay", Canadian Geotechnical Journal, Vol. 26, pp. 122-131.

Lefebvre, G., LeBoeuf, D., and Hornych, P. (1992). "Slope Failures Associated with the 1988 Saguenay Earthquake, Quebec, Canada." Canadian Geotechnical Journal, Vol. 26, No 1, pp. 162-164.

Mesri, G. (1989). " A reevaluation of $S_{u(mob)} = 0.22\,\sigma'_p$ using laboratory shear tests." Canadian Geotechnical Journal, Vol. 26, pp. 162-164.

Newmark, N.M. (1965). "Effects of Earthquakes on Dams and Embankments", Geotechnique, Vol. 15, No.2, June, pp.139-159.

Seed, H.B., and Wilson, S.D., (1967). "The Turnagain Heights Landslide, Anchorage, Alaska", J. Soil Mechanics and Foundation Div., ASCE, Vol. 93, No SM4, July, pp.325-353.

Seed, H.B. (1967). "Slope Stability During Earthquake",J. Soil Mechanics and Foundation Div.,ASCE, . Vol. 93,No. SM4,July,pp.299-255.

Seed, H.B., and Idriss, I. M., (1981). "Evaluation of Liquefaction Potential of Sand Deposits Based on Observation of Performance in Previous Earthquakes", In-situ Testing to evaluate Liquefaction Susceptibility, ASCE, St Louis, Missouri, Oct.

Seed, H. B. (1968). "Landslide During Earthquakes due to Soil Liquefaction." American Society of Civil Engineers, Journal of the Soil Mechanics and Foundations Division, Vol. 94, No. SM 5, pp. 1055-1122.

Shannon and Wilson, (1964)."Report on Anchorage Area Soil Studies, Alaska", Report prepared for the U.S. Army Engineer District, Anchorage, Alaska, Contract No. DA-95-507-CIVENG-64-18.

Stark, T.D. and Eid, H.T. (1994). "Drained Residual Strength of Cohesive Soils." J. Geotechnical Eng. Div., ASCE, Vol. 120, No. 5, pp. 856-871.

Stark, T.D. and Contreras, I.A. (1995). "Constant Volume Ring Shear Apparatus." Paper Submitted for review and possible publication in the ASTM Geotechnical Testing Journal.

Steinbrugge, K.V. (1970). "*Earthquake Engineering*", Robert L. Wiegel, Editor, Prentice-Hall, pp. 167-226.

Updike, R.G. (1988). "A model for earthquake-induced translatoey landslides in Quaternary sediments." Geological Society of America Bulletin, Vol. 100, pp.183-792.

Wilson, S.D. (1967). "Investigation of Embankment Performance", J. Soil Mechanics and Foundation Div., ASCE, v. 93, No. SM4, July, pp. 135-156.

Woodward-Clyde Consultants (1982). "Anchorage Office Complex, Geotechnical Investigation, Anchorage, Alaska", Report to Alaska Department of Transportation and Public Facilities, Design and Construction, Anchorage, Alaska.

A CORRELATION BETWEEN LANDSLIDES AND RAINFALL DATA ALONG A-LI SAN HIGHWAY OF TAIWAN

S. T. Chen, S. G. Kao and C. C. Chen

Department of Mineral and Petroleum Engineering,
National Cheng Kung University, Tainan, Taiwan

ABSTRACT

A study was carried out to see the possibility of building a warning system for landslides along A-Li San Highway by the real-time rainfall data. A survey was done on the relationship between the rainfall data and landslide record. The stability of the rock slopes was evaluated by the SMR method . If the limiting thresholds of rainfall are expressed in the way of intensity-duration curves suggested by Caine, the rock slopes with higher SMR values would have a higher intensity-duration limiting threshold. It also suggests that a warning system with probability of occurrence is more probable.

KEYWORDS

Slope failure, rainfall data, warning system, SMR values, limiting threshold, occurrence rate.

INTRODUCTION

It is a common experience that landslides,especially natural slope failures, occur during the period of heavy rainfall. Indeed, rainfall is one of the major factors which can trigger the occurrence of landslides. This phenomenon may be explained by the following reasons:

1.Rainfall increases the pore pressure of the earth material and then reduce the effective normal stress acting on the potential sliding surfaces. Finally, it will reduce the resistance force for slope failure.

2.Rainfall will increase the water content of the earth material. As a result, the strength of the earth material is reduced and the weight of the sliding mass is increase.

Since rainfall plays an important role in landsliding, it is logical to use the rainfall data to predict the occurring time of landslide. Although some landslide warning systems based on rainfall data have been established (keefer *et al.*, 1987), a clear theoretical consideration for the idea are still under developed,

The Central Weather Bureau of R.O.C. (Taiwan) is going to set up some real-time rainfall telestations in Chia-Nan region (Chia-Nan region includes Chiayi and Tainan counties and cities). It is planning to use the rainfall data obtained from these telestations and radar system to build a flood warning system for that area. The first author of this paper wish to use these data to build a landslide warning system in future if it is feasible. Since all of the urban area of Chia-Nan region are located on a plain (Chia-Nan plain), a mountainous highway must be chosen for this study.A-Li San is a famous and crowded resort with elevation of approximately 2200m. A-Li San Highway which connects Chiayi and A-Li San is the busiest mountainous highway in Chia-Nan Area, so it was chosen for this research. Then a study on the correlation between landslide occurrence and rainfall was carried out to investigate the possibility of building a reasonable accurate landslide warning system in future.

GEOGRAPHY AND GEOLOGY

Figure l is a simple map of the mountainous part of A-Li San Highway. This map shows the locations of the larger villages along the Highway and their elevations. The lowest elevation of the study area is 280m (Chuko)and the highest elevation is 2660m (Tatacha). The average annual rainfall is about 3100mm per year, average raining days are around 126 days. The rains are concentrated between April and September.

Most of the rocks exposed to surface are sandstones and shales, many rock slopes consist of interbed sandstones and shales. These rocks are highly fractured and folded.Of course many slope surfaces are covered by weathered materials, colluvium and alluvium. For a quick classification of the failure slopes according to their stability, these slopes were first

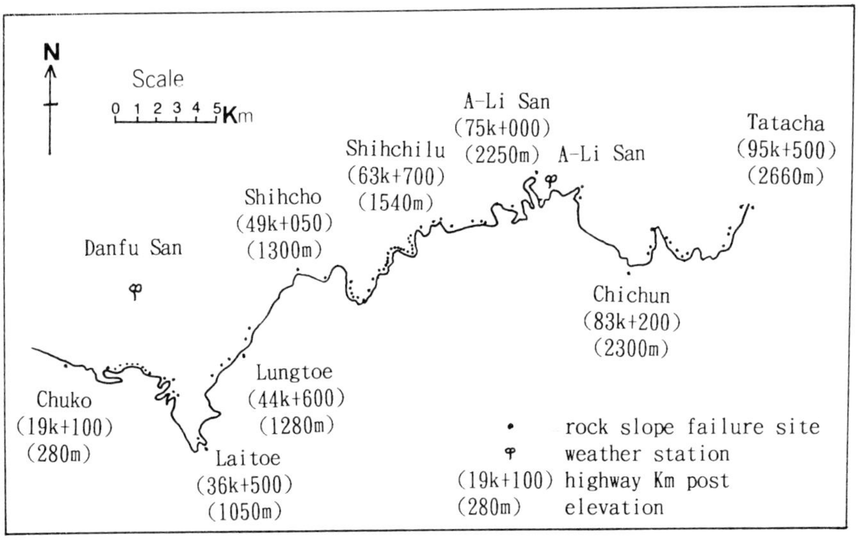

Fig. 1. The location of major villages and rock slope failure sites along A-Li San Highway

classified into soil and rock slopes judging by their sliding material. The rock slopes were further classified with SMR (Slope Mass Rating) method which is described in detail by Romana (1993). The SMR rating of a slope can be described by the following equation

$$SMR = RMR + (F_1 \cdot F_2 \cdot F_3) + F_4 \quad (1)$$

In Equation 1,RMR is rock mass quality rated by the method suggested by Bieniawski (1979).F_1,F_2,and F_3, are adjustment factors for the attitudes of the geological weak planes and their relationship with the attitudes of the slope surfaces. F_4 is the adjustment factor for the excavation method of the slopes. Each rock slope can be given a grade between 0 and 100 by SMR rating method, and the slopes with higher grades should be more stable than the slopes with lower grades. Though the absolute value of the grades of the slopes may be rather meaningless, but the relative stability of the rock slopes should be able to be expressed by their SMR values. The reason that the authors adopted this rating method is that it is a convenient, economical and fast method for investigating a large number of rock slopes in a very short time.

RAINFALL AND LANDSLIDE DATA

Since only two weather stations along A-Li San Highway have existed long enough to record all the rainfall data since the highway was built and have landslide data (started from 1984). All the rainfall data of these two weather stations were obtained in hourly basis. Figure 1 also shows the location of the two weather stations. When the landslide data were analyzed ,the slides occurred west of Shicho used the data from Dafu San weather station and the other slides used the data from A-Li San station. Since the density of the weather station is rather low, the accuracy of the rainfall data would not be very good, however the result obtained from statistical analysis work should be still valid for their trend.

Landslide data were provided by Taiwan Provincial Highway Bureau. These data include the occurring time, amount of sliding material, location, and remedial treatment of each landslide. The occurring time was recorded when the blockade of highway was reported by the local residents or workers. Figure 2 shows an example of the combination of rainfall and landslide occurring time data. One rainstorm may have many reports of highway blockade time, and one highway blockade time report may contain many landslide occurrence.

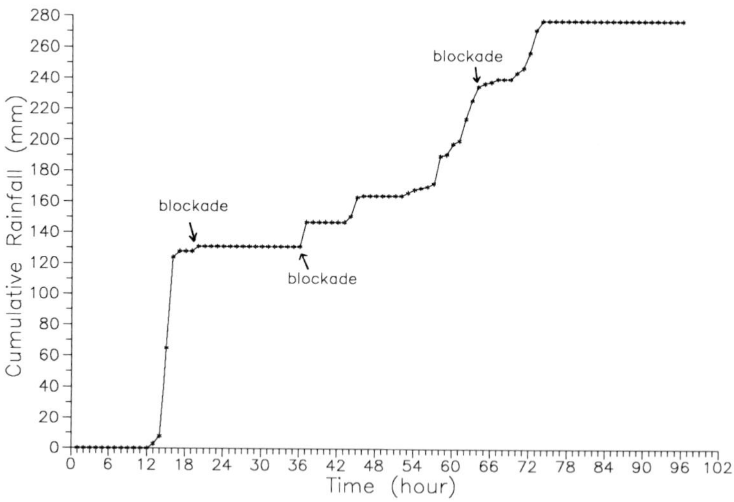

Fig. 2. An example of highway blockade (landslide occurring) time for a rainstorm

Between September of 1984 and September of 1994, 561 landslides were recorded for the 75 Km long mountainous part of A-Li San Highway, and above 95 % of these landslides were triggered by heavy rainfall. Since too many landslides occurred , only 184 landslides which are larger and more recently occurred were studied in more detail way. Among these landslides, 78 are rock slides and 106 are earth or debris slides. These 78 rock slides were classified further by SMR method and obtained their SMR values.

CORRELATION BETWEEN RAINFALL DATA AND THE SMR VALUES OF THE FAILED ROCK SLOPES

The rainfall data and SMR values of the 78 rock slopes were analyzed by statistical methods. Since more than one rock slides may occur for each landslide occurring time record (highway blockade time), only the highest SMR value of the failed rock slopes was used to correlate with the rainfall data. Therefore, the SMR value of each correlated rainfall data means that the highest SMR value slope (the best stability quality slope) which can be failed by the specified rainstorm.

After statistical analysis, it is found that the SMR values of the failed rock slopes have a close relationship with the cumulative rainfall before slides occurred but have a much less relationship with the intensity of rainfall. However it is best expressed by the following equation

$$SMR = 3.67 + 2.6 \cdot \ln I_6 + 5.46 \cdot C \qquad (2)$$

Where:

SMR: the highest SMR values of the failed rock slopes

I_6: the average intensity of rainfall for the period 0~6 hrs before occurring of landslide, in mm/hr.

C: the cumulative rainfall before occurring of landslide, in mm.

Figure 3 shows the data points of the highest SMR values of the failed rock slopes and the related rainfall data. It seems that Equation (2) shows the trend quite well for the relationship between rainfall data and SMR values of the failed rock slopes.

Keefer, *et al*(1987) used rainfall intensity-duration concept to predict the landslide occurring time in San Francisco Bay region. By simulate this concept, the rainfall intensity-duration

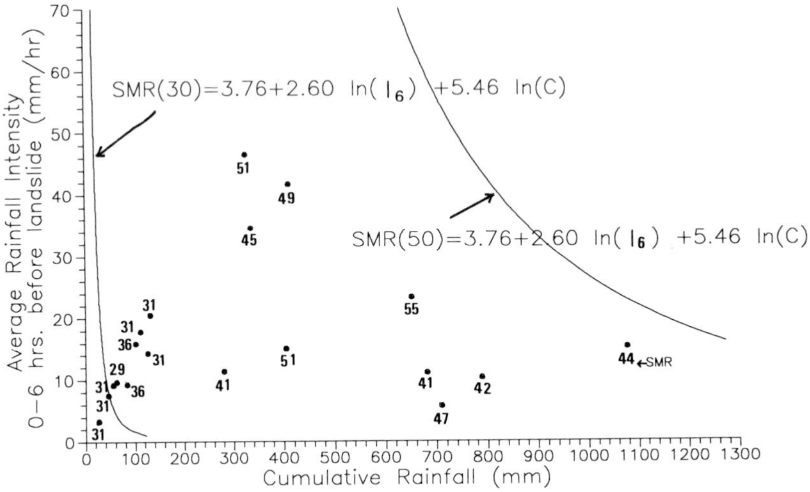

Fig. 3. Relationship between the SMR values of the failed rock slope and the rainfall which triggering failure

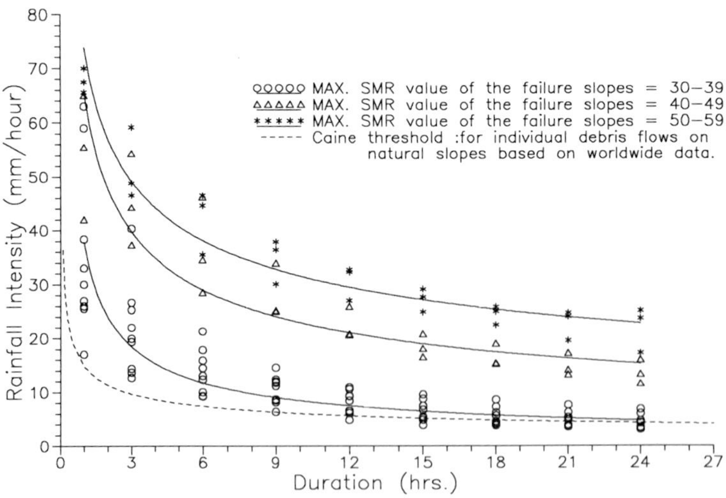

Fig. 4. Rainfall intensity-duration curves for the failed rock slopes along A-Li Highway

curves for each rock slope failure can be plotted. After averaging the rainfall intensity -duration curves for each range of SMR values of the failed rock slopes, Fig. 4 was obtained.

Figure 4 shows that the rock slopes with higher SMR values need higher intensity of rainfall (for a given rainfall period) to fail them. Figure 4 also shows that Caine's threshold (1980) for debris flow is lower than the rock slopes of poor stability quality (SMR=30~39). It is reasonable, since the debris accumulated slopes are generally less stable than the poor quality rock slopes. Unfortunately, all the studied landslide occurring time contain both of the rock and soil slope failures.In this study no rainfall intensity-duration curve of failed soil slopes can be used to compare with Caine's threshold curve.

PREDICTION OF LANDSLIDE OCCURRING TIME BY RAINFALL DATA

Since landslides have a close relationship with rainfall, and rainfall data could be obtained in an economical and convenient way for a large region, an useful warning system could be built if the relationship between landslide occurring time and rainfall data can be established. After analyzing the word-wide data, Caine (1980) suggested a limiting threshold for shallow landslides and debris flow, as the following equation

$$I= 14.82D^{-0.39} \quad (3)$$

Here :

I : rainfall intensity, in mm/hr

D: duration of rainfall, in hr.

Caine suggests that once the intensity-duration of a rainstorm is higher than the value predicted by Equation (3), it may provoke the shallow instability on undisturbed slopes. The duration of rainfall time is best defined between 10 minutes to 10 days. However Equation (3) is the minimum requirement for triggering a shallow landslide or debris flow, it may need a higher intensity-duration rainfall for triggering a shallow landslide or debris flow for a specified slope.

Keefer *et al*. (1987)derived an equation similar to Equation (3) in the form of :

$$I=aD^{b} \quad (4)$$

This equation is based on the following assumption:

1.The landslides are caused by the increase of pore pressure which is induced by infiltrated rainfall.
2.There is a critical level of the pore pressure for triggering landslides.
3.All the rainfall that falls on the slopes will infiltrate into the saturated zone.
4.The rate of duainage is proportional to the thickness of the saturated zone.

For the rock slope failures along A-Li San Highway, Equation (4) seems to be work also. As shown by Fig. 4, the intensity-duration curves of the failed rock slopes were plotted by regression analysis of the data points with Equation (4). These data fit quite well to these curves, and these curves also are quite smiliar to Caine's threshold, which is plotted according to Equation (3).

There are many other methods to estimate the occurring time of landslides by rainfall data, especially for the debris flow (Mark and Newman, 1985; Cannon and Ellen, 1985). Chiang and Lin (1991) collected many different methods proposed by Japanese and English literatures and used them to estimate the occurring time of the debris flow in Taiwan. In general, all these methods are derived by experience method, in other words by statistical treated of the collected dada. Therefore, these methods are good only for the landslides occurred in a specified way (such as debris flow) and in a specified region.

Of course, prediction of landslide occurring time by measurement of pore pressure (Yatabe *et al* ., 1986) and sliding mass displacement rate (Hoek and Bray (1981)) would be much more accurate. However, the authors of this paper find that after putting many instruments on the possible failure slopes judged by experience, the failure may still occur on the slopes without any instrument. The authors feel that these measurement methods may not fit the purpose of landslide warning for a large area.

For A-Li San Highway area, there are mixtures of rock and earth slopes with difference stability qualities, Fig. 3 and Fig. 4 are designed to estimate the landslide occurring time for the rock slopes with different stability qualities, it must use the lowest standard to set up the limiting threshold for landslide and debris flow warning system for whole region. Therefore , the curves of the rock slopes with 30 SMR values may be used as a warning system basis. If debris flow is considered, Caine's limiting threshold may be chosen for warning basis. The authors of this paper have tried many other methods to estimate the occurring time of the landslides along A-Li San Highway, however these methods may be too complicated for practical application and would not be discussed in this paper.

CORRELATION BETWEEN THE LANDSLIDE OCCURRENCE RATE AND RAINFALL DATA

Since the factors triggering the failures of slopes are very complicated and rainfall is just one of the very important factors, prediction of landslide occurring time by rainfall data alone should not be expected to be very accurate. However, we do know the fact that more landslides may occur during a heavier rainstorm period. Table 1 shows the data obtained in this study, it indicates that higher landslide occurrence rate will be associated with higher rainfall intensity. The authors of this paper feel that a landslide warning system accompanied with probability of occurrence is a more reasonable one to use. However the true statistical meaning of Table 1 is still to be analyzed,the probability of landslide occurrence may not be just expressed by the landslide occurrence rate as shown in Table 1.

Table 1. The relationship between the rate of landslide occurrence and rainfall data in A-Li San Highway region during 1985～1993 period.

Duraton (hrs)	Average Rainfall Intensity (mm/hr)	Tatal number of Rainfall (1)	No. of Rainfall Triggering Slope Failure(2)	Landslide Occurrence rate (2)/(1)×100%
3	> 10	187	45	24.1
	> 15	151	40	26.5
	> 20	77	29	37.7
	> 25	53	23	43.4
	> 30	39	19	48.7
	> 40	20	14	70
	> 50	9	8	88.9
6	> 10	124	42	33.9
	> 15	55	23	41.8
	> 20	32	18	56.3
	> 25	19	15	78.9
	> 30	16	14	87.5
	> 40	12	12	100

The landslide occurrence data in Table 1 can be further divided into two period, the period of earlier stage after the operation of the highway (1985～1988)and the period of later stage (1989～1993).Table 2 shows that the highway slopes in later period are more stable than in the earlier period.This phenomenom can be simply explained by the fact that more additional protection measures were put on the highway slopes each year, the landslide occurrence rate for same rainfall condition will be then decreased gradually. Therefore, the warning system set up according to the past landslide records may be too conservative for a developing region.

Table 2. The difference between the rates of landslide occurrence in the earlier and later stage of A-Li San Highway operation period.

Duraton (hrs)	Average Rainfall Intensity (mm/hr)	Landslide Occurrence Rate (%)	
		Earlier Stage (1985～1988)	Later Stage (1989～1993)
3	> 10	24.7	23.5
	> 15	26.6	26.4
	> 20	48.5	29.5
	> 25	52.2	36.7
	> 30	61.1	38.1
	> 40	90	50
6	> 10	39.6	29.6
	> 15	52.2	34.4
	> 20	60	52.9
	> 25	88.9	70
	> 30	100	75

CONCLUSION

By the study on the relationship between landslide occurrence and rainfall data for the slopes along A-Li San Highway of Taiwan, the following conclusions have been made.

1.The intensity-duration curves of real time rainfall data may be useful for landslide warning system for A-Li San Highway.

2.The SMR values of rock slopes might be useful for predicting landslide occurring time, usually the rock slopes with higher SMR values need higher rainfall intensity and accumulated rainfall to fail them from a statistical point of view.

3.The rainfall data may only be used to predict the landslide occurring time for a large area with numerous slopes. It is difficult to use them to predict the landslide occurring time for a specific slope.

4. It is more appropriate to use probability concept to predict the failure of slopes by rainfall data. A larger rainfall will have a higher probability to cause a slope to fail in a region, but it is not necessary that a larger rainfall will certainly fail a slope in a region when the smaller rainfall failed a slope in the same region before.

ACKNOWLEDGEMENT

This research was supported by National Science Council of R.O.C. with project number NSC 83-0414-P006-010B,the authors deeply appreciate it.

REFERENCES

1. Bieniawski, Z.T.(1979). The Geomechanics Classification in Rock Engineering Applications. *Proc. 4th Intl. Cong. Rock Mechanics Montreux* 2,40-48.
2. Caine, N. (1980). The Rainfall Intensity-Duration of shallow Landslides and debris flow. *Geogr. Ann.* 62A,23-27.
3. Cannon, S. H. and S.D. Ellen, (1985). Rainfall that Resulted in Abuntant Debris-Flow Activity During the Storm, *U.S. Geol. Surv. Prof. Pap.1434*,27-33.
4. Chiang Y. C. and C.Y.Lin (1991). An Analysis of Rainfall Characteristics for Debris-Flow Events in Taiwan. *Jour. Chinese Soil & Water Cons.*, 22,21-37.
5. Hoek, E. and J. W. Bray,(1981). *Rock Slope Engineering*. The Institution of Mining and Metallurgy,London.
6. Keefer, D. K. *et al.*(1987). Real-Time Landslide Warning During Heavy Rainfall. *Science.* 238, 921-925.
7. Mark, R. K. and E.B. Newman (1985). Rainfall Totals Before and During the Storm : Distribution and Correlation with Damaging Landslides. *U.S. Geol. Surv. Prof. Pap. 1434*,17-26.

8. Romana, M. R. (1993). A Geomechanical Classification for Slope:Slope Mass Rating. in: ,*Comprehensive Rock Engineering* (J.A. Hudson, ed.), Vol.3,pp.575-599.Pergamon Press, Oxford.
9. Yatabe, R.,N. Yagi,and M. Enoki,(1986). Prediction of Slope Failure Based on the Amount of Rainfall. *Japanese Society of Civil Engineers*, 376,p.297-305.

REDUCING LANDSLIDE RISK IN URBAN AREAS – EXPERIENCE IN THE UNITED STATES

R. L. Schuster

U.S. Geological Survey, Box 25046, Mail Stop 966, Denver, CO 80225, U.S.A.

ABSTRACT

Reducing the landslide risk in urban areas, and thus mitigating economic and social losses due to landslides, can be achieved by one or more of four approaches: (a) restricting development in landslide-prone areas (i.e., avoidance); (b) using codes in excavation, grading, landscaping, and construction; (c) using engineered physical measures (drainage, slope-geometry, modification, structures, and vegetation) to prevent or control landslides; and (d) developing landslide warning systems. Optimal approaches to reduction of landslide risk usually involve a carefully assembled mix of the above hazard-reduction strategies and techniques. In addition, landslide insurance can provide financial protection for individual property owners.

KEYWORDS

Landslide; hazard; risk; mitigation; insurance, multiple hazards.

INTRODUCTION

In addition to killing people, landslides destroy or damage residential and commercial properties and developments in urban areas, as well as agricultural and forest lands, and impair water quality of rivers and streams. (In this paper, I will use the broad definition of *landslide* to include all gravity-related slope failures of earth materials (Varnes, 1978; Cruden and Varnes, in press). Landslides constitute a major socioeconomic problem in the United States. The areas most seriously affected are the West Coast, the Rocky Mountains, and the Appalachian Mountains (Fig. 1), but landslides occur in every state and are a significant factor in more than one half of the states. In the U.S., landslides annually cause an estimated $1-2 billion dollars in economic losses and about 25-50 deaths (Committee on Ground Failure Hazards, 1985). Most of these losses occur in urban areas, particularly in the State of California.

This paper will discuss reducing social and economic losses due to landslides by means of effective planning and management. The concepts presented and examples of this use are based primarily on experience in the United States; however, the same techniques are proving to be successful in other industrial nations as well as in developing countries.

Careful development of hillside slopes can reduce losses due to slope failure by avoiding the hazards or

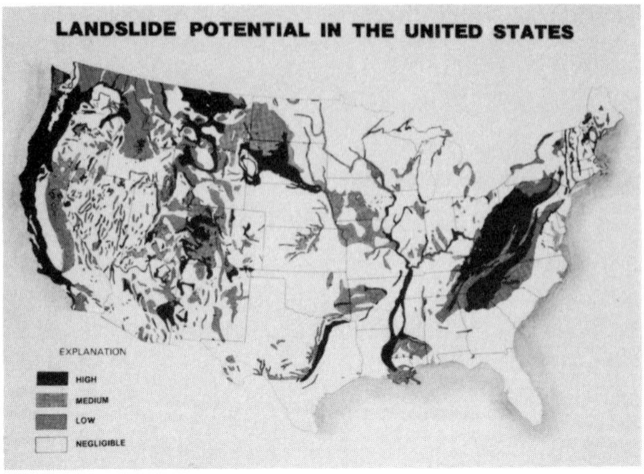

Figure 1. Map showing relative potential for landsliding in different parts of the conterminous United States (after Radbruch-Hall *et al*,, 1982).

by reducing the damage potential. Reducing landslide risk can be achieved by one or more of four approaches: (1) restricting development in landslide-prone areas, (2) using excavation, grading, landscaping, and construction codes, (3) using engineered physical measures (e.g., drainage, slope-geometry modification, restraining structures, and vegetation) to prevent or control landslides, and (4) developing warning systems (Kockelman, 1986). These methods of hazard mitigation, when used with modern technology, can greatly reduce losses due to landslides. Schuster and Leighton (1988) have estimated that these methods can reduce landslide losses in California by more than 80 percent. Slosson and Krohn (1982) have stated that enactment of these approaches *has already* reduced landslide losses in the City of Los Angeles by 92-97 percent.

This paper also briefly reviews the states of development of landslide insurance programs as mitigative measures for landslide hazards. While insurance may not reduce overall hazards or costs directly, non-subsidized landslide insurance does offer promise as a means of wider distribution of landslide costs and of reducing landslide losses for individual property owners. In addition, advice from insurance organizations can influence in a positive manner the owners of land subject to landslide hazards.

Landslides often occur as elements of interrelated multiple natural-hazard processes in which an initial event activates secondary events, or in which two or more natural hazard processes occur at the same time. Examples are combinations of earthquakes, volcanic eruptions, and landslides. The resulting multiple-hazard problems require a shift in perspective from mitigation of individual hazards, such as landslides, to a broader systems framework that takes into account the characteristics and effects of all the processes involved.

MAJOR POLICY OPTIONS FOR LANDSLIDE HAZARD REDUCTION

Alternative options of management policy are available to decisionmakers who are concerned with natural hazard reduction (Petak and Atkisson, 1982; Olshansky and Rogers, 1987; Olshansky, 1990). The three basic options are: (1) take no action at all, (2) provide relief and rehabilitation assistance after disasters occur, and (3) take action to contain or control hazards before serious damage occurs (Rossi *et al.*, 1982). Prior to about 1950, the first two options dominated United States practice. However, as a result of technical advances and social change, the concept of prevention of landslide disasters by appropriate land-use development or engineered physical mitigative measures is becoming more common in the United States, particularly in urban areas.

PREREQUISITES FOR REDUCING LANDSLIDE HAZARDS

Successful landslide-hazard reduction efforts in the United States commonly utilize four basic factors: (1) an adequate base of technical information on the hazards and risks, (2) a technical community able to apply, and enlarge upon, this data base, (3) an able and concerned local government, and (4) a citizenry that realizes the value of and supports a program that promotes the health, safety, and general welfare of the community (U.S. Geological Survey, 1982). The key to a successful landslide hazard-reduction program is the understanding of the landslide problem within the geographic entity involved. The work of Varnes (1978) and Cruden and Varnes (in press) in classifying and describing landslides, and of Varnes and the International Association of Engineering Geology Commission on Landslides and other Mass Movements on Slopes (1984), has been very helpful in this regard. Recognizing and identifying landslides have been discussed in detail by Rib and Liang (1978), Hansen (1989), and Soeters *et al.* (in press).

Reliable landslide-hazard maps are of significant value in establishing hazard-reduction programs. These maps commonly indicate where landslides have occurred in the past, the locations of landslide-susceptible areas, and the probability of future occurrences. Brabb (1984) described and presented examples of various types of landslide maps: inventory, susceptibility, loss-evaluation, and risk-determination. Of particular interest in increasing the efficiency of landslide hazard mapping is the use of computer techniques to produce digital maps. In the United States, the Geographic Information Systems (GIS) approach is the most commonly used technique for producing digital maps and associated digital data.

Governmental organizations in the United States have reported on landslide-hazard mitigation approaches that are of value to others who are attempting to develop plans for their own areas. For example, the State of Colorado has prepared two such reports (Jochim *et al.*, 1988; Wold and Jochim, 1989) that aim to reduce landslide losses by:

(1) identifying local government resources, plans, and programs that can assist in loss reduction;

(2) identifying unmet local needs that must be addressed to reduce losses;

(3) identifying state- and local-agency capabilities and initiatives that can deal with unmet local needs;

(4) developing cost-effective mitigation projects that can be expected to reduce landslide losses;

(5) educating state and local officials and emergency-response personnel on landslide hazards and on potential methods for loss reduction; and

(6) establishing means to provide a long-term governmental process to reduce losses.

Similar plans have been prepared for the City of Cincinnati, Ohio (Hamilton County Regional Planning Commission, 1976), and Portola Valley, California (Mader *et al.*, 1988). In addition, Hamilton County, Ohio, has prepared a report on the duties of the County's Earth Movement Task Force (Hamilton County, 1982). This report provides advice to those wishing to establish similar working groups.

The State of Colorado also has set a national precedent in dealing with natural hazards (including landslides) by means of the Colorado Natural Hazards Mitigation Council, an official statewide 300-member group comprised of earth scientists, engineers, planners, and local and state policy makers whose goal is to formulate new state policies regarding natural hazards (Colorado Natural Hazards Mitigation Council, 1992). A prime strategy of the Council is to unify technical experts and policymakers on issue-directed, hazards-reduction teams. These teams deal with hazards in areas that are politically responsive to innovative solutions. They prepare statewide plans based on these solutions, and the plans are used to develop policy decisions for future state hazards legislation.

An important aspect of reducing landslide hazards is collecting and disseminating information on landslides to scientists, engineers, policy makers, and the public. An excellent United States example is the U.S. Geological Survey's National Landslide Information Center (NLIC) in Golden, Colorado (Brown, 1992). The NLIC maintains a multiple-entry database to foster national and international exchange of landslide information with scientists, engineers, and decisionmakers, and with other stakeholders who need such data.

METHODOLOGY FOR LANDSLIDE-HAZARD MITIGATION

Reducing landslide hazards in the United States has been achieved by (1) avoidance, i.e., reduction of damages and loss of life by restricting development in landslide-prone areas, a function assisted by the mapping of landslide susceptibility, (2) requiring that excavation, grading, landscaping, and construction be conducted so as to not contribute to instability of slopes, and (3) engineered physical control measures, such as drainage, slope-geometry modification, and protection of property and structures from landslides by erecting protective barriers or systems of ground anchors (Erley and Kockelman, 1981; Kockelman, 1986). These techniques are used individually or in various combinations to reduce losses due to existing or potential slope failures. The first two methods can be promoted by public legislation. For urban areas in the United States, such legislation commonly is under the jurisdiction of city or county governments. However, most other countries with major landslide losses have incorporated a strong federal or provincial role in dealing with all aspects of landslide hazard reduction to ensure consistent standards of practice and application (Swanston and Schuster, 1989). In the United States, the federal government functions primarily as a source of technical expertise, research support, and funding of state and local control works.

Restricting Development in Landslide-susceptible Areas

One of the most effective and economical ways to reduce landslide losses is to locate developments on stable ground and to dedicate landslide-prone areas to open space or low-intensity uses. This procedure, which is commonly known as "avoidance," can be accomplished by (1) removing or converting existing development, and/or (2) discouraging or regulating new development in unstable areas (Kockelman, 1986).

Removing or converting existing development. Landslide damage to existing development can be reduced or eliminated by evacuating the area or by converting existing facilities or structures to uses less vulnerable to slope failure (Erley and Kockelman, 1981; Kockelman, 1986). Permanent evacuation of the distressed property commonly requires public acquisition of the land. Converting existing facilities and structures to uses that are less vulnerable to slope failure can be undertaken by individual property owners, by developers, or, in the case of public land, by the government. The feasibility of successful conversion depends on the value and criticality of the facilities, whether they can be successfully altered so that they can resist slope movement, and the level of concern of their owners.

Discouraging or Regulating New Development. Where feasible, the most effective and economical method of reducing slope-failure losses is to discourage new development in landslide-prone areas (U.S. Geological Survey, 1982). Methods that have been successful in discouraging new development in the United States are:

(1) Public-information programs. -- Any successful program of landslide-hazard mitigation requires the support of the people involved. When the public is properly informed of the existence of hazards, it generally will support land-use controls that will reduce losses due to the hazards.

(2) Warning signs. -- Warning signs posted by local governments can be used to alert prospective property buyers or developers to potential landslide hazards.

(3) Recording the hazard. -- Local government offices should ensure that public records on land ownership include available information on slope-failure hazards. Such records alert land purchasers, developers, lenders, and tax assessors to potential hazards.

(4) Disclosure of hazards to potential property purchasers. -- Local government entities can discourage development in hazardous areas by enacting hazard-disclosure laws that alert potential property buyers to hazards (Kockelman, 1986). In some parts of the United States, such disclosure is required by law. For example, Santa Clara County, California, requires all sellers of real estate within the County's landslide, flood, and fault-rupture zones to provide prospective purchasers with written statements of the geologic hazard (Santa Clara County Board of Supervisors, 1978).

(5) Exclusion of public facilities. -- To discourage development in landslide-prone areas, local governments can prohibit construction of public facilities, such as streets and water supply, electric power, and sewage disposal utilities.

(6) Special assessments and tax credits. -- Local governments can assess added taxes on properties scheduled for development in hazardous areas. Conversely, tax credits can be awarded for lands left in an undeveloped state.

(7) Financing policies. -- Lending institutions can discourage development in landslide-prone areas by denying loans for development in these areas.

(8) Insurance costs. -- High costs of insurance can discourage development in hazardous areas.

(9) Government acquisition. -- Local government agencies can promote avoidance by acquiring landslide-prone properties by purchase, condemnation, tax foreclosure, dedication, devise (will), or donation (Kockelman, 1986). Local governments commonly dedicate such properties to use as parks or "open space."

Government Regulation of Development. It is unrealistic to assume that developing landslide-prone lands can be discouraged indefinitely by the non-regulatory methods noted above. Thus, government regulation often is necessary. In the United States, restrictions on land use because of natural hazards are generally imposed and enforced by local governments by means of zoning districts and regulations. Examples of regulations of land use in areas subject to landslide activity are:

(1) Land-use zoning regulations -- Land-use zoning provides direct benefits by controlling development of vacant lands in areas prone to slope failure and by limiting expansion of existing development in those areas. Under zoning ordinances, land uses with the least danger of activating landslides include open space, parks, non-irrigated agriculture, woodlands, wildlife refuges, and recreation. Regulations can include provisions that prohibit specific land uses or operations that could cause mass movement or that might be vulnerable to slope failure, such as irrigation, construction of roads and buildings, construction of reservoirs, storage or disposal of liquid wastes, and operation of off-road vehicles. Zoning regulations also can control the location and density of development in hillside areas.

To assist counties and municipalities in designing land-use regulations in hillside areas in the State of Colorado, the Colorado Geological Survey has prepared model regulations (Rogers *et al.*, 1974) that permit the following land uses in designated landslide-prone areas:

(a) Recreational uses that do not require permanent structures for human habitation, including parks; wildlife and nature preserves; golf courses; picnic grounds; and hiking, fishing, hunting, and skiing uses that do not result in high population concentrations.

(b) Low-density agricultural uses, such as grazing, truck-crop farming, and forestry.

(c) Low-density and temporary industrial and commercial uses, such as parking areas and storage yards for portable equipment.

(2) Subdivision regulations -- Local government regulation of the design of subdivisions (planned local units of land designed with streets, sidewalks, sewer systems, etc., in preparation for building homes) is another means of controlling the development of landslide-prone areas. Subdivision design and zoning regulations must be coordinated carefully with adequate technical information regarding local landslide problems.

(3) Sanitary regulations -- Local sewage-disposal systems utilizing ground absorption (septic tanks, leaching fields, and seepage beds and pits) can saturate the surrounding ground and cause landslides. Thus, the design, engineering, and construction of these installations must be controlled by local government regulations.

Success of Development Restriction. In the United States, implementing avoidance procedures has met with mixed success. In some areas, and particularly in California, land-use programs instituted by local governments have been extensive and have been generally successful in reducing landslide losses. However, in many parts of the nation having landslide-susceptible areas, there are no widely accepted procedures or regulations for considering landslides as part of the land-use planning process (Committee on Ground Failure Hazards, 1985).

Land-use zoning has probably been the most effective means of discouraging or regulating development. For example, in San Mateo County, California, a landslide-susceptibility map (Brabb *et al.*, 1972) has been in use since 1975 as a zoning map that controls the density of development in a county with varying degrees of landslide hazard. For areas in the highest category of landslide susceptibility, only one residence is permitted on each 16 ha of land (San Mateo County, 1973). From 1972 until 1982, all of the new landslides (mostly slips, slides, and slumps) that occurred were either in areas already mapped as landslides or in areas judged highly susceptible to landsliding. Thus, the zoning procedure was an outstanding success at that point. However, in 1982, an unusual combination of meteorological events caused exceedingly high rainfall intensities, and thousands of debris flows occurred in areas where few had been observed previously (Brabb, 1984). Thus, the 1972 landslide-susceptibility map had been successful in predicting the locations of future deep-seated landslides, but had not been successful for debris flows. The new debris flows had not been expected because the landslide-susceptibility map had been based on airphoto examination of only deep-seated landslides, without consideration of the special circumstances causing debris flows.

Another approach to zoning has been used in rapidly developing Fairfax County, Virginia, where maps used for zoning purposes outline various degrees of hazard in different geologic materials (Obermeier, 1979). Developers are required to obtain professional engineering advice for sites to be developed in specific geologic materials. The County requires investigation reports and development plans; those are reviewed by a Geotechnical Review Board retained by the County. The result has been a "drastic reduction in landslides" (Dallaire, 1976).

The author has found little information in the literature testifying to the success of removing or converting existing development as a tool in the reduction of losses due to landslides. The most obvious examples are those in which existing developments have been wholly or partly destroyed by slope failures, and, as a reaction to those losses, a decision has been made to replace the original development with a usage less prone to slope-failure damage. A noteworthy example of this type of land-use planning was the moving of the town of Yungay, Peru, from its original site, which had been overrun by the 1970 Huascaran debris avalanche, to a new site that is located a few hundred meters away and is topographically protected from a repeat of the 1970 event.

More common are instances where removing or converting an existing facility has been only partially successful because of the resistance of property owners, developers, or communities. An excellent example is provided by the City of Anchorage, Alaska, which received heavy damage from landslides activated by the 1964 Alaska earthquake. As a result of the earthquake, a Scientific and Engineering Task Force was established by the United States government to assess the damage, to evaluate future hazards, and to make recommendations that would minimize the impact of any future earthquake or landslide activity. The panel recommended that the 1964 Anchorage landslide areas be converted to non-intensive usage, such as parks, recreation areas, car parking areas, and light structures no more than two stories high (Hansen *et al.*, 1966; Mader *et al.*, 1980). The recommended grading and building restrictions were incorporated into an urban renewal plan that was relatively successful for government properties, but proved to be unsuccessful for private properties because owners felt that property compensation at post-disaster values was not sufficient inducement to relocate. In the L Street landslide area, the recommendations of the Task Force were ignored even by the city government. About a year after the earthquake, the Anchorage City Council rezoned the area to permit higher residential densities, and between 1964 and 1980, extensive new construction, including five new high-rise buildings, took place on or adjacent to the landslide (Mader *et al.*, 1980)

Excavation, Grading, and Construction Regulations

Excavation, grading, and construction ordinances have been developed to ensure that construction in landslide-prone areas is planned and conducted in a manner that will not impair the stability of hillside slopes. These codes commonly: (1) regulate, minimize, or prohibit excavation and fill activities in landslide-prone areas, (2) require a permit prior to scraping, excavating, filling, or cutting any lands, (3) require proper engineering design, construction, inspection, and maintenance of cuts, fills, and drainage systems. They are particularly applicable to expanding urban areas.

In the United States, there is no "uniform" code that is applied nationwide to ensure standardization of the above criteria. Instead, in dealing with stability of hillside slopes on private lands, agencies at state and local levels apply design and construction criteria that fit their specific needs. The federal government usually has not participated in formulating and enforcing such codes. However, it has exerted influence on the engineering profession by means of the codes used by government agencies in their own construction programs on federal projects. Federal standards for excavation and grading -- such as those developed by the Bureau of Reclamation, the U.S. Army Corps of Engineers, and the Soil Conservation Service, each of which is in charge of major construction activities in the United States -- are often used by other organizations in both the public and private sectors (Committee on Ground Failure Hazards, 1985).

The development of United States excavation and grading codes as related to geologic hazards began in California shortly after World War II. At that time, the spiraling demand for housing in southern California due to a rapidly expanding population intensified the development of foothill and mountain slopes for residences (Scullin, 1983). In addition, improved earth-moving technology made the development of hillside slopes economically feasible. The resulting poorly organized development, combined with unusually heavy rainfall in the early 1950's, led to significant landslide activity and major economic losses (Jahns, 1969). As a result, in 1952 the City of Los Angeles adopted the first grading ordinance in the United States. This code was far from perfect, and, during the period 1956-61, governmental agencies and hillside-development industries of southern California suffered through multitudinous problems in applying the original code and its early modifications (Scullin, 1983). However, this original code formed the basis for all subsequent codes adopted by local governments throughout the United States.

The early work in California also became the basis for the Grading and Excavation Code, Chapter 70 of the Uniform Building Code established by the International Conference of Building Officials (1985). This code has been adopted directly or used as a model by local governments in many countries. One of the basic tenets of Chapter 70 is that professional judgments are required in dealing with excavation and grading:

> "When required by the building official, each application for a building permit shall be accompanied by two sets of plans and specifications, and supporting data consisting of a soils engineering report and engineering geology report. The plans and specifications shall be prepared and signed by a civil engineer when required by the building official."

<u>Success of Excavation, Construction, and Grading Codes.</u> The best example in the United States of a successful program to reduce landslide losses has been that developed by the City of Los Angeles over the past 40 years. The Los Angeles loss-reduction program relies heavily on grading regulations that serve as a model for urban areas throughout the world. These regulations require specific evaluations of landslide potential by engineering geologists and geotechnical engineers before construction, as well as inspection of grading operations during construction. In 1952, following severely damaging winter storms, a grading code was adopted that instituted procedures for safe development of hillsides; these grading regulations were significantly improved in 1963. The benefits resulting from these regulations were illustrated by the extent of landslide damages that occurred in Los Angeles during severe storms of 1968-69 and 1978. During the storms of 1968-69, for a comparable number of building sites, the damages to sites developed before institution of grading codes in 1952 were nearly 10 times as great as damages to those developed after 1963 (Slosson, 1969). As shown in Table 1, similar results occurred for the 1978 storms.

Table 1. Relationship between modern grading codes and slope failures for Los Angeles building sites from the catastrophic February 1978 southern California storm (Slosson and Krohn, 1979).

Building code in effect	Number of sites developed	Number of site failures	Percentage of site failures	Damage costs
Pre-1963 (pre-modern code)	37,000	2,790	7.5	$40-49 million
Post-1963 (modern code)	30,000	210	0.7	$1-2 million

In 1979, Slosson and Krohn (1979) noted that damages to private property constructed in Los Angeles since the 1963 enactment of stringent grading regulations were reduced by more than 90 percent as compared to property developed before enactment of the grading code.

Protecting Existing Development

In spite of the avoidance and regulatory procedures listed above, there will continue to be development in areas subject to landslides. Thus, land-use planning programs for landslide-prone areas should include measures to protect facilities, structures, and people. Erley and Kockelman (1981) divided protective measures in such areas into: (1) physical methods of control of unstable slopes, and (2) monitoring and warning systems.

Physical Control Measures. The most commonly used physical measures for control of unstable slopes are: (1) drainage of surface and ground water, (2) modification of the slope geometry by excavation, (3) construction of toe earth buttresses as counterforts, (4) use of restraining structures, such as retaining walls and ground anchors, (5) hardening of the unstable mass, and (6) use of vegetation ("biomass") to either strengthen the soil or act as a retaining structure. All of these physical control methods have been discussed extensively in the landslide literature (e.g., Baker and Marshall, 1958; Veder, 1981, pp. 73-173; Holtz and Schuster, in press; Schuster, in press; Wyllie and Noorish, in press). If correctly designed and constructed, physical control measures are excellent site-specific management tools for landslides. The principal negative factor in their use is their high cost. Thus, they are used only if avoidance is not feasible or economically desirable.

Because of its high stabilization efficiency in relation to cost, drainage is the most widely used physical stabilization method in most countries. In the United States, drainage is used for slope stabilization both by government agencies and by private landowners and developers. Underground drainage systems and pumping wells remove ground water; surface water is drained from unstable areas by ditches.

Slope modification by removal of all or part of the earth driving the landslide is commonly used to increase stability. Stability also is increased by placing earth-buttress counterforts at the toes of potential failures. In California, this is the most common mechanical (as contrasted to hydrologic) method of landslide control (Committee on Ground Failure Hazards, 1985).

Mud flows and debris flows can sometimes be diverted around structures or facilities by means of deflection walls, i.e., retaining walls placed at an angle other than 90° to the direction of the slope

(Hollingsworth and Kovacs, 1981). In addition, check dams and catchment basins, constructed in gullies, canyons, and valleys of mud-flow-prone areas, can intercept flows before they reach developments.

Physical protection is most efficient for the control of small landslides. It requires costly engineering design, quality construction, and continuing maintenance. Landslide control in this manner by government entities can be self-defeating in that remedial public-works measures may encourage development in landslide-prone areas by leading the public to believe that problems have been eliminated, not just reduced.

Monitoring and Warning Systems. Hillside areas subject to landslides can be monitored so that downslope residents can be warned of slope movement, and, if necessary, be evacuated. Monitoring techniques traditionally have included field observation and the use of extensometers, inclinometers, tiltmeters, electrical fences, and trip wires (Wilson and Mikkelesen, 1978). Recent innovations in monitoring devices include acoustic instrumentation, television, guided radar, laser beams, and vibration meters. Data from any of these devices can be telemetered to central receiving stations.

One of the most important areas of landslide research involves the development of real-time warning systems for issuing warnings of landslides in urban areas during major storms. Such a system has been developed for the San Francisco Bay region, California, by the U.S. Geological Survey in cooperation with the U.S. National Weather Service. This system is based on: (1) empirical and theoretical relations between rainfall intensity/duration and landslide initiation, (2) geologic determination of areas susceptible to landslides, (3) real-time monitoring of a regional network of telemetering rain gages, and (4) National Weather Service precipitation forecasts (Keefer *et al*, 1987).

The Territory of Hong Kong also relies on a rainfall-monitoring system for identifying periods of high landslide potential. This system is maintained by the Geotechnical Control Office (GCO) within the Engineering Department of the Hong Kong government (Geotechnical Control Office, 1985). During heavy rainstorms, the GCO operates on an emergency basis to provide advice on remedial measures for landslides that may occur.

Success of Physical Controls and Warning Systems. If designed and constructed satisfactorily, physical control methods provide effective, but generally expensive, means of landslide hazard mitigation. For this reason, they often are used where landslide costs are very high because of high population densities and property values. An excellent example of such application is in urban areas of Japan where costs of landslides have historically been very high. Because of almost yearly landslide disasters, the Japanese government has dedicated large sums of money to the construction of check dams ("Sabo works) and other physical controls, resulting in a general reduction of deaths and damages due to major landslide disasters.

A successful prototype of a real-time warning methodology is the aforementioned system developed for the San Francisco Bay region by the U.S. Geological Survey and the U.S. National Weather Service (Keefer *et al.*, 1987). The system was used to issue the first regional public warnings in the United States during the storms of February 12-21, 1986, which produced 800 mm of rainfall in the San Francisco Bay region. These warnings were issued over public radio and television stations in the area. According to eyewitness accounts of landslide occurrence, the warnings accurately predicted the times of major landslide events. At present, this system still functions in protection of the Bay Area populace. It is continually being improved, and is intended as a prototoype for other heavily populated, landslide-prone urban areas.

LANDSLIDE INSURANCE -- PROTECTION FOR PROPERTY OWNERS

Insurance programs can reduce the impact of landslide losses to individual property owners by spreading these losses over a larger base (Schuster and Fleming, 1986). The use of insurance as a mitigative measure has the following advantages over other strategies (Olshansky and Rogers, 1987):

(1) In theory, insurance provides equitable distribution of costs and benefits. If property owners in landslide-prone areas were to pay premiums reflecting their actual risk and if insurance were to fully compensate victims, costs and benefits would be equitably distributed.

(2) Landslide insurance encourages hazard reduction if premium rates reflect not only the degree of natural hazard but also the quality of physical control measures.

(3) Using insurance to reduce landslide hazards appeals to those opposed to government regulation because, as compared to the other approaches, it depends more on the private market than on government intervention.

The most successful application of insurance to mitigation of losses due to natural hazards has been in New Zealand, where a national insurance program assists home owners whose dwellings have been damaged by landslides or other hazards that cannot within reason have been prevented or controlled by the home owners (O'Riordan. 1974). A disaster fund, accumulated from a surcharge to the national fire-insurance program, reimburses property owners for losses. This natural-hazard insurance program is an outgrowth of New Zealand's Earthquake and War Damage Act of 1944.

Landslide insurance can be divided into two types, public and private. In the United States, public landslide insurance is available in certain circumstances through the National Flood Insurance Program, which was created by the Housing and Urban Development Act of 1968. An amendment to this act extended its application to "mudslides" in 1969. However, the range of phenomena defined by the term "mudslide" was not made clear. As presently worded, the regulations include landslides that are proximately caused or precipitated by accumulations of surface water or ground water (Committee on Methodologies for Predicting Mudflow Areas, 1982). The insurance on these "water-caused" landslides is provided by private insurance companies, but it is underwritten and subsidized by the federal government.

However, the Federal Emergency Management Agency (FEMA), which administers the NFIP, has been unable to effectively implement this "mudslide" insurance program, largely because of technical difficulties in defining "mudslide" and in mapping "mudslide" hazard zones (Olshansky and Rogers, 1987). A possible solution to this dilemma would be to add all types of landslides to the NFIP. A bill proposing this solution was introduced in the U.S. House of Representatives in 1981. This bill would have immediately added landslide coverage to the nearly two million existing NFIP policies, and a new landslide mapping program would have been undertaken to provide data for the underwriters. However, FEMA opposed this bill because of its high cost and difficulty in administration. Partly as a result of FEMA opposition, the bill was killed before reaching a vote by the House of Representatives.

Yelverton (1973) has reviewed United States experience in landslide insurance as a basis for proposing a national landslide insurance program. However, other than the NFIP, landslide insurance generally is not available in the United States. While the concept of private landslide insurance is an appealing one, it has certain drawbacks in practice, and the private sector does not appear to be interested in offering this coverage. This reluctance to offer landslide coverage is a long-standing one partially based on several costly and highly publicized landslides that have occurred along the California coast. In addition, private insurers hesitate to offer landslide coverage because of the problem of "adverse selection," which is the tendency for only those who are in hazardous areas to purchase insurance (Olshansky and Rogers, 1987; Olshansky, 1990).

Olshansky and Rogers (1987) have summarized the need and general requirements for landslide insurance in the United States as follows: "insurance can equitably provide funds to compensate for landslide damage that will inevitably occur even when there are strict land use and grading controls. For insurance to be an effective solution, though, a comprehensive government landslide insurance fund is needed, or, alternatively, some other form of government intervention is needed to induce or require private insurers to cover landslides. Controls on building, development, and property maintenance would need to accompany mandatory insurance. Insurance and appropriate government intervention can operate together, each filling a need not served by the other, and each improving the performance of the other in reducing landslides and compensating victims."

MULTIPLE-HAZARD REDUCTION

Typically, landslide-hazard reduction is undertaken as an individual exercise. However, landslides often occur as elements of interrelated multiple natural-hazard processes in which an initial event triggers secondary events (Advisory Board on the Built Environment, 1983; Advisory Committee on the International Decade for Natural Hazard Reduction, 1987). In other cases, two or more natural hazard processes, not directly related to each other but triggered by a common cause, may occur at the same time in the same or adjacent localities. United States examples are the 1964 Alaska earthquake, which triggered tsunamis, local flooding, and many landslides, and the 1980 Mount St. Helens eruption, which led to major landslides, floods, and wildfires that consumed large tracts of timber.

Multiple-hazard problems require a shift of perspective from mitigation of individual hazards, such as landslides, to a broader systems framework that takes into account the characteristics of more than one hazard (Advisory Committee on the International Decade for Natural Hazard Reduction, 1987). Therefore, in planning landslide-hazard reduction strategies, attention should be paid to possible interrelationships between landslides and other hazards. For example, a building moved from a floodplain to a hillside to avoid floods may be at increased risk from earthquakes or landslides. At present, United States land-use regulations commonly deal individually with hazards such as floods and landslides. Future codes should treat all hazards that threaten an area.

CONCLUSIONS

In the United States, a wide range of landslide-hazard mitigation strategies and techniques has been developed and some are in extensive use. Optimal approaches to reduction of landslide hazards generally involve a carefully assembled mix of these hazard-reduction strategies and techniques. To plan a coordinated and successful reduction program requires input and cooperation from geologists, engineers, planners, land owners and developers, lending organizations, government entities, and insurance companies. This methodology has been relatively successful in landslide-prone areas, particularly in California.

REFERENCES

Advisory Board on the Built Environment (1983). *Multiple Hazard Mitigation*, Report of a Workshop on Mitigation Strategies for Communities Prone to Multiple Natural Hazards, 60. National Research Council, National Academy Press, Washington, D.C.

Advisory Committee on the International Decade for Natural Hazard Reduction (1987). *Confronting Natural Disasters--An International Decade for Natural Hazard Reduction*, 60. U.S. National Academy of Sciences, National Academy Press, Washington, D.C.

Baker, R.F., and H.E. Marshall (1958). Control and correction. In *Landslides and Engineering Practice* (E.B. Eckel, ed.), Highway Research Board Special Report 29, pp. 150-188. National Academy of Sciences/National Research Council, Washington, D.C.

Brabb, E.E., (1984). Innovative approaches to landslide hazard and risk mapping. In: *Proceedings of the 4th International Symposium on Landslides*, Toronto, 1, 307-323.

Brabb, E.E., E.H. Pampeyan, and M.G. Bonilla (1972). *Landslide Susceptibility in San Mateo County, California*. U.S. Geological Survey Miscellaneous Field Studies Map MM-360, scale 1:62,500.

Brown, W.M., III (1992). Information for disaster reduction: The National Landslide Information Center, U.S. Geological Survey. In *Proceedings of the Sixth International Symposium on Landslides*, Christchurch, New Zealand, 2, 891-892.

Colorado Natural Hazards Mitigation Council (1992). *Annual Report, 1991-92,* 55 pp. Colorado Office of Emergency Management, Golden, Colorado.

Committee on Methodologies for Predicting Mudflow Areas (1982). *Selecting a Methodology for Delineating Mudslide Hazard Areas for the National Flood Insurance Program*, 35 pp. National Research Council, Washington, D.C.

Committee on Ground Failure Hazards (1985). *Reducing Losses from Landsliding in the United States,* 41 pp. Commission on Engineering and Technical Systems, National Research Council, Washington, D.C.

Cruden, D.M., and D.J. Varnes (in press): Landslide types and processes. In *Landslides: Investigation and Mitigation* (A.K. Turner and R.L. Schuster, eds.), Transportation Research Board Special Report. National Academy of Sciences/National Research Council, National Academy Press, Washington, D.C.

Dallaire, Gene (1976). Consultants reviewing plans for other consultants in Fairfax County, Va.; landslides greatly reduced. *Civil Engineering*, American Society of Civil Engineers, 46, No. 9, 77-79.

Erley, D., and W.J. Kockelman (1981). *Reducing Landslide Hazards--A Guide for Planners.* American Planning Association Planning Advisory Services Report No. 359, 29 pp.

Geotechnical Control Office (1985). *Description of the Geotechnical Control Office Engineering Development Department*, 8 p. Hong Kong.

Hamilton County (1982). *Earth Movement Task Force,* 38 pp. plus appendix. Hamilton County, City of Cincinnati, Ohio.

Hamilton County Regional Planning Commission (1976). *Hillside Development Study--Hillside Development Plan and Strategy, 86 pp.* Regional Planning Commission, Hamilton County, Cincinnati, Ohio.

Hansen, A. (1989). Landslide hazard analysis. In *Slope Instability* (S. Brunsden and D.B. Prior, eds.), pp. 523-595. John Wiley & Sons, New York.

Hansen, W.R., E.B. Eckel, W.E. Schaen, R.E. Lyle, W. George, and G. Chance (1966). *The Alaska Earthquake of March 27, 1964--Investigations and Reconstruction Effort*, U.S. Geological Survey Professional Paper 541, 111 pp.

Hollingsworth, R., and C.S. Kovacs (1981) Soil slumps and debris flows. *Bulletin of the Association of Engineering Geologists,* 18, No. 1, 17-28.

Holtz, R.D., and R.L. Schuster (in press). Stabilization of soil slopes. In *Landslides: Investigation and Mitigation* (A.K. Turner and R.L. Schuster, eds.). National Academy of Sciences/National Research Council, National Academy Press, Washington, D.C.

International Conference of Building Officials (1985). *Uniform Building Code*, 817 pp. Whittier., California.

Jahns, R.H. (1969). Seventeen years of response by the City of Los Angeles to geologic hazards. In *Geologic Hazards and Public Problems* (R.A. Olson and M.M. Wallace, eds.), Conference Proceedings, pp. 283-296. Office of Emergency Preparedness, Washington, D.C.

Jochim, C.L., W.P. Rogers, J.O. Truby, R.L. Wold, Jr., G. Weber, and S.P. Brown (1988) *Colorado Landslide Hazard Mitigation Plan,* 149 pp. Colorado Geological Survey, Department of Natural Resources, Denver, Colorado.

Keefer, D.K., R.C. Wilson, R.K. Mark, E.E. Brabb, W.M. Brown, III, S.D. Ellen, E.L. Harp, G.F. Wieczorek, C.S. Alger, and R.S. Zatkin (1987). Real-time landslide warning during heavy rainfall. *Science*, 238, 921-925.

Kockelman, W.J. (1986). Some techniques for reducing landslide hazards. *Bulletin of the Association of Engineering Geologists*, 23, No. 1, 29-52.

Mader, G.G., W.E. Spangle, and M.L. Blair (1980). *Land Use Planning after Earthquakes,* 24 pp. William Spangle and Associates, Inc., Portola Valley, California.

Mader, G.G., T.C. Vlasic, and P.A. Gregory (1988). *Geology and Planning: the Portola Valley Experience*, 75 pp. William Spangle and Associates, Inc., Portola Valley, California.

Obermeier, S.F. (1979). *Slope Stability Map of Fairfax County, Virginia.* U.S. Geological Survey Miscellaneous Field Studies Map MF-1072, scale 1:48,000.

Olshansky, R.B. (1990). *Landslide Hazard in the United States--Case Studies in Planning and Policy Development*, 176 pp. Garland Publishing, Inc., New York

Olshansky, R.B., and J.D. Rogers (1987). Unstable ground: Landslide policy in the United States. *Ecology Law Quarterly*, 13, No. 4, 939-1006.

O'Riordan, T. (1974). The New Zealand natural hazard insurance scheme: Application to North America. In *Natural Hazards--Local, National, Global* (G.F. White, ed.), pp. 217-219. Oxford University Press, New York.

Petak, W.J., and A.A. Atkisson (1982). *Natural Hazard Risk Assessment and Public Policy--Anticipating the Unexpected*, 489 pp. Springer-Verlag, New York.

Radbruch-Hall, D.H., R.B. Colton, W.E. Davies, I. Lucchita, B.A. Skipp, and D.J. Varnes (1982). *Landslide Overview Map of the Conterminous United States*, U.S. Geological Survey Professional Paper 1183, 25 pp.

Rib, H.T., and Ta Liang (1978). Recognition and identification. In *Landslides: Analysis and Control* (R.L. Schuster and R.J. Krizek, eds.), Transportation Research Board Special Report 176, pp. 34-80. National Academy of Sciences, Washington, D.C.

Rogers. W.P., L.R. Ladwig, A.L. Hornbaker, S.D. Schwochow, S.S. Hart, D.C. Shelton. D.L. Sroggs, and J.M. Soule (1974). Model geologic hazard area control regulations, Appendix. In *Guidelines and Criteria for Identification and Land-use Controls of Geologic Hazard and Mineral Resource Areas*, pp. 135-146. Colorado Geological Survey, Denver.

Rossi, P.H., J.D. Wright, and E. Weber-Burdin (1982). *Natural Hazards and Public Choice--The State and Local Politics of Hazard Mitigation*, 337 pp. Academic Press, New York.

San Mateo County (1973). *Ordinance No. 2229*, Board of Supervisors, San Mateo County, California.

Santa Clara County Board of Supervisors (1978). Geological ordinance no. NS-125.35, *Santa Clara Code*, Section C-12-600 and following, San Jose, California.

Schuster, R.L. (in press). Recent advances in slope stabilization. In *Proceedings*, Sixth International Symposium on Landslides, Christchurch, New Zealand, 1992.

Schuster, R.L., and R.W. Fleming (1986). Economic losses and fatalities due to landslides, *Bulletin of the Association of Engineering Geologists*, 23, No. 1, 11-28.

Schuster, R.L., and F.B. Leighton (1988). Regulations in California, U.S.A. In *Landslides and Mudflows* (E.A. Kozlovskii, ed.), 2, 116-122. UNESCO/UNEP, Moscow.

Scullin, C.M. (1983). *Excavation and Grading Code Administration, Inspection and Enforcement*, 405 pp. Prentice-Hall, Inc., Englewood Cliffs, New Jersey.

Soeters, R., K. Van Westen, and K.A. Turner (in press). Recognition and identification. In *Landslides: Identification and Mitigation* (A.K. Turner and R.L. Schuster, eds.), Transportation Research Special Report. National Academy of Sciences/National Research Council, Washington, D.C.

Slosson, J.E. (1969). The role of engineering geology in urban planning. In *The Governor's Conference on Environmental Geology*, Colorado Geological Survey Special Publication No. 1, pp. 8-15.

Slosson, J.E., and J.P. Krohn (1979). AEG Building Code Review--Mudflow/debris flow damage; February 1979 storm--Los Angeles area. *California Geology*, 32, No. 1, 8-11.

Slosson, J.E., and J.P. Krohn (1982). Southern California landslides of 1978 and 1980. In *Storms, Floods, and Debris Flows in Southern California and Arizona, 1978 and 1980*, Proceedings of a Symposium, National Research Council, and Environmental Quality Laboratory, California Institute of Technology, Pasadena, California, September 17-18, 1980, pp. 291-319. National Academy Press, Washington, D.C.

Swanston, D.N., and R.L. Schuster (1989). Long-term landslide hazard mitigation programs: Structure and experience from other countries. *Bulletin of the Association of Engineering Geologists*, 26, No. 1, 109-133.

U.S. Geological Survey (1982). *Goals and Tasks of the Landslide Part of a Ground-failure Hazards Reduction Program.* U.S. Geological Survey Circular 880, 49 pp.

Varnes, D.J. (1978). Slope-movement types and processes. In *Landslides--Analysis and Control* (R.L. Schuster and R.J. Krizek, eds.), Transportation Research Board Special Report 176, pp. 11-33. National Academy of Sciences/National Research Council, Washington, D.C.

Varnes, D.J., and the International Association of Engineering Geology Commission on Landslides and Other Mass Movements on Slopes (1984). *Landslide Hazard Zonation--A Review of the Principles and Practice*, United Nations Economic, Scientific and Cultural Organization, Natural Hazards series, No. 3, 63 pp.

Veder, C. (1981). *Landslides and Their Stabilization*, 247 pp. Springer-Verlag, New York.

Wilson, S.D., and Mikkelsen, P.E. (1978). Field instrumentation. In *Landslides: Analysis and Control* (R.L. Schuster and R.J. Krizek, eds.), Transportation Research Board Special Report 176, pp. 112-138. National Academy of Sciences/National Research Council, Washington, D.C.

Wold, R.L., and C.L. Jochim (1989). *Landslide Loss Reduction: A Guide for State and Local Government Planning*, 50 pp. Colorado Geological Survey Special Publication, Denver, Colorado.

Wyllie, D.C., and I.N. Noorish (in press). Rock slope stabilization and control. In *Landslides: Investigation and Mitigation* (A.K. Turner and R.L. Schuster, eds.), Chapt. 18. Transportation Research Board Special Report, National Academy of Sciences/National Research Council, National Academy Press.

Yelverton, C.A. (1973). Land failure insurance. In *Geology, Seismicity and Environmental Impact*, pp. 15-20. Association of Engineering Geologists, Brentwood, Tennessee.

MECHANISM OF INITIATION OF DEBRIS FLOW

R. H. Chen,* M. L. Lin* and H. Chen**

** Department of Civil Engineering, National Taiwan University, Taipei, Taiwan*
*** Department of Geology, National Taiwan University, Taipei, Taiwan*

ABSTRACT

The initiation condition and failure mode of debris flow have been studied by many researchers, and several possible mechanisms of initiation of debris flow were proposed. However, due to the difficulty of observing the initiation condition of debris flow in the field before and while it occurred, model study was performed in this research. Mobilization of debris flows was also studied and simulated by inducing the pore water pressure in samples during triaxial tests. The rate of the increase in pore water pressure was based on rainfall intensity and the coefficient of storage of the soil. The results show that the strength parameters of soils and the factors of safety obtained in slope stability analysis from conventional tests are higher than those from this proposed method.

KEYWORDS

Debris flow; mechanism; model test; strength; failure.

INTRODUCTION

The Ofelia typhoon dropped a large quantity of rain on the Hualien area on the 23rd of June, 1990. This storm had a high wind speed of 45 m/s and a maximum precipitation of 106 mm/hr from 2 PM to 3 PM. The severe storm produced an enormous flood and rapid geomaterial movement suddenly without any warning. Thus the debris flow was formed from the upper part to the lower part of the gully on the mountain facing the back of the Tungmen village. The debris flow delivered a mixture of rock fragments and fine particle fractions to the village, located at the exit of the gully.

The elevation of Tungmen village is around 170 meters and is on an alluvium delta. Around 1300 people live in this deltaic region. The hazard caused the deaths of 29 people, injured 68 people, collapsed 24 houses, and damaged 11 buildings. Nearly half of the village was covered by the geomaterial of the debris flow of more than 55,000 m^3.

This paper will study the mechanism of initiation of this debris flow. According to Johnson and Sitar (1990), Fleming, *et al.* (1989), and Ala and Mathewson (1990), the initiation of debris flow was triggered by the sudden increase in pore water pressure in the soil mass, which caused the soil mass to become unstable. Such rapid increase in pore water pressure can not be explained by the infiltration of rain water, instead, the pore water pressure increase may result from the upwelling of ground water in the fractured base rock. Thus, the "field stress path" concept was proposed for the initiation of debris flow.

Another failure mechanism used by many researchers to analyze the debris flow initiation was the infinite slope failure mode with various forms of modification for ground water condition (Harris, 1977, Yu, 1987, Lin, *et al.*, 1989). Takahashi (1978, 1981) suggested that the surface runoff was a significant factor for triggering debris flow. He assumed the seepage in the slope as parallel to the slope face and the shear force induced by the surface flow was considered in the derivation of solution. This solution would reduce to the infinite slope condition when the thickness of the surface runoff was zero. However, in order to observe the initiation condition of debris flow, model study was performed in this research.

SITE INVESTIGATION

The Hualien area is located on the plate boundary of the Philippian Sea plate and the Eurasian Continental plate. These plates have been very active recently. The study area is located to the west of Hualien city (Fig. 1). It is geologically composed of the Tananao schist formation formed on the Paleozoic.

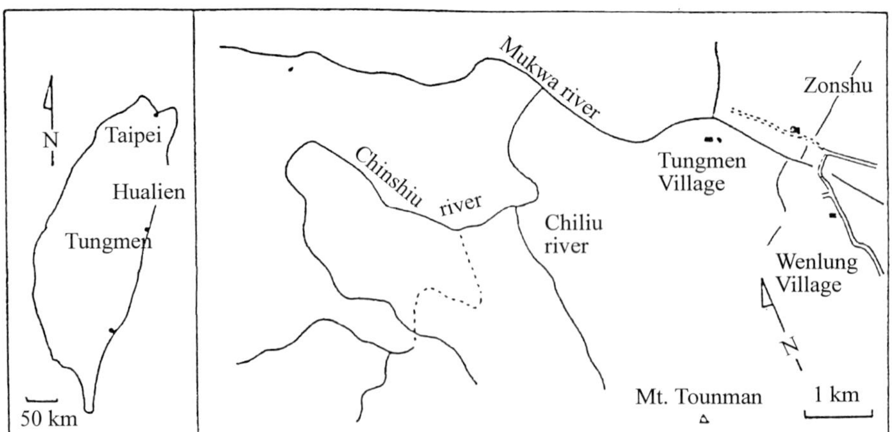

Fig. 1. Location of the study area.

Geology

The rock types in this area mainly consist of quartz-mica schist, phyllite, mica schist, mica quartzite and a bit lesser portion of scattered amounts of chlorite schist and metabasite. The metabasite is normally distributed on the bottom of all the above rock types in the study area. In general, the quartz-mica schist, phyllite mica schist and mica quartzite have all been called the black schist because they have an overall black colouration. These black schists are well cemented and have more of a black colour with a greater concentration of mica in the composition.

The discontinuities of the schistorsity and joint are well developed. In Tungmen village, the attitude of the schistosity is N40°E and the dip is 11° toward west. There is a 2 mm to 4 cm spacing distributed on the schistosity. The attitudes of three sets of joints are: (a) N80°W with a dip of 74° toward the north; (b) N10°W with a dip of 78° toward the south; and (c) N67°W with a dip of 88° toward the south. The spacing of the different discountinuities ranges from 10 cm to 1 meter. The dip direction of these joints is nearly vertical to the schistosity. These discontinuities and slope surfaces formed into variations of dip and wedge of slope.

The discontinuities are densely distributed and the spacing between the discontinuities is always less than one meter. This means that the geological structure contains a number of the discontinuities distributed across a broad area. Due to the shallow slope, the debris flow tends to attenuate downstream. However, it still maintains enough speed, mass and energy to collect additional debris from the gully walls and floor as the mass marches inexorably to the gully exit. The irregular spatial distribution of partial size and the concentration of boulders in the flow changes greatly. These changes affect not only mobility (in terms of speed and distance covered) but also the resulting degree and type of devastation. It appears that large concentrations of sizable boulders in the moving mass can obliterate even solid concrete houses and other structures more easily than fine particles in mud flows without such boulder concentration.

Geomorphology

Tungmen village is located on the south bank of the Mukwa River. Lower parts of the village are located right at the exit of a large gully into the Mukwa river. The back of Tungmen village is at the base of a series of steep mountain slopes. The trees and various vegetation have developed very well. The slope grade averages from around 30° to 60°. The upper part of the slope on south of Tungmen village has a stream flowing down the slope toward the village. The stream bed is normally dry.

The composition of the deposited material includes rock fragments and fine particles. The size of the big boulders is around 1 meter to 4 meters. The length of the fan is around 120 meters and its width around 70 meters. Its thickness is around 6 meters. The slope grade is around 25° on the gully after the debris flow. Most of the gully wall has been cleanly eroded from a "V" shape into more of a "U" shape leaving the walls fairly smooth and featureless.

MODEL TEST

In order to study the initiation and failure conditions of debris flow in the laboratory, model tests were performed. Material used in the model test was reconstituted with scaled-down particles and laws of similitude were taken into account when forming the model.

Material

The soil material used in the model was obtained from the site of Tungmen debris flow. The soil was majorly composed of gravels with nonplastic fines. The grain size distribution of the soil is as shown in Fig. 2. The soil is classified as GM according to the Unified Soil Classification System. Material used in the model test was reconstituted using the field material with scaled-down particle size. The scaling down of particles and reconstitution of material was conducted using the "modified weight replacement method" developed in the previous research (Chen, *et al.*, 1993), which provided satisfactory results in shearing and permeability properties. The grain size distribution of the reconstituted material is also shown in Fig. 2 with maximum grain sizes of 1.5 in. (38.1 mm), 1.0 in. (25.4 mm), and 0.5 in. (12.7 mm) each. The material was compacted in three layers with a controlled dry unit weight of 18.64 kN/m^3, which was the field dry unit weight.

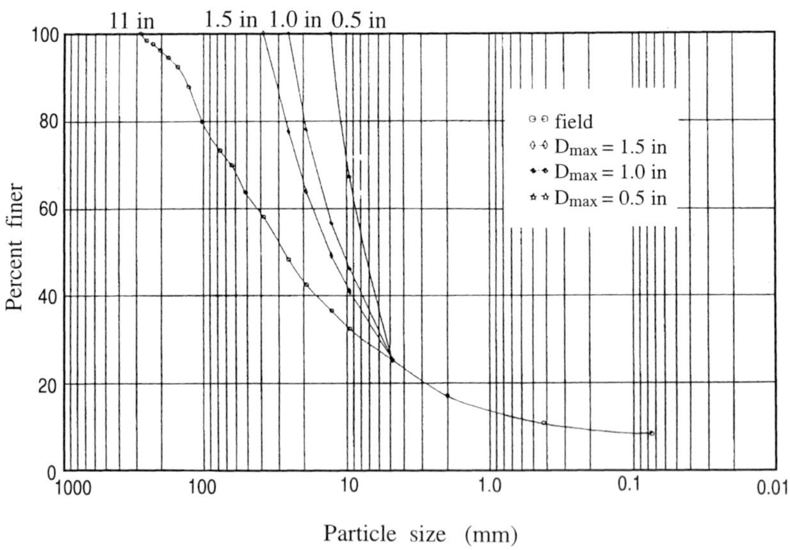

Fig. 2. Grain size distribution of field and reconstituted materials.

Water Tank

The water tank used for the model test was sized according to the laws of similitude. The ratio of geometry of model to the prototype in field was kept constant as 1:50. The width, height, and length of the water tank was 0.42 m by 0.55 m by 2.5 m. Considering the similitude in geometry, the thickness of the specimen was determined to be 0.1 m with a maximum grain size of 0.5 in. The length of the specimen was limited by the tank and its peripheral devices, and was determined to be 1.15 m. The water tank can be raised to rising angles ranged from zero to 30 degree. In order to provide proper roughness and permeability for the boundary of the tank, a layer of permeable material was fixed to the side walls and base of the tank. The permeable material was composed of the same soil material of the specimen but with grain sizes larger than #40 sieve size, and mixed with proper amount of glue. Thus, it will provide about the same roughness as in the soil layer but still be relatively permeable. Based on previous discussions, water was supplied to the specimen in two different ways. In the first method, water flowed from a reservoir through an energy dissipation zone to the upper face of the specimen, which was likely to generate seepage in the soil parallel to the slope surface. In the second method, water was supplied to the specimen through the outlets buried in the pervious material on the base of the tank, which was to simulate the condition of upwelling of ground water. The rate of the water flowing to the specimen was controlled by regulators and could vary from 0.5 $\sim$ 50 ℓ/min. The layout of the tank is illustrated in Fig. 3.

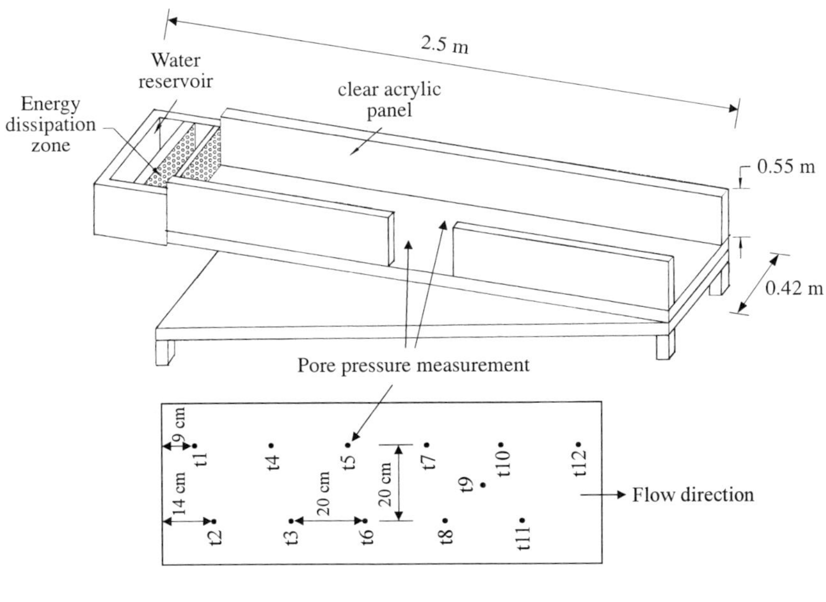

Fig. 3. Layout of water tank for the model test.

Model Test and Testing Results

Two rising angles were used in the model test. The rising angle of 20 degree was selected for it was the slope angle observed in the Tungmen debris flow, a smaller rising angle of 13.5 degree was also tested for comparison. The sequence of water supply was also separated into two ways: a slow step-wise increase in flow rate, and a continuous rapid increase in flow rate. The purpose of the two distinctive ways of flow rate increase was to verify the influences of peak flow rate versus accumulative flow quantity. The compacted soil specimen was submerged overnight for saturation, then drained immediately before the test. The water tank was raised slowly to the designated height and the rising angle of the tank was measured using slant rule. The water flow was supplied from either the top face or the bottom face of the specimen with given increase sequence of flow rate. The process of the debris flow development was recorded by video camera, and pore water pressures at the base of the specimen were measured at 12 different locations as shown in Fig. 3.

For the tests performed with water supplied from the top face of the specimen, erosion ditch developed first, and as the flow rate continued to increase, water flow concentrated in the ditch and ditch was cut down and small-scaled slope failure would occur at the lower slope surface. If the flow rate was increased further, the slope failure would enlarge and proceed uphill, and eventually cause a massive movement of the material. The thickness of the surface runoff was about 3 mm, regardless of the variation of flow rate. The values of pore water pressure measured at the base of the specimen increased with increasing water flow, and reached the peak values before the failure occurred. The pore water pressure decreased as soon as the failure surface started to develop. The peak value of the pore water pressure were about less than 0.1 m of water head. The progress of the failure was about the same for both rising angles and both flow rate sequences, except that the failure developed more rapidly with higher rising angle and larger flow rate. A typical model test after failure is shown in Fig. 4.

For the test performed with water supplied from the bottom face of the specimen, the pore water pressure increased with increasing flow rate, and cracks started to develop on the surface of the soil mass. With further increase of flow rate, sliding occurred in the soil mass and more cracks developed on the surface of soil mass, and finally a massive soil flow was induced. The peak values of pore water pressure measured in this set of test were much higher than those measured in the test with water supplied from the top face of the specimen, and the soil moved as a whole mass in the final status instead of being concentrated in a channel. For the flow rate sequence of step-wise increase, failure would not occur if the flow rate was maintained at a low value for a long period of time (large flow quantity). However, if a large increase of flow rate was applied, the sliding would start to develop fairly rapidly. The resulting pore water pressure in this condition would be larger than those from the step-wise increase of flow rate. It appeared that the effects of the peak flow rate might be more significant than the accumulative quantity of water flow. The peak value of pore water pressure was larger than 0.1 m of water head in both flow rate sequences, which implied that the water head was higher than the surface of the specimen. The pore water pressure for the model test with lower rising angle would reach higher values than those of the larger rising angle. Thus the resulting effective stresses at the base of the specimen were very low and failure would occur fairly rapidly in a massive way. A typical set of model test after failure is shown in Fig. 5.

Discussion of Testing Results

For all the tests performed in this study, the thickness of the surface runoff was all about 3 mm regardless of the manners and sequences of water flow, thus the failure mechanism proposed by

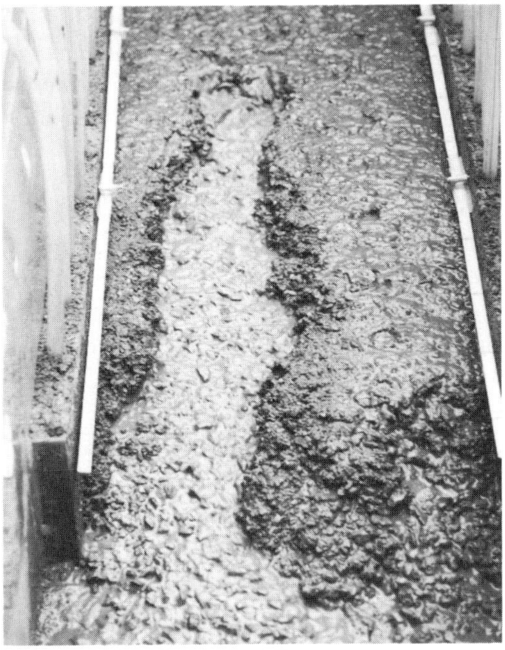

Fig. 4. A typical set of model test with water supplied from the top surface.

Fig. 5. A typical set of model test with water supplied from the base surface.

Takahashi (1978, 1981) can not be verified. For the test performed with water supplied from the top face of the specimen, failure was induced by surface erosion of the specimen, further increase in flow rate worsened the erosion and caused a small scale slope failure, which later triggered a massive soil movement. The debris flow occurred in this circumstances would be of a smaller scale and its development was confined in the channel formed by erosion. For the test performed with water supplied from the bottom face of the specimen, the pore water pressure would reach a higher value and failure occurred fairly rapidly with soil moved as a whole mass within the tank. The measured pore water pressure in this condition was much higher than pore water pressure measured when water supplied from the top face of the specimen, and the soil at failure was very close to the liquefied condition. A sudden increase in flow rate had more significant effects than the accumulative flow quantity. The model test with smaller rising angle would withstand larger flow rate or pore water pressure without inducing failure in both ways of water supply. Both types of water supply could induce massive soil movement, but with water supplied from the base of the soil mass (upwelling), the failure occurred in a large scale and proceeded more rapidly.

DEBRIS FLOW INDUCED BY PORE WATER PRESSURE

Anderson and Sitar (1991) studied the mechanism of slope failure induced by rainfall. They adopted the field stress path (FSP) concept in the triaxial compression tests for analyzing the critical condition when mud flow was initiated. Figure 6(a) shows an infinite slope with seepage. The effective principal stresses acting on the soil element as shown in Fig. 6(b) can be represented by Mohr's circle. In the field, when the soil is saturated, its Mohr's circle shifts horizontally to the left with the same radius (Fig. 7). If seepage force continues to increase, the increase in pore water pressure will decrease the effective stress. Thus, the effective stress circle continues to move to the left until it touches the failure envelope and the flow initiates. Anderson and Sitar stated that the pore pressure measured by tensiometers showed the above phenomenon. Pore pressure would change from negative value to positive value if rainfall would continue for a certain period of time. Hence this field stress path concept assumed failure of soil is caused by the increase of pore water pressure and the reduction in effective stress. However, there is no change in shear stress.

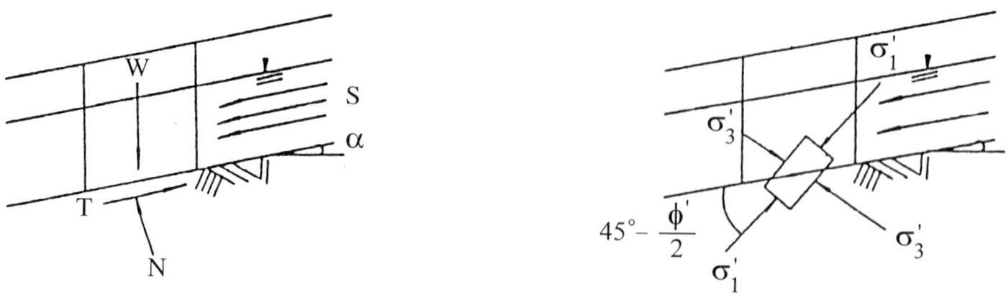

(a) acting forces (b) principal stresses

Fig. 6. The soil element in an infinite slope.

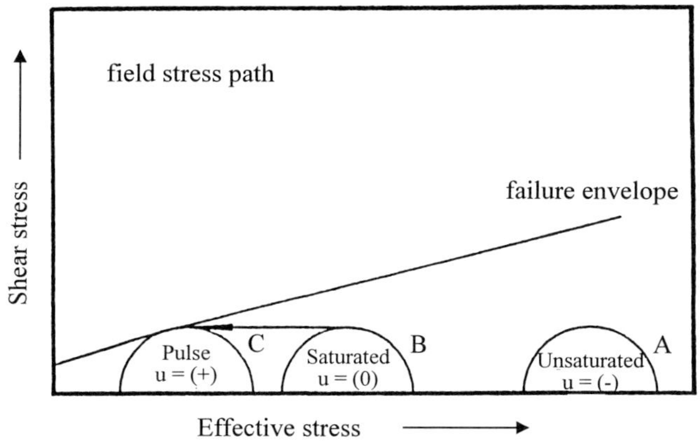

Fig. 7. The simulation of the effective stress path.

Pore Pressure Increment

To simulate the pore pressure change in the field, it was done by changing the back pressure inside the soil sample. The pore pressure increment shown in Eq. (1) was obtained from Boussinesq's equation in which the aquifer was assumed homogeneous and isotropic:

$$\frac{\partial u}{\partial t} = \gamma_w \frac{\partial h}{\partial t} = \gamma_w \frac{I}{S} \tag{1}$$

where u is the pore water pressure, h is the hydraulic head, I is the rain intensity, S is the coefficient of storage, t is time, γ_w is the unit weight of water.

Based on site investigation, the following values were chosen and substituted into Eq. (1): $I = 246$ mm/hr, $S = 0.02$. Hence, the pore pressure increment is about 2 kPa per minute.

$$\frac{\partial u}{\partial t} = \gamma_w \frac{I}{S} = \frac{9.8 \times 246 \times 10^{-3}}{3600 \times 0.02} = 0.033 \text{ kPa/sec} = 2 \text{ kPa/min} \tag{2}$$

FSP Triaxial Compression Test

For field-stress-path (FSP) triaxial compression tests, the grained size distribution of soil was also modified by the weight replacement method as described previously. Soils with different maximum grain sizes of 0.75 in., 0.5 in., and 0.375 in., were tested.

The main part of testing equipment is the same as that for conventional test except that the load was applied by a dead weight hanging on the loading frame. At the beginning, samples were saturated and then were consolidated isotropically or anisotropically. After the sample having been fully consolidated, then back pressure was applied into it by an increment of 1 kPa/30 sec.

The shear strain rate versus time curve is shown in Fig. 8(a). It can be seen that initially the curve was close to the horizontal axis. After some time, the slope of the curve increased gradually

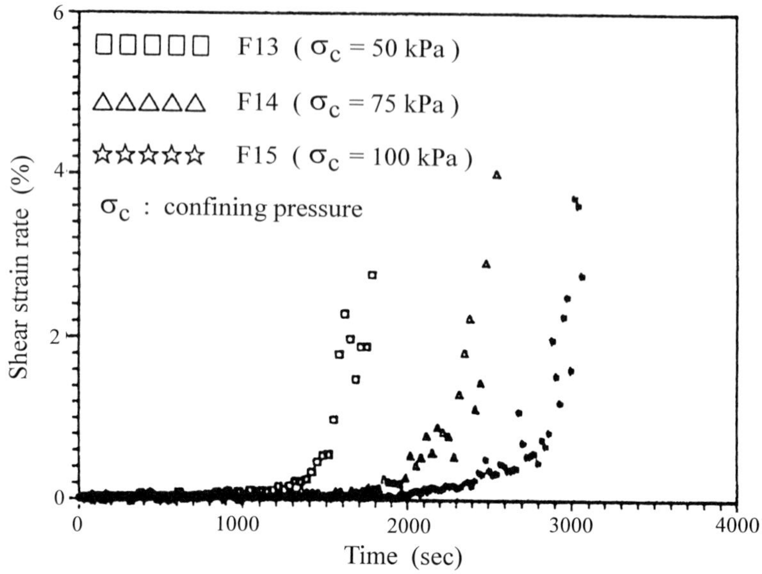

(a) shear strain rate

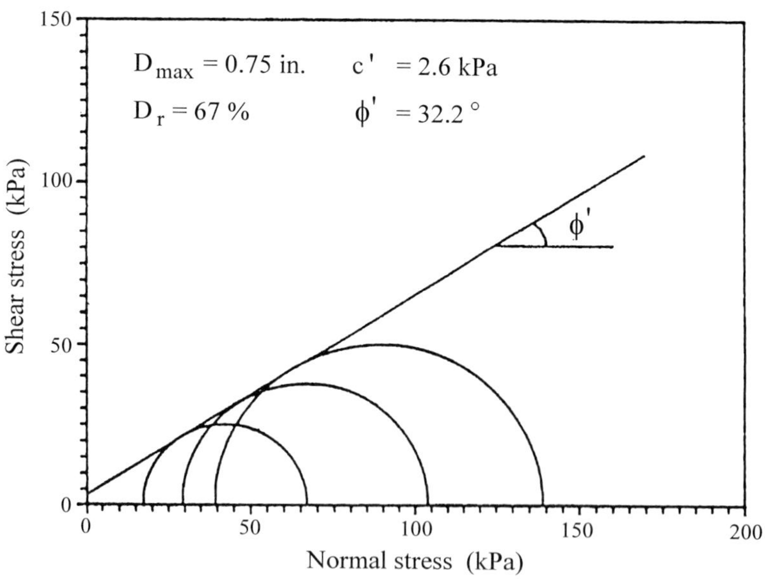

(b) Mohr's circles at failure

Fig. 8. Results of the FSP tests.

and the curve would become very steep eventually. The failure point is determined as at the end of the linear portion.

The Mohr's circle of effective stresses at failure state can be plotted as shown in Fig. 8(b). In order to compare the results from conventional and FSP triaxial compression tests, data are presented in Table 1. It shows that the effective friction angles obtained from FSP tests are about $6^\circ \sim 15^\circ$ lower than those from conventional tests. This phenomenon is due to different failure mechanisms. In conventional tests, samples were compressed by increasing vertical load, hence soil particles were in close contact during the test. However, in FSP tests, the ambient pressure of sample was kept constant. Samples were failed by increasing back pressure that caused soil particles to separate. By increasing back pressure continuously, sample would gradually dilate until it failed suddenly as a flow.

Table 1. Comparison of the factors of safety

(a) Isotropically consolidated undrained test

Maximum grain size (in.)	Relative density (%)	Strength parameters	Factor of safety
0.375	67	$c' = 2.8$ kPa $\phi' = 33.4^\circ$	0.96
0.375	25	$c' = 0$ $\phi' = 30.3^\circ$	0.85
0.5	67	$c' = 0$ $\phi' = 38.6^\circ$	1.16
0.5	25	$c' = 0$ $\phi' = 37.2^\circ$	1.11
0.75	67	$c' = 0.6$ kPa $\phi' = 46.5^\circ$	1.54
0.75	25	$c' = 1.9$ kPa $\phi' = 44.7^\circ$	1.44

(b) Field stress path test

Maximum grain size (in.)	Relative density (%)	Strength parameters	Factor of safety
0.375	67	$c' = 3.8$ kPa $\phi' = 27.6^\circ$	0.76
0.375	25	$c' = 4.3$ kPa $\phi' = 24.0^\circ$	0.65
0.5	67	$c' = 3.8$ kPa $\phi' = 29.4^\circ$	0.82
0.5	25	$c' = 5.8$ kPa $\phi' = 26.3^\circ$	0.72
0.75	67	$c' = 2.6$ kPa $\phi' = 32.2^\circ$	0.92
0.75	25	$c' = 4.8$ kPa $\phi' = 29.4^\circ$	0.82

Besides, samples having more uniform grain size distribution and lower relative density will have smaller friction angles as shown in Table 1.

Stability Analysis

The cohesion of soil is fairly small as shown in Table 1, therefore it will not be considered in this analysis. In stability analysis, the factor of safety for an infinite slope with the ground water table at surface is defined as follow,

$$F_s = \frac{\gamma_{sat} - \gamma_w}{\gamma_w} \frac{\tan\phi}{\tan\theta} \tag{3}$$

where γ_{sat} is the saturated unit weight of soil, γ_w is the unit weight of water, ϕ is the friction angle of soil, θ is slope angle. In this analysis, $\gamma_{sat} = 2.09$ kN/m^3, $\theta = 20°$.

The results from stability analysis are also shown in Table 1. It can be seen that the factors of safety using the strength parameters obtained from conventional tests are generally greater than 1.0. On the other hand, when the strength parameters of FSP tests are used, the factors of safety obtained are all less than 1.0. Consequently, from the results of this analysis and the failure conditions, it can be concluded that Tungmen debris flow was more likely induced by the pore water pressure increase in the soil mass.

CONCLUSIONS

1. From site investigation, highly fractured rock on site is very likely the source of debris materials.
2. In model test, water was supplied either from the top face or from the bottom face of the specimen. Both types of water supply induced massive soil movement, but with water supplied from the bottom of the soil mass, the failure occurred in a large scale and proceeded more rapidly.
3. The field stress path test can simulate the failure mechanism of debris flow better than conventional triaxial test. The former obtained lower frictional angle of the debris than the latter, and would trigger debris flow more easily.
4. In the stability analysis on debris flow, cares should be taken when the strength parameters obtained from conventional tests are used, because it may overestimate the factor of safety.

REFERENCES

Anderson, S. A. and N. Sitar (1991). Influence of stress path on soil strength parameter and analysis of rainfall-induced slope failure. *Engineering Geology and Geotechnical Engineering*, 27, 21-1–21-16.

Ala, S., and C. C. Mathewson (1990). Structural control of ground-water induced debris flow. *Hydraulics/Hydrology of Arid Lands*, 590–595.

Chen, R. H., M. L. Lin, and H. Chen (1993). Study on debris flow in Hualien Area (I). Report No. NSC81-0414-P002-16B, National Science Council, Taiwan, R. O. C.

Fleming, R. W., S. D. Ellen, and M. A. Algus (1989). Transformation of dilative and contractive landslide debris into debris flows – an example from Martin County, California. *Engrg. Geol.*, 27, 201–223.

Harris, C. (1977). Engineering properties, ground water condition and the nature of soil movement on a solifluction slope in north Norway. *Quart. J. Engrg. Geol.*, 10, 27–43.

Johnson, K. A., and N. Sitar (1990). Hydrologic conditions leading to debris flow initiation. *Can. Geotech J.*, 27(6), 789–801.

Lin, P. S., S. C. Fang, K. R. Lin, and S. Y. Fong (1989). Debris flow in lateritic gravelly deposits. *Proc., Japan-China Joint Seminar on Natural Hazard Mitigation*, Kyoto, Japan.

Takahashi, T. (1978). Mechanical characteristics of debris flow. *J. Hydraulics Div., ASCE*, 104(8), 1153–1169.

Takahashi, T. (1981). Debris flow. *Ann. Rev. Fluid Mech.*, 13, 57–77.

Yu, F. C. (1987). Basic study on the debris flow (I), study on the mechanism of the occurrence of debris flow. *J. Chinese Soil and Water Conservation.*, 18(2), 28–40.

Flood Engineering

URBAN FLOOD HAZARD AND ITS MITIGATION

B. C. Yen

Hydrosystems Laboratory, Department of Civil Engineering, University of Illinois, 205 N. Mathews Avenue, Urbana, IL 61801, U.S.A.

ABSTRACT

Urban flood disasters are caused by natural geophysical events as well as manmade incidents. Mitigation means can be technical and nontechnical, structural and nonstructural. The most frequent urban floods are rain-induced catchment flooding. Structural means of constructing sewers and channels were the main approach of flood mitigation. Today, ponding and nonstructural measures, particularly real-time operation, evolve to become viable alternatives. However, the traditional concept of return period for rainfall prediction and the associated design peak discharge is inadequate for the nonstructural operation and real-time forecasting. Information on temporal distribution of rain is needed to produce runoff hydrograph and flood volume. Furthermore, reliability analysis is a better alternative to the return period for urban stormwater management. Finally, new method is now available to determine channel/sewer capacity under backwater conditions, allowing separation of hydraulics from hydrology in flood analysis, providing the potential for improvement in flood analysis for stormwater management and flood mitigation.

KEYWORDS

Flood hazards; urban drainage; stormwater management; sewers; urban hydrology; return period; risks; design rainfall; flood prediction; real-time forecasting; storm drainage.

CAUSES OF URBAN FLOOD HAZARDS

Flooding of urban land is a frequent, natural or manmade disaster in large or small scales. It may occur locally covering only a small area such as a local catchment flooding. It may cover a large area as in the case of flooding from a large river passing through a city. It occurred in ancient time and it will continue to occur in the future. But with proper mitigation measures urban flood hazard can be much minimized.

The causes of urban flood hazard may be natural or manmade. Sources of excessive water in the natural hydrologic cycle that could cause urban flooding include the following:

- Rain;
- Snow melting;
- River flood;
- Coastal storm surge and tsunamis.

Manmade causes of urban flooding abound, including

- Failure of water carrying facilities such as pipes, hydrants, pumps and other components of water supply network;
- Failure of water storage facilities such as water towers, reservoirs and ponds, embankments and dams;
- Blockage of drainage facilities such as inlets, sewers, ditches and culverts;
- Geophysical events such as earthquakes and high winds that induce spillage of water or breakage of water facilities;
- Human error that causes spillage of water or failure of water facilities.

The 1992 Chicago downtown flood is a recent example of manmade urban flood disaster whereas the 1993 floods in St. Louis, Des Moines and other cities is an example of natural urban flood disaster. Among the various possible causes, natural rain-produced flood is the most frequent one in cities.

MITIGATION MEANS

Urban flood mitigation means can be technical or nontechnical. They are complementary rather than mutually exclusive. For manmade flood hazards, the best mitigation means are through regular maintenance of the facilities and proper operational procedures, which in turn require adequate training of the operators. Of course, properly prepared enforceable standards and codes for the design and construction of the facilities that could cause manmade flooding are also prerequisites for hazard mitigation.

For natural flood hazards, effective mitigation requires reliable estimation of the flood water and appropriate design and construction of the drainage facilities, in addition to proper operation and maintenance. In other words, it involves both structural and nonstructural measures. Structures such as sewers, channels, pumping stations and ponds may be built to drain or control the flood water. On the other hand, nonstructural measures may be used to manage the flood water and to minimize flood damages. These nonstructural measures include:

- Land use zoning;
- Flood insurance;
- Regulations to encourage local detention or retention storage of flood water;
- Flood warning system;
- Real time forecasting of floods;
- Real time operation on runoff to reduce flood peak.

Urban water disasters caused by river floods or coastal storm surges and tsunamis should be mitigated through nonstructural measures such as land-use zoning, flood insurance, real-time forecasting and warning in addition to structural measures of levees, floodwalls and pumps. in the United States the most commonly used design and construction standards for levees and seawalls are those of the Army Corps of Engineers.

Since rain produced flood is the most frequent water disaster in urban areas, infrastructures such as roadside gutters, sewers, and pumping stations have been built to mitigate the rain induced flood hazards. In other words, traditionally structural approach is the main approach to deal with urban flood hazards by draining the excess water away from urban area as fast and as much as possible and feasible. The structure is designed and built to protect the area up to a threshold level of rainfall or flood, and it is assumed that rains or floods below the threshold level will be handled adequately without flooding. The basic design procedure is first to determine the design rainfall; second, to transform the design rainfall to the design flood discharge; third, compute the required size of the structure; and fourth decide on the actual size used based on commercial availability, economic and other considerations.

However, modern urban stormwater hazards go beyond flooding. Urban nonpoint pollution hazards have caught considerable attention in recent years. Besides, many urban floods are caused by inadequate sewer capacities to handle increasing flow due to additional urbanization beyond the original design condition; and feasible remedies often involve local storage. For these two kinds of stormwater management problems the runoff information needed includes not only the peak discharge as in the traditional case but also the runoff volume and its time distribution. Technically the basic procedure is first to determine the project rainfall; second, to transform the rainfall to runoff; and third, develop the management or operational strategy on the stormwater.

PREDICTION VS. FORECASTING IN URBAN STORMWATER MANAGEMENT

In the technical approach to manage urban stormwater and mitigate flood disasters, two types of problems of different time scales are involved. They are:

- Long-term prediction for planning and design;
- Short-term forecasting for real-time operation and warning.

The first type of prediction can be applied to both structural and nonstructural measures. The prediction is usually based on statistical analysis of past rainfall or flood records and the results are used for design of sewers, ponds and other structures, or for planning of future management strategy, including budget planning for the future. It predicts future occurrence probability but not the time or sequence of occurrences. The return period in the intensity-duration-frequency (IDF) relationship is a typical example. Example of more detailed rainfall-occurrence-probability-prediction technique that can be used in urban runoff control is described by Chow and Yen (1976).

The second type of flood forecasting is mostly for nonstructural measures of stormwater management. It usually forecasts in the near future in a relatively short time scale the magnitude as well as the time of occurrence of the flood to be used for real-time operation. The forecasting is most likely done through simulation or time series analysis.

Unlike the first type of prediction for which the methodologies have been developed for nearly a century, the second type of real-time forecasting is relatively new and new techniques are still being developed. For example, use of remote sensing such as radar and GIS for real-time stormwater control is still in experimental stage. The rainfall data requirement for the second type is different from that for the first time. In fact, even for the first type of runoff prediction, the rain data requirement and analysis are considerably more complicated for planning and design involving the volume of runoff than traditional design of sewers requiring only the peak discharge of runoff.

MODELING RAINFALL-RUNOFF RELATION FOR URBAN FLOODS

The objective of modeling runoff from rainfall for urban flood mitigation is to determine the hydrograph (i.e. discharge as a function of time), Q(t), or its characteristic values such as peak discharge at the outlet of the catchment. As discussed in Yen (1986b) the runoff can be expressed symbolically and nondimensionally as a function of the catchment characteristic, rain input, and fluid property parameters, π_w, π_r and π_f, respectively,

$$\frac{Q(t)}{iL^2} = F_1\left(\frac{t}{t_d}, \pi_w, \pi_r, \pi_f\right) \tag{1}$$

in which
F = a function;
i = a characteristic rain intensity, e.g.,average intensity;
t_d = a characteristic rain time, e.g.,duration;
L = a characteristic length of catchment, e.g., $\sqrt{A}$;
A = catchment area.

The catchment parameter π_w in terms is a function of the characteristics of the catchment and sewer network,

$$\pi_W = F_2\left(\frac{L_X}{L}, \frac{L_Z}{L}, n_s, n_d, S_x, S_z, n_k\right) \tag{2}$$

where

L_x = catchment length;
L_z = catchment width;
S_x,S_z = representative catchment slope components;
n_s = a symbolic dimensionless parameter representing catchment shape, in addition to L_x/L and L_z/L, e.g.,it could be a roundness factor, circumference $/\sqrt{A}$, L_c/L where L_c is the distance of catchment centroid from outlet, or any combination of them;
n_d = a symbolic dimensionless parameter representing the pattern of catchment drainage or component;
n_k = a symbolic dimensionless parameter representing catchment land use which affects both surface roughness and infiltrability.

Similarly, the rain input parameter can be represented dimensionlessly and symbolically as

$$\pi_r = F_3\left(\frac{t_d i}{L}, r_t, r_s, r_m, \Phi\right) \tag{3}$$

where

r_t = a symbolic dimensionless parameter representing the temporal distribution of rain;
r_s = a symbolic dimensionless parameter representing the spatial distribution of rain over the catchment;
r_m = a symbolic dimensionless parameter representing the rainstorm movement;
Φ = a dimensionless parameter representing the antecedent soil moisture or initial losses.

The fluid properties are the viscosity, density and surface tension. Nondimensionally they can be replaced by characteristic catchment flow Reynolds number, $\mathbb{R}$, Froude number $\mathbb{F}$, and Weber number $\mathbf{W}$, i.e.,

$$\pi_f = F_4\,(\mathbb{R},\ \mathbb{F},\ \mathbf{W}) \tag{4}$$

Existing rainfall-runoff models applicable to urban storm flood mitigation can be classified into four groups: (i) the lumped-system models, (ii) the hydrologic system conceptual models, (iii) the physical-process based distributed system models, and (iv) time series models.

Lumped-System Models

In the lumped-system models, the catchment is treated as a black box. Spatial variations of the catchment and of the input are ignored and the flow process inside the catchment is not considered. Thus F_2 in Eq. 2 is reduced to a coefficient depending only on some gross property of the catchment such as its land use and mean slope; whereas r_t, r_s, and r_m, in Eq. 3 are all ignored; and π_f (Eq. 4) is also ignored. The coefficient values are mostly determined from regression of observed past data. In the lumped-system group the models can further be subclassified into those giving only the peak discharge Q_p for storm events and those giving the entire hydrograph, Eq. 1. Models of the first subgroup in the form of equations or graphs abound. Among them the best known and most widely used is the rational formula. These formulas and graphs were developed for design of sewers, culverts or channels requiring only a design discharge but not the entire hydrograph. In the second subgroup the best known model is the unit hydrograph. Another example is the SCS (1975) method which is essentially the rational method applied to time-varying rainfall to produce runoff hydrographs. Descriptions of a number of the lumped-system models can be found in Chow (1964).

A third subgroup of the lumped-system models is the correlation models that correlate the flood stage or discharge at the location of interest to the same at an upstream location. This is a commonly used method for river flood forecasting. This method works satisfactorily if there is sufficient data to establish the correlation, if the lateral flow from tributaries and other sources or losses are relatively insignificant, if there is no significant storage or flow control (such as gates and locks) between the locations and if the flood-response characteristics of the river do not change with time, i.e., a stationary system.

Conceptual Models

In the conceptual hydrologic system models, instead of expressing π_w and π_f in Eq. 1 in terms of the real physical parameters they are replaced by hypothetical parameters. A typical example of conceptual models in rainfall-runoff process is the Nash (1957) linear reservoir model. In Nash's model the catchment, irrespective of its geometric properties, is hypothetically regarded as behaving like a series of identical reservoirs in sequence. At any time the outflow from the reservoir is linearly proportionally to the water stored in the reservoir. The final output from this series of reservoirs is the catchment outflow. In a similar conceptual way, a linear channel is simply a hypothetical stretch that delays the input by a constant length of time without changing its magnitude (Dooge 1959). Many linear and nonlinear reservoir and channel models have been proposed. Recently tanks, orifices, weirs, and even a pipe (Jensen, 1984) have also been proposed to hypothetically simulate the catchment. The best known conceptual model application to urban catchments is the Australian nonlinear-reservoir RORB model (Laurenson and Mein, 1985). Common to all these conceptual models are two basic equations, the continuity equation relating at any time t the system input I(t) to the outflow, Q(t), and the storage in the system, s(t),

$$I - Q = \frac{ds}{dt} \tag{5}$$

and the storage equation at any time t

$$s = F_5 (I,Q) \tag{6}$$

in which F_5 is a function.

Different assumptions on F_5 representing the catchment classify the different conceptual models. A relative general form of the storage equation is (Chow, 1972)

$$s = \sum_i a_i \frac{d^i I}{dt^i} + \sum_j b_j \frac{d^j Q}{dt^j} \tag{7}$$

in which the coefficients a_i and b_j are functions of both I and Q. Those interested in conceptual models may refer to, e.g., Chow (1972) and Yen (1989).

Physical-Process Based Models

Physically based models simulate the catchment transformation process on the basis of the physical process, i.e., considering π_w and π_f in Eq. 2 and 4 to some degree of reality. This approach requires the consideration of the processes in the elements or components and their relative distribution within the catchment.

In formulating a physically based model, the following phases are considered after the rainfall input has been determined.

(a) Decomposition of the catchment. In this phase one determines the level of detail of subcatchments and components such as the pervious and impervious surfaces, street gutters and inlet catch basins to be used in the model. Figure 1 shows schematically some components of a urban storm drainage system. The less the details, the simpler the model, but the greater the loss of physical reality.

(b) Methods for rainwater abstractions. In this phase the methods to calculate the losses of rainwater due to interception, depression storage and infiltration are selected. One should consider if the abstraction values are assigned catchment wide or if they are allowed to

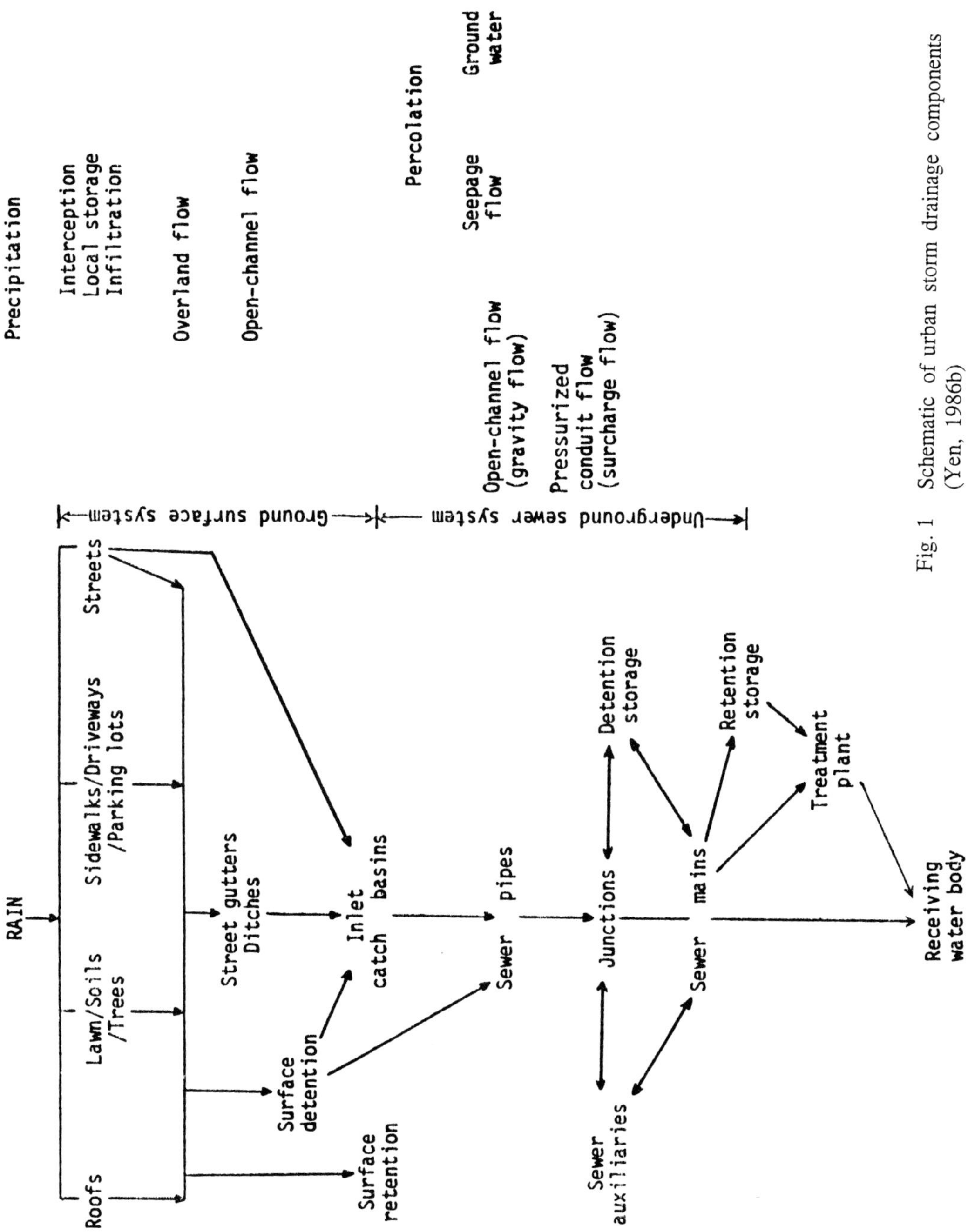

Fig. 1 Schematic of urban storm drainage components (Yen, 1986b)

vary for different locations and different types of surfaces. The latter is more physically satisfactory but it requires more input information.

(c) Runoff from subcatchments. In this phase the methods of transforming rain excess water to runoff and the routing of runoff on the surfaces, sewers and channels are selected in accordance with the level of decomposition of the catchment. One major difference between the physically based models and the conceptual models is the use of the equations in this phase. In the physically based models for routing runoff in a catchment, in addition to the continuity equation, a flow momentum equation is used. The continuity equation can be based on the controlled reach form, Eq. 5, or in the cross-sectional averaged form, as follows

$$\frac{\partial A}{\partial t} + \frac{\partial Q}{\partial x} = \int q_1 \, d\sigma \tag{8}$$

in which

t = time;
x = flow longitudinal direction;
a = flow cross-sectional area normal to x;
Q = discharge through the cross section A;
σ = perimeter bounding the cross section A;
q_1 = lateral flow rate per unit length of perimeter σ along unit length of x, being negative for outflow.

The momentum equation can be the full dynamic wave equation or any of its simplifications,

$$\frac{1}{gA}\frac{\partial Q}{\partial t} + \frac{1}{gA}\frac{\partial}{\partial x}\left[\frac{Q}{A}\right]^2 + \frac{1}{gA}\int_\sigma U_x q_1 d\sigma + \frac{\partial Y}{\partial x} - S_o + S_f = 0 \tag{9}$$

Dynamic wave

Noninertia

Kinematic wave

in which

Y = depth of flow of the cross section, measured vertically;
S_o = channel or overland slope;
S_f = friction slope;
V = Q/A, cross-sectional average velocity along x direction;
g = gravitational acceleration;
U_x = x-component velocity of q_1.

It is not necessary to use the same routing method for the different types of surfaces in a catchment. For example, for an aggregated pervious surface, the time-area method with an assumed velocity usually suffice. Conversely, for a system of street pavement, gutter and inlet catch basin, a kinematic wave routing may not be sufficient because of the mutual backwater effects.

(d) Routing of flow in sewer and channel system. The outflow from a subcatchment enters the inlet catch basin that connects the subcatchment to the sewer or channel system. The inlet hydrographs of the subcatchments serve as the input to the sewer systems. The routing is computed using the same continuity and momentum equations given in Eqs. 8 and 9. Those interested in details may refer to Yen (1986a).

Time-Series Simulation Models

Time series stochastic simulations can be used for forecasting as well as prediction. The catchment or river discharge at a given moment is determined from the time series established based on past runoff events. They are seldom used for urban catchment flood mitigation because usually there is a lack of adequate runoff data for such simulation. They are more suitable for urban river flooding where the required data are more likely available. A number of stochastic process simulation techniques have been suggested for flood forecasting problems (Yevjevich, 1972; Salas et al., 1980; Bras and Rodriguez-Iturbe, 1984). Nevertheless, application of time-series simulation for urban flood mitigation is still at an initial research stage.

Urban Runoff Routing Models

There exist many distributed simulation models that can be used in urban flood mitigation. They route the rainwater through overland surface and sewer and channel networks. All of them use a higher hydraulic level of routing for the sewer/channel network than for the overland flow The better known models that use dynamic-wave-sewer routing include the British model SPIDA, Danish model MOUSE, French model CAREDAS, as well as U.S. models UNSTDY, IUSR and SWMM-Extran. There are only a few noninertia models. Contrarily, there are numerous nonlinear-kinematic-wave and lower-level models. Details of most of the models are given in the user's manual of the models. Summary description of selected models can be found, e.g., in Chow and Yen (1976), Yen (1986a, b) and Bedient and Huber (1993).

INADEQUACY OF RAINSTORM DESIGN RETURN PERIOD IN URBAN FLOOD MITIGATION

Traditionally, urban storm flood mitigation structures are designed based on the rainstorm of certain occurrence frequency, i.e., return period of the design rainfall, rather than the design flood. The implicit logic behind this approach is that rainfall is practically unaffected by urban development whereas runoff changes with land condition and degree of urbanization.

In designing urban flood mitigation infrastructures there are two different types of failure. The first type is the structural failure which involves damage or change of the structure, hence hindering the ability of the structure to function as originally desired. Examples are failure of a pump, silting of a pond, collapse of a sewer, scour of a culvert, and damage of a treatment plant. The second type is the performance failure wherein the structure is not damaged but the performance threshold is exceeded and undesirable consequences occurred. Examples include flooding of streets due to inadequate sewer capacity, litter blocking of catch-basin inlets, and excessive combined sewer overflow into receiving water body. Yen and Ang (1971) called the first type of failure a catastrophic type failure and the second type of performance failure a property damage type failure.

Traditionally, urban sewers are designed with rainstorm return period of 2, 3 or 5 years because they belong to the performance type failure; whereas pumping stations are designed using rainstorm return period of 50 or 100 years because they belong to structural type failure. This design-storm return period approach is compatible with the concept of intensity-duration-frequency (IDF) relationship for rainfall prediction. The IDF relation is derived from the envelope curves covering the maximum rainfalls for given durations and frequencies. It is regarded as conservative for design of structures against possible future rainfalls of specified frequency. However, the design rainstorm from the IDF relationship is hypothetical and it does not represent any one of the actual storms in the future. It provides no information on the temporal variation of the rainfall within the specified rain duration. The design-storm return periods appeared in many city drainage codes were mostly suggested by experienced lead engineers in the past who had observed that such a design level provided a reasonable balance between investment and the protection from flood damage losses in the past. In recent years, the design return period for sewers has been rising worldwide due to increasing damage costs of floods, especially for developed countries.

In storm drainage it is the runoff, not the rainfall, that is to be managed. If the stochastic temporary characteristics of the storm runoff, such as its probability distribution, return period and hydrograph are known, there will be little interest in the rainfall. However, information on storm runoff is far less readily available than rainfall. Furthermore, even in the rare case that observed runoff data is available at specific points of a catchment, the data may not be directly applicable for the same catchment at different times because the observed record reflects a watershed of changing past conditions which are different from those in the future having different degrees of urbanization. Thus, the result of a frequency analysis of the observed runoff data of a catchment is inapplicable to the future of different land use conditions.

Contrarily, rainfall may be regarded as practically unaffected by urbanization. Therefore, it is traditional in storm drainage design to consider rainfall frequency as a reference scale of the level of protection. In doing so, it is implicitly assumed that the abstractions from rainfall are similar or relatively negligible for different rainstorms. This assumption is an acceptable approximation for heavy rainstorms having multi-year return periods, but not for the frequent small rainstorms now also considered in stormwater pollution control. Obviously, this implicit assumption of rainfall return period being the same as peak runoff return period is closer to reality for heavy rain storms of large return periods than for much lighter, frequent rainstorms.

However, urban drainage using only the design-rainstorm return period is totally inadequate and outdated. First, urban storm drainage problems have expanded beyond the traditional structural approach of design and build to drain the water to include also nonstructural measures of stormwater management for pollution control as well as flood mitigation. Particularly, real-time flood routing and control is evolving as a practical means in urban stormwater operation and management. Second, design and operation of urban drainage facilities is becoming more sophisticated and often involves a significant amount of water storage in the drainage network, hence requiring information on the temporal and spatial distributions of the rainfall in addition to the IDF relationship to determine the hydrograph. Third, today the public often desire to know the risk they are exposed to. The rainstorm return period constitutes only a portion of the risk.

Incidently, in the frequency analysis to determine the rainfall magnitude-return period relationship different types of probability distributions have been proposed to fit the rainfall data. However, any attempt to find the "true" distribution is futile because all the distributions are human-imposed. Rather, the question should be: Which distribution is most representative to the set of data and not contradicting to the physical reality? It should also be noted that for a limited data sample the best-fit distribution is not necessarily the best distribution for prediction.

RISKS IN URBAN DRAINAGE DESIGN

Urban drainage, like all engineering projects, is always subject to uncertainties in its planning, design and operation. Therefore the best design protection level should be an optimal trade-off between the costs of construction, operation and maintenance and the damage costs. There are numerous sources of uncertainties. The major ones are:

- Natural randomness of rainfall, soil condition and other geophysical events;
- Modeling uncertainty;
- Parameter uncertainties;
- Data uncertainties due to measurement errors, inadequate data, and error in data handling and transcription;
- Operational uncertainties due to construction and manufacture deviations, deterioration and maintenance deficiency, and procedural errors.

Flooding due to inadequate sewer or channel capacity belongs to the failure of performance type rather than structural failure. Hence, there are no "right" or "wrong" design return periods for the sewers or channels. A large sewer provides a high level of protection and hence low average annual damage cost while its construction cost is high. On the other hand, a small sewer designed with a low return period is less costly to construct but has a comparatively high average annual damage cost. Table 1 is an example taken from Yen and Jun (1984) for the design of the 28-acre Fourth Street catchment in Champaign, Illinois.

The flooding probability, or risk, given in Table 1 accounts for not only the natural randomness (return period) of the rainfall but also the other uncertainties including those on the surface conditions related to conversion of rainfall to runoff, those on modeling reliability, those on data adequacy and accuracy, and those on the deviation of construction from the ideal design conditions: the slope may be off; the size, shape and material of the pipes may not be exact; the pipe roughness changes with time due to sliming; the pipes may deteriorate or settle. There exist several methods to combine the uncertainties of the various factors to yield the total risk (Ang and Tang, 1984; Yen and Tung, 1993). A first-order second-moment analysis (Yen et al., 1986) is employed to compute the values listed in Table 1.

In Table 1 the results were obtained using the Illinois Least-Cost Sewer System Design Models ILSD-1&2 (Yen et al., 1984). Parts (a) and (b) are ILSD-1 least-cost optimal designs considering only the construction (material, excavation and installation labor) costs using a design rainfall return period of 2 years and 25 years, respectively. No risk-damage cost is considered in the optimization. Expectedly, the design using 25-year return period yields a system with bigger sewer pipes and lower risks than the 2-year return period design. The construction costs are $63,850 and $48,210, respectively, for the 25-year and 2-year designs.

Part (c) of Table 1 is the result of the ILSD-2 risk-based least cost design for a service period of 25 years. In addition to the costs considered in ILSD-1 designs, the risk-damage costs of insufficient capacity of each of the sewers are also included in the optimization and the risks are allowed to vary without constraint for all the sewers. The result is a larger system than cases (a) and (b) with a construction cost of $92,818, risk-damage cost of $19,065 and a total cost of

Table 1. Risks of sewers for example designs

Sewer	Sewer diam (in.)	Sewer slope (%)	Sewer capacity (cfs)	Risk	
				In 25 years	In one year
(a) Design for rainfall of 2-year return period					
11	21	0.454	9.91	0.45	0.10
21	24	0.696	17.53	0.65	0.17
22	15	0.825	5.45	0.20	0.04
23	12	1.160	3.56	0.41	0.09
31	24	1.319	24.13	0.67	0.18
32	21	0.356	8.77	0.66	0.18
33	15	1.778	8.00	0.40	0.09
34	18	0.452	6.56	0.65	0.18
41	27	3.235	51.72	0.63	0.16
42	21	2.295	22.29	0.76	0.22
51	36	2.417	96.28	0.62	0.16
(b) Design for rainfall of 25-year return period					
11	27	0.338	16.73	0.031	0.005
21	27	1.013	28.94	0.092	0.014
22	15	0.947	5.84	0.145	0.026
23	15	0.878	5.62	0.042	0.006
31	27	1.979	40.46	0.093	0.015
32	24	0.356	12.53	0.202	0.037
33	18	1.778	13.01	0.030	0.005
34	18	1.088	10.17	0.134	0.023
41	36	1.907	85.54	0.083	0.013
42	27	1.620	36.60	0.145	0.023
51	48	1.389	157.19	0.089	0.014
(c) Risk-damage-cost-included ILSD-2 optimal design for 25-year service period					
11	36	0.338	36.0	All less than 0.00012 (sewers 11–42)	All less than 0.00004 (sewers 11–42)
21	36	0.934	59.8		
22	21	0.825	13.4		
23	21	0.935	14.2		
31	36	1.979	87.1		
32	30	0.550	28.3		
33	21	2.264	22.1		
34	24	1.111	22.1		
41	48	1.969	187.2		
42	36	1.420	73.8		
51	54	1.694	237.7	0.002	0.0004

$111,883. Despite their lower construction costs, cases (a) and (b) have a higher total cost than case (c) because of their high risk-damage costs.

The risk-based optimal design of case (c) offers not only a lower total cost than the design without considering the risk cost, it also gives lower risks for the service of the sewers as shown in the table, thus, a better design. If there is any constraint on the maximum acceptable risk level for any sewer, this constraint can easily be accounted for in the risk-based optimal design. Conversely, for the non-risk based models, special treatment must be made to indirectly account for the risk level constraint.

In all three cases the risk, and hence the return period, is different for different sewers in a design, and different from the original design return period. In fact, from the viewpoint of risk-based design and decision, the return period is no more than an intermediate parameter.

CHANNEL CAPACITY VS. FLOOD TO BE DRAINED

Urban flood problems actually consists of two different problems; namely, the hydraulic problem of the flow capacity of the sewer or channel, and the hydrology problem of how much water to be drained. The water to be drained need not be from rainfall; it can be from failure of structures, or, as an extreme example, from fire hydrants. The water to be drained can be viewed as the demand to be estimated hydrologically, whereas the sewer/channel capacity can be regarded as the supply to be determined hydraulically. A comparison of the sewer/channel capacity with the peak discharge of the flood to be drained determines if there is a flooding problem. Unfortunately, in urban as well as rural drainage analysis these two different problems have not been dealt separately. One reason is that on the supply side there exists no method to determine satisfactorily the sewer/channel capacity other than the just-full, steady, uniform flow value. In a sewer/channel network there is mutual backwater effect, and hence the capacity of each sewer/channel should be determined by considering the interacting sewers/channels in the network as a system. In unsteady-flow routing the peak discharge is a function of the inflow hydrograph and there is no single unique value of the sewer/channel capacity.

Conversely, on the demand side there are a number of hydrologic methods that could be used to estimate the amount of water to be drained, either as a peak discharge or as an entire hydrograph. Some of these methods are presented in Chow (1964), Bedient and Huber (1993) and Yen (1986b, 1989).

Recently, a hydraulic-performance-graph method has been developed that could be used to estimate the capacity of sewers/channels under backwater conditions (Yen and Gonzalez, 1994). The hydraulic performance graph is essentially a summary of all the backwater profile information for a channel. A typical graph is shown nondimensionally in Fig. 2 for example sewers B10, B20 and B30 in Sevuk and Yen (1982). In the figure Q_f is the steady, uniform, full-pipe discharge; S_o, L and D are the slope, length and diameter of the sewer pipe; the subscripts ds and us of depth Y denote the depths at the downstream and upstream ends of the sewer pipe, respectively; and Y_n is the normal flow depth for the specified discharge Q. The N-line represents normal flow condition. The C curves denote critical flow condition at the downstream end of the sewer. Using this graph the sewer/channel capacity under different backwater conditions can be estimated easily requiring no complicated routing or backwater profile computations. The graph can also be used for direct identification of the flooding bottleneck locations in a sewer/channel system. Thus, it offers a new approach for relatively simple and

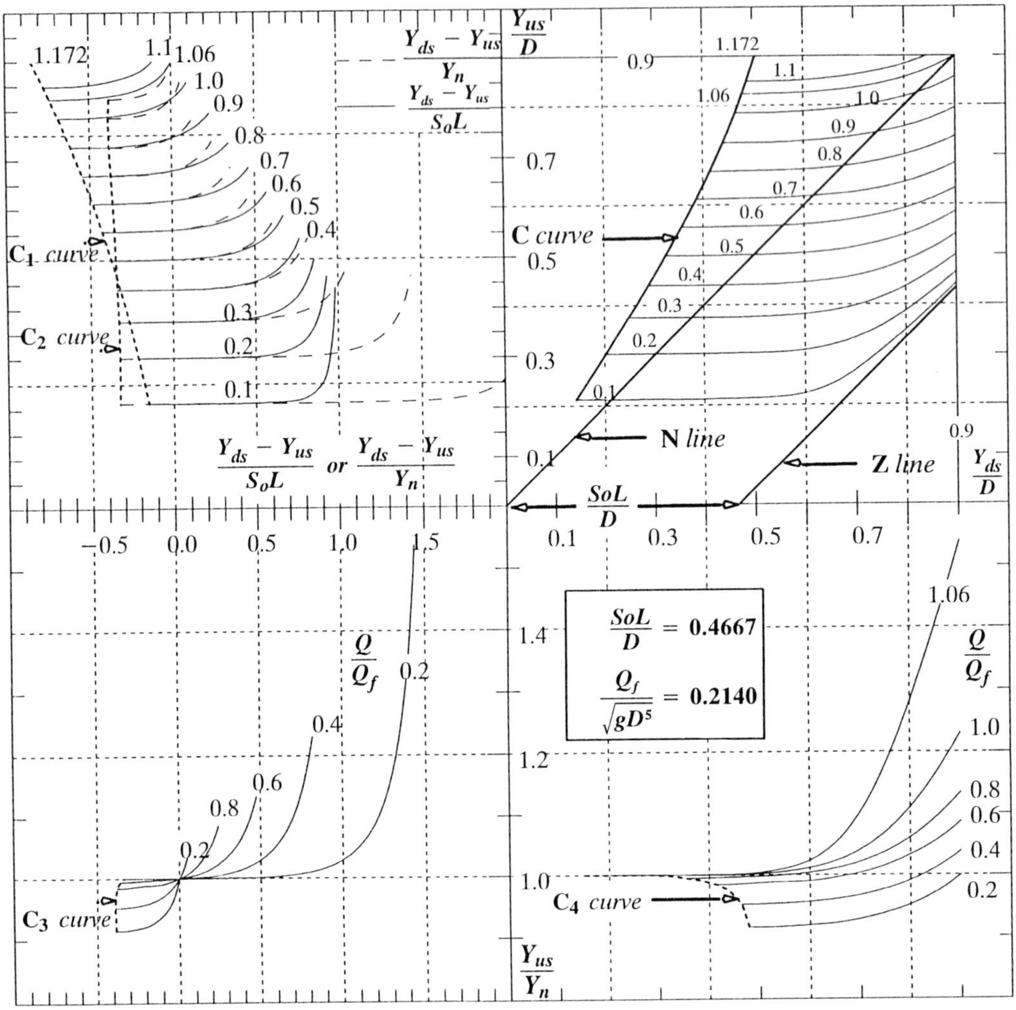

Fig. 2 Example nondimensional hydraulic performance graph for sewers.

efficient analysis of flood mitigation separating the hydrologic "demand" determination from the hydraulic "supply"determination.

REFERENCES

Ang, A.H.-S.and W.H. Tang (1984). *Probability Concepts in Engineering Planning and Design, Vol II: Decision, Risk, and Reliability.* John Wiley & Sons, Inc., New York.

Bedient, P.B. and W.C. Huber (1993). *Hydrology and Floodplain Analysis.* 2ed., Addison-Wesley Publishing Col, Reading, Massachusetts.

Bras, R. and I. Rodriguez-Iturbe (1984). *Random Functions and Hydrology.* Addison-Wesley Publ. Co., Reading, Massachusetts.

Chow, V.T. (ed.) (1964). *Handbook of Applied Hydrology.* McGraw-Hill Book Co., New York.

Chow, V.T. (1972). Hydrologic modeling (a Freeman lecture). *J. Boston Soc. Civ. Engrs.,* 60(5), 1-27.

Chow, V.T. and B.C. Yen (1976). Urban stormwater runoff: Determination of volumes and flowrates. *Environmental Protection Technology Series, EPA-600/2-76-116,* U.S. EPA, Cincinnati, Ohio.

Dooge, J.C.I. (1959). A general theory of the unit hydrograph. *J. Geophys. Res.* AGU, 64 (1), 241-256.

Jensen, M. (1984). A method for simplified urban catchment description. *Proc. 3rd IAHR/IAWQ Intern. Conf. Urban Storm Drainage,* Vol. 1, pp. 213-221. Chalmers Univ. of Tech., Gothenburg, Sweden.

Laurenson, E.M. and R.G. Mein (1985). RORB-Version 3 runoff routing program user manual. *Report,* Dept. of Civ. Eng., Monash Univ., Clayton, Vic. Australia.

Nash, J.E. (1957). The form of instantaneous unit hydrograph. *Pub. No. 45,* Vol. 3, pp. 114-121, Intern. Assoc. Hydrol. Sci., Paris.

Salas-La Cruz, J.D., J.W. Delleur, V. Yevjevich and W.L. Lane (1980). *Applied Modeling of Hydrologic Time Series.* Water Resources Publications, Highlands Ranch, Colorado.

Sevuk, A.S. and B.C. Yen (1982). Sewer network routing by dynamic wave characteristics. *J. Hydraulic Div.,* ASCE, 108 (HY3), 379-398.

Soil Conservation Service (1975). Urban hydrology for small watersheds. *Tech. Release 55.* U.S. Dept. of Agriculture, Washington, D.C.

Yen, B.C. (1986a). Hydraulics of sewers. In: *Advances in Hydroscience Vol. 14* (B.C. Yen, ed.), pp. 1-122. Academic Press, Orlando, Florida.

Yen, B.C. (1986b). Rainfall-runoff processes on urban catchments and its modeling. In: *Urban Drainage Modeling* (C.Maksimovic and M. Radojkovic, ed), pp. 3-26. Pergamon Press, Oxford.

Yen, B.C. (1987). Urban drainage hydraulics and hydrology: From art to science. In: *Urban Drainage Hydraulics and Hydrology* (B.C. Yen, ed.) pp. 1-24. Water Resources Publications, Highlands Ranch, Colorado.

Yen, B.C. (1989). Flood forecasting and its reliability. In: *Taming the Yellow River: Silt and Floods* (L.M. Brush et al., ed), pp. 163-196, Kluwer Academic Publishers, Dordrecht, The Netherlands.

Yen, B.C. and A. H.-S. Ang (1971). Risk analysis in design of hydraulic projects. In: *Stochastic Hydraulics,* Proc. of the 1st Internat. Symp. (C.L. Chiu, ed.), pp. 694-701. University of Pittsburgh.

Yen, B.C. and J.A. Gonzalez (1994). Determination of Boneyard Creek flow capacity by hydraulic performance graph. *Research Report.* Water Resources Center, Univ. of Illinois at Urbana-Champaign.

Yen, B.C. and B.-H. Jun (1984). Risk consideration in design of storm drains. *Proc. 3rd IAHR/IAWQ Intern. Conf. Urban Storm Drainage,* Vol. 2, pp. 695-704. Chalmers Univ. of Tech., Gothenburg, Sweden.

Yen, B.C. and Y.K. Tung (eds.) (1993). *Reliability and Uncertainty Analyses in Hydraulic Design.* Am. Soc. Civ. Engrs, New York.

Yen, B.C., S.-T. Cheng and C.S. Melching (1986). First order reliability analysis. In: *Stochastic and Risk Analysis in Hydraulic Engineering* (B.C. Yen, ed.), pp. 1-36, Water Resources Publications, Highlands Ranch, Colorado.

Yen, B.C., S.-T. Cheng, B.-H. Jun, M.L. Voorhees, H.G. Wenzel, Jr. and L.W. Mays (1984). Illinois least-cost sewer system design model: ILSD-1 & 2 user's guide. *Research Report* 188, Water Resources Center, Univ. of Illinois at Urbana-Champaign.

Yevjevich, V. (1972). *Stochastic Processes in Hydrology.* Water Resources Publications, Highlands Ranch, Colorado.

INUNDATION MODELS FOR PA-CHANG CREEK BASIN IN TAIWAN

M. -H. Hsu, F. -C. Wu, W. -H. Teng and
S. -H. Lin

Department of Agricultural Engineering,
National Taiwan University, Taipei, Taiwan

ABSTRACT

Two inundation models are developed in this study, namely, two-dimensional finite difference model and cell inundation model. The simulated results by 2-D finite difference model are in good accuracy, however, this model is not satisfactory for urgent forecasting as used in flood warning system due to its long simulation time. The cell model is developed to simulate the inundation within the limited pre-forecasting time. The many parameters of the cell model are calibrated in accordance with the simulated results of the 2-D finite difference model, and two case studies are performed for verification. The cell model is then applied to the simulations of inundation in Pa-chang Creek Basin, Taiwan. The simulated results reveal that cell model is both efficient and accurate. The computational time needed by cell model is approximately 1/26 of that needed by 2-D finite difference model.

KEYWORDS

Inundation model, finite difference model, cell model, flood warning system, digital terrain model (DTM), geographic information system (GIS).

INTRODUCTION

Pa-chang Creek, located in the southern part of middle Taiwan, discharges into Taiwan Straits with total length of 81 kilometer and drainage basin of 475 square kilometer (see Fig. 1). The slope of the upstream reach is very steep, while the midstream and downstream drainage basins are plains with mild slope, as shown in Fig. 2. The bountiful precipitation in upstream area, and the mild slope as well as the river meandering in the midstream and downstream reaches are the main reasons that caused severe inundation in the past. Therefore, there is a very urgent demand of an inundation forecasting system for damage mitigation. In this forecasting system, the runoff model and inundation model suiting for the Pa-chang Creek Basin have to be developed for estimating flood discharge, inundation area and depth. This study concerns numerical simulations of the inundation in Pa-chang Creek Basin, and also intends to find out the inundation area for the rainfall of various return periods. Two inundation models have been developed in present study. They are two-dimensional finite difference model and cell inundation model.

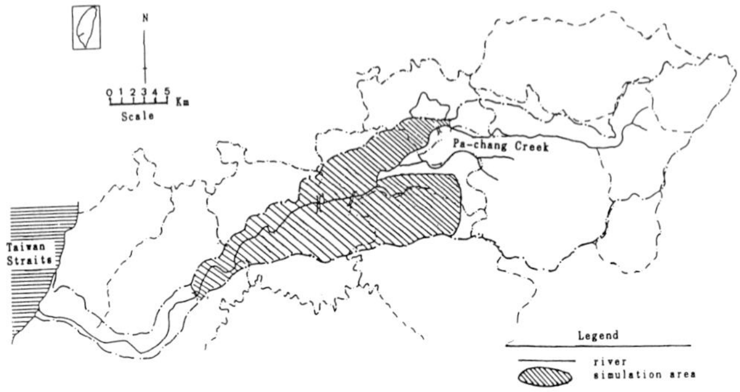

Fig. 1. Schematic location map of mid and downstream Pa-chang Creek simulation area

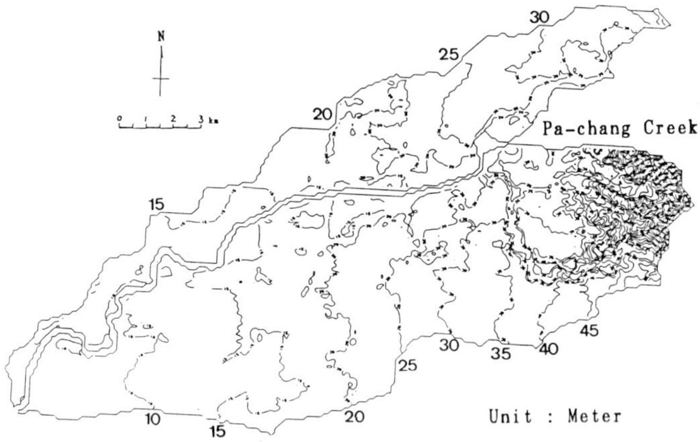

Fig. 2. Digital terrain map of mid and downstream Pa-chang Creek simulation area

Hromadka et al. (1989) applied the two-dimensional diffusion hydrodynamic model to the evaluation of floodplain depths resulting from an overflow of a leveed river. Xanthopoulos and Koutitas (1976) investigated numerically the propagation of a flood wave on a two dimensional dry plain. Vongvisessomjai et al. (1985) developed a cell model in computing flood water stage and discharge of the Bangkok flood plain. Yen and Hsu (1984), Hsu et al. (1990), Yen et al. (1990), Hsu (1992), Hsu and Teng (1993), and Hsu et al. (1994) have been working on simulation of inundation and flood routing over the past decade and obtained considerable results on these issues.

NUMERICAL MODELS

Two-Dimensional Finite Difference Model

On the assumption that accelerations of water flow on the land surface are small compared to gravity and friction, the inertial terms in motion equations are neglected. Then, the depth-average shallow water equations on land surface can be written as:

$$\frac{\partial d}{\partial t}+\frac{\partial (du)}{\partial x}+\frac{\partial (dv)}{\partial y}=q \tag{1}$$

$$-\frac{\partial h}{\partial x}=S_{fx}+\frac{qu}{dg} \tag{2}$$

$$-\frac{\partial h}{\partial y}=S_{fy}+\frac{qv}{dg} \tag{3}$$

in which d= depth of flow; h= water stage; u, v= velocity component in x- and y-direction respectively; t = time; g= gravitational acceleration; q= effective precipitation per unit area; $S_{fx}=nu|u|/d^{4/3}$, friction slope in x-direction; $S_{fy}=nv|v|/d^{4/3}$, friction slope in y-direction; n= Manning's roughness. In the above equations, d, u, and v are three dependent variables to be solved. For the present study, a two-step alternating direction explicit scheme (ADE) has been employed. Three finite difference equations derived from Eqs. (1) to (3) in each time step are as follows:

(1) For the first time step (m + 1/2):

$$d_{i,j}^{m+1/2}=d_{i,j}^{m}-\frac{1}{2}\Delta t\left\{\frac{\left[\left(d^m u^{m+1/2}\right)_{i+1/2,j}-\left(d^m u^{m+1/2}\right)_{i-1/2,j}\right]}{\Delta x}+\frac{\left[(dv)_{i,j+1/2}^{m}-(dv)_{i,j-1/2}^{m}\right]}{\Delta y}-q_{i,j}^{m+1/2}\right\} \tag{4}$$

$$\frac{\left[(d+z)_{i,j}^{m+1/2}-(d+z)_{i+1,j}^{m+1/2}\right]}{\Delta x}=\left\{u^{m+1/2}\left[\frac{(n_x)^2|u|^{m+1/2}}{(d^m)^{4/3}}+\frac{q^{m+1/2}}{\left(d^m g\right)}\right]\right\}_{i+1/2,j} \tag{5}$$

$$\frac{\left[(d+z)_{i,j}^{m+1/2}-(d+z)_{i,j+1}^{m+1/2}\right]}{\Delta y}=\left\{v^{m+1/2}\left[\frac{(n_y)^2|v|^{m+1/2}}{(d^{m+1/2})^{4/3}}+\frac{q^{m+1/2}}{\left(d^{m+1/2}g\right)}\right]\right\}_{i,j+1/2} \tag{6}$$

where m is time index; i and j are x- and y-direction spatial indices; $\Delta t=t^{m+1}-t^{m}$; $\Delta x=x_{i+1}-x_i$; $\Delta y=y_{j+1}-y_j$. Using Eqs. (4) and (5), $u^{m+1/2}$ and $d^{m+1/2}$ along x-direction can be solved simultaneously. Then, $v^{m+1/2}$ can be found from Eq. (6).

(2) For the second time step (m+1):

$$d_{i,j}^{m+1}=d_{i,j}^{m+1/2}-\frac{1}{2}\Delta t\left\{\frac{\left[(du)_{i+1/2,j}^{m+1/2}-(du)_{i-1/2,j}^{m+1/2}\right]}{\Delta x}+\frac{\left[\left(d^{m+1/2}v^{m+1}\right)_{i,j+1/2}-\left(d^{m+1/2}v^{m+1}\right)_{i,j-1/2}\right]}{\Delta y}-q_{i,j}^{m+1}\right\} \tag{7}$$

$$\frac{\left[(d+z)_{i,j}^{m+1}-(d+z)_{i,j+1}^{m+1}\right]}{\Delta y}=\left\{v^{m+1}\left[\frac{(n_y)^2\left|v^{m+1}\right|}{(d^{m+1/2})^{4/3}}+\frac{q^{m+1}}{\left(d^{m+1/2}g\right)}\right]\right\}_{i,j+1/2} \tag{8}$$

$$\frac{\left[(d+z)_{i,j}^{m+1}-(d+z)_{i+1,j}^{m+1}\right]}{\Delta x}=\left\{u^{m+1}\left[\frac{(n_x)^2\left|u^{m+1}\right|}{(d^{m+1})^{4/3}}+\frac{q^{m+1}}{\left(d^{m+1}g\right)}\right]\right\}_{i+1/2,j} \tag{9}$$

In the second time step, Eqs. (7) and (8) are used to solve v^{m+1} and d^{m+1} along y-direction, and then u^{m+1} is solved by Eq. (9).

Cell Inundation Model

In addition to the two-dimensional finite difference model, the cell model proposed by Cunge (1980) is adopted in this study and modified by considering the geographic characteristics of the drainage basin such as roads, ditches, and hydraulic structures. The drainage basin is divided into storage cells interconnected by a number of hydraulic links such as channel, weir, and orifice. The schematization of the basin is not arbitrary but is based on the physical boundaries such as roads, dikes, and natural levels, etc. The change in the cell storage is described by the continuity equation. Each cell is connected with the neighboring cells by the hydraulic links which carry flow exchange between them. The flow is considered to concentrate in the link in which its overbank portions serve only the storage effects. According to the real situation, there is no overspill across the physical boundaries, the flow exchange between the cells are therefore determined by the discharge equation. The numerical solution in terms of water levels at the cells and discharges along the hydraulic links are obtained by solving the two governing equations.

Continuity Equation. Derivation of the continuity equation for a cell in the differential form leads to:

$$A_i\frac{dH_i}{dt}=P_i+\sum Q_{i,k} \tag{10}$$

where H_i= water level of cell i ; A_i= surface area of cell i ; P_i= external discharge into cell i , i.e. effective precipitation; $Q_{i,k}$= flow exchange between cell i and cell k. Positive values of $Q_{i,k}$ mean that flow is from cell k to cell i , negative values from cell i to cell k. The discharge $Q_{i,k}$ in Eq. (10) can be represented by any appropriate discharge formula as described below.

Discharge Between Cells. (1) Channel-link discharge formula:

$$Q=\Phi\left[(H_H-Z_H)+(H_L-Z_L)\right]^{5/3}\left[H_H-H_L\right]^{1/2} \tag{11}$$

where Φ is the flow parameter; H_H, H_L= high and low water levels of two neighboring cells respectively; Z_H, Z_L= ground elevations of the cells with high and low water levels respectively.

(2) Weir-link discharge formula:

for a free overflow weir, $(H_L-Z_W)<\frac{2}{3}(H_H-Z_W)$,

$$Q = \mu_1 b(2g)^{1/2} \cdot \frac{2}{3} \cdot (H_H - Z_W)^{3/2} \tag{12}$$

for a drowned weir, $(H_L - Z_W) \geq \frac{2}{3}(H_H - Z_W)$,

$$Q = \mu_2 b(2g)^{1/2}(H_L - Z_W)(H_H - H_L)^{1/2} \tag{13}$$

where μ_1 and μ_2 are weir discharge coefficients; b is effective width of the weir; Z_W is the elevation of weir crest; and other notations are previously defined.

(3) Submerged orifice discharge formula:

$$Q = C_d bw[2g(H_H - 0.63w)]^{1/2} \tag{14}$$

where C_d is the orifice discharge coefficient; b is the width of orifice; w is the height of orifice.

In these discharge formulas, parameters such as weir discharge coefficients, effective width of weir, orifice discharge coefficient, and width as well as height of orifice are determined according to the simulation results of the two-dimensional finite difference model, which will be discussed later.

Numerical Method

The finite difference approximation for Eq. (10) is used as follows:

$$A_i \frac{H_i^{n+1} - H_i^n}{\Delta t} = \left[\theta P_i^{n+1} + (1-\theta)P_i^n\right] + \theta \sum_k Q_{i,k}\left(H_i^{n+1}, H_k^{n+1}\right) + (1-\theta)\sum_k Q_{i,k}\left(H_i^n, H_k^n\right) \tag{15}$$

in which θ is a weighting factor and $0 \leq \theta \leq 1$. It is explicit scheme for $\theta = 0$ and implicit scheme for $\theta \geq 0.5$. $\Delta t = t^{n+1} - t^n$. With Eqs. (11) to (14) and appropriate boundary and initial conditions, the water level H_i^{n+1} of cell i at time step t^{n+1} in Eq. (15) can be solved.

The most important part of a complete flood warning system is the inundation-forecasting model due to the fact that it must overcome the time and spatial limits to simulate the inundation in the shortest computational time so as to preserve sufficient time for damage mitigation. According to the previous studies listed in the first section (Xanthopoulos and Koutitas, 1976; Hromadka et al., 1989; Hsu et al., 1990), the simulation results from the two-dimensional finite difference model are fairly reasonable and in good accuracy. However, this model is not satisfactory for urgency owing to the long simulation time. The cell inundation model is developed because it bears the advantages of efficiency and accuracy.

DETERMINATION OF PARAMETERS FOR CELL MODEL

Previous studies on the simulation of inundation were carried out using two-dimensional finite difference model. The advantage of this model is the capability of precisely reflecting the geometry of the study area and the ground surface resistance. The only parameter to be determined is Manning's roughness coefficient, which, most of the case, is within a definite range of values in consideration of the actual ground surface conditions and the numerical convergence. For the example of the inundation model for Taipei city (Hsu et al., 1990), the simulation of inundation can be performed in very good accuracy after the values of Manning's n have been calibrated and verified with the measured data. However, the long simulation time is the shortcoming of this model.

In order to simulate the inundation more efficiently, the cell model is developed to overcome the limit of the forecasting time. This conceptual model needs much shorter time for simulation because the inundation flow in equation of motion is simplified to the idealized channel, weir, and orifice flows. However, there are many discharge coefficients to be determined, which are crucial factors for successful simulations. These parameters are calibrated according to the simulation results of the two-dimensional finite difference model. In other words, the inundation area and depth computed from these two models are required to be identical to some extent. By doing so, the cell model used in this study will bear the advantages of accuracy and efficiency.

A fictitious square simulation area and the Gia-yi area (shown in Figs. 5 and 6) are two case studies for calibration of the cell model by the 2-D finite difference model. For the fictitious square area, Figures 3 and 4 are the simulated contours of inundation depth for 24 hours by the 2-D finite difference model and the cell model respectively. For Gia-yi area, Figures 7 and 8 are the simulated contours of inundation depth for 24 hours by the 2-D finite difference model and the cell model respectively. Table 1 lists the simulation results for 24 hours by these two models. These simulation results reveal that the calibrated cell model is satisfactory in terms of accuracy.

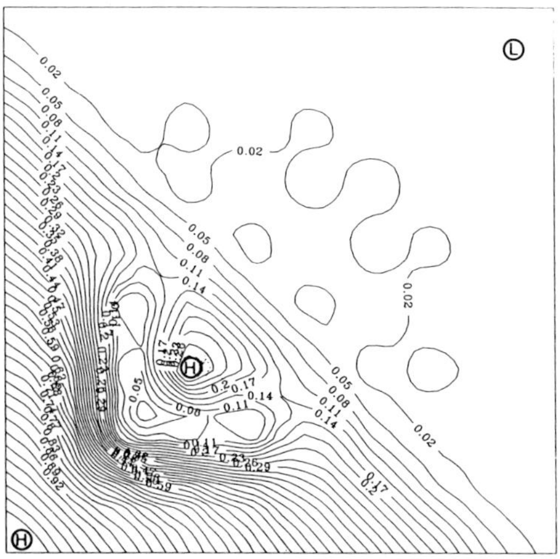

Fig. 3. Simulated contours of inundation depth at 24 hours by 2-D finite difference model (Fictitious square simulation area)

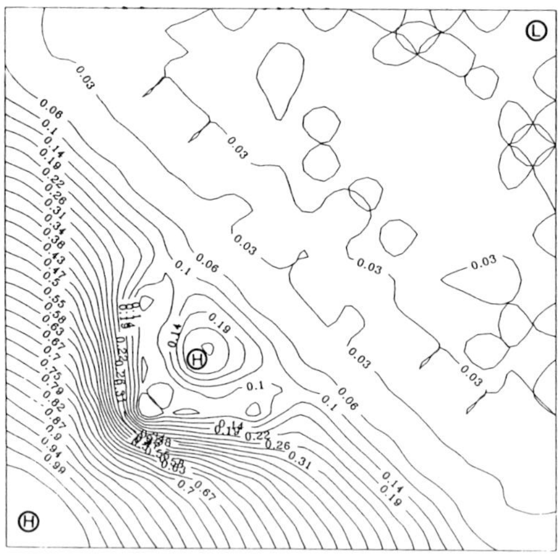

Fig. 4. Simulated contours of inundation depth at 24 hours by cell model (Fictitious square simulation area)

For the simulation of inundation in Pa-chang Creek Basin, all the parameters of cell model are calibrated according to the simulation results from 2-D finite difference model under the rainfall of 10-year return period. The calibrated cell model is then applied to the simulations of inundation in the same area under the rainfall of 50- and 100-year return periods.

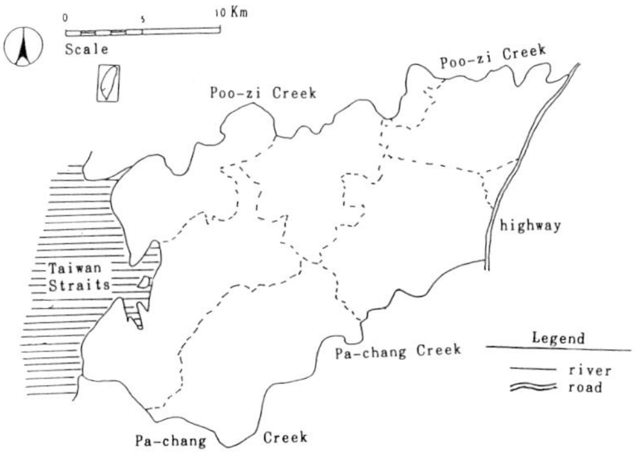

Fig. 5. Schematic location map of Gia-yi area

Fig. 6. Topographic contour map of Gia-yi area

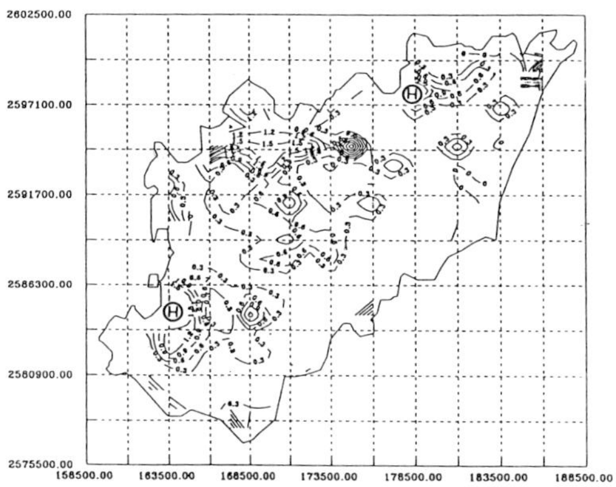

Fig. 7. Simulated contours of inundation depth at 24 hours by 2-D finite difference model (Gia-yi area)

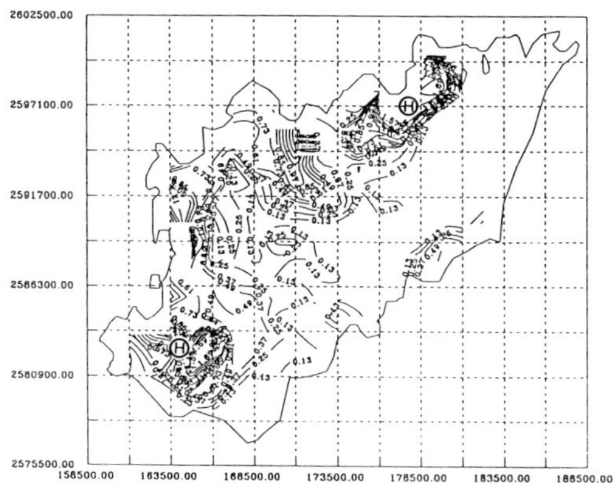

Fig. 8. Simulated contours of inundation depth at 24 hours by cell model (Gia-yi area)

Table 1. Comparison of inundation depth-area relationships in Gia-yi area* by 2-D finite difference model and cell model (Time = 24 hrs)

Inundation	Inundation Area (ha.)	
Depth, d (m)	Finite Difference Model**	Cell Model***
$0.0 < d < 0.5$	26,226	26,100
$0.5 < d < 1.0$	4,612	6,019
$1.0 < d$	4,928	3,647

* Total area = 35,766 ha.
** Total number of grids = 8,939
*** Total number of cells = 133

APPLICATIONS

Simulation Area

The simulation area for the application of inundation models is the mid and downstream drainage basin of Pa-chang Creek. Figure 2 is the digital terrain map based on the 120 m × 120 m resolution topographic data supported by the digital terrain model (DTM). Figure 9 shows the 4,493 grids used in 2-D finite difference model. For cell model, the simulation area is divided into 86 storage cells, as shown in Fig. 10.

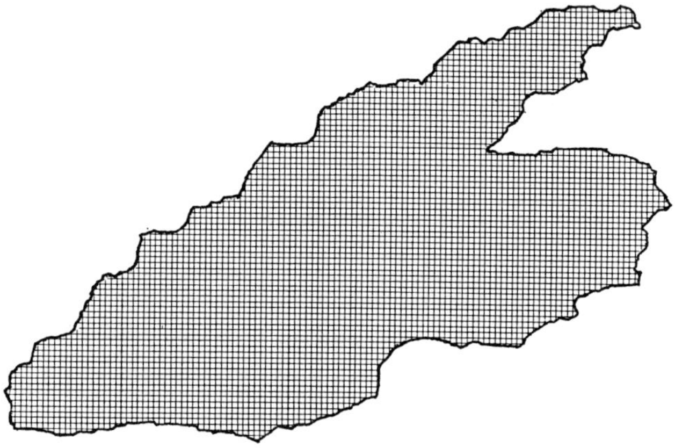

Fig. 9. Grids used in the simulation by 2-D finite difference model

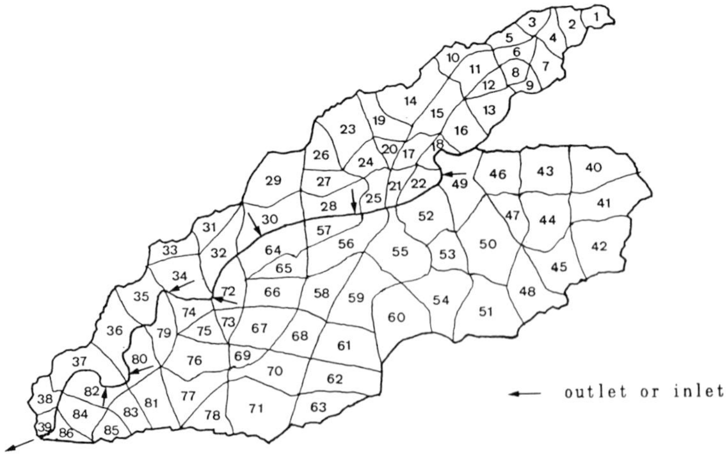

Fig. 10. Cells used in the simulation by cell model

Results

The simulations of inundation in Pa-chang Creek Basin by the two numerical models have been completed. The simulated contours of inundation depth under 10-, 50-, and 100-year return periods rainfall for 12, 16, and 24 hours are illustrated in Figure 11 through 16, and results listed in Table 2 through 4. From the results of the 2-D finite difference model listed in Table 2 and shown in Figures 11 and 12, the area of the portion that the water depth exceeds 0.5 m is 43 hectares at 16 hours; while for the cell model, it is 44 hectares, which illustrates good agreement between the results from these two models. Figure 12 through 14 as well as Table 3 illustrate the increasing of water depth with time. In consideration of the ground surface elevation, it can be concluded that the inundation is concentrated in the outlet part of the drainage system and some local lowlands. It is clearly observed from Figure 12 through 14 that the range of inundation expands from the lowlands along the Pa-chang Creek to higher areas through the drainage system. From Figures 12, 15, 16 and Table 4, it is found that the range of inundation is indeed increasing with the length of return period.

The computational times needed for the simulation in the Pa-chang Creek mid and downstream basin are 14.3 minutes and 369.8 minutes for the cell model and the 2-D finite difference model respectively, as listed in Table 5. Therefore, the computational time needed by the cell model is approximately 1/26 of that needed by the finite difference model.

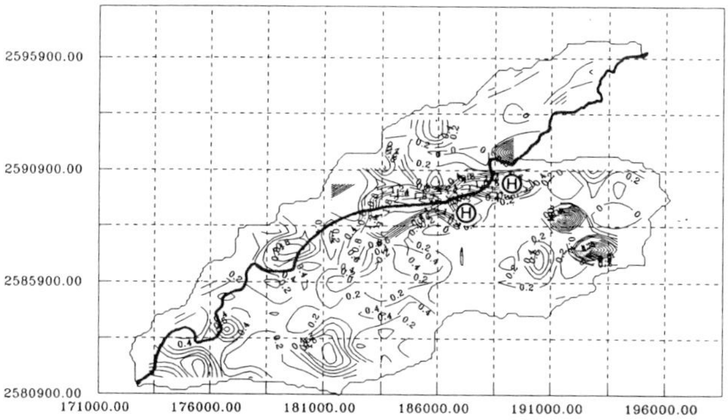

Fig. 11. Simulated contours of inundation depth in Pa-chang Creek Basin at 16 hours by 2-D finite difference model (Return period = 10 years)

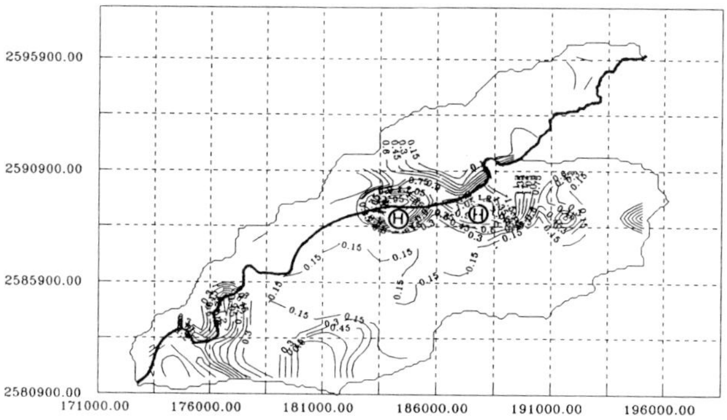

Fig. 12. Simulated contours of inundation depth in Pa-chang Creek Basin at 16 hours by cell model (Return period = 10 years)

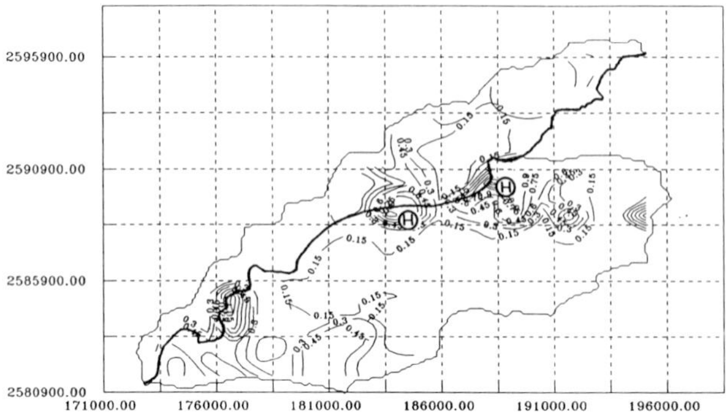

Fig. 13. Simulated contours of inundation depth in Pa-chang Creek Basin at 12 hours by cell model (Return period = 10 years)

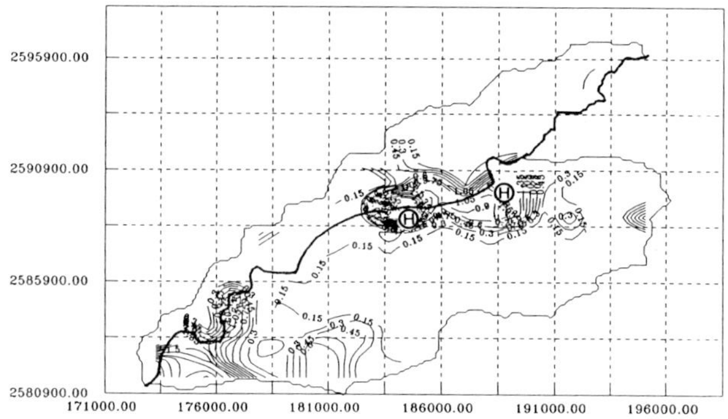

Fig. 14. Simulated contours of inundation depth in Pa-chang Creek Basin at 24 hours by cell model (Return period = 10 years)

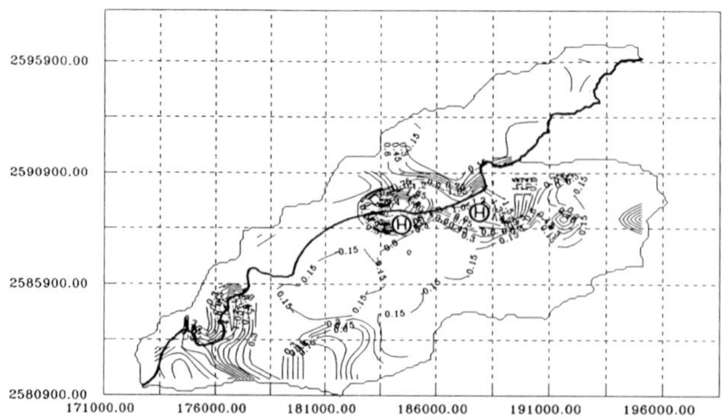

Fig. 15. Simulated contours of inundation depth in Pa-chang Creek Basin at 16 hours by cell model (Return period = 50 years)

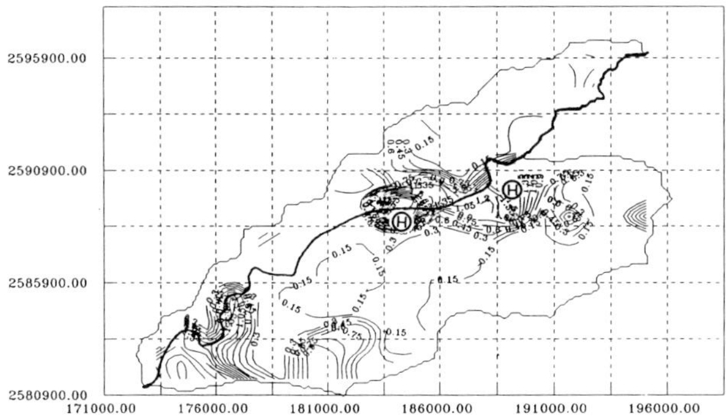

Fig. 16. Simulated contours of inundation depth in Pa-chang Creek Basin at 16 hours by cell model (Return period = 100 years)

Table 2. Comparison of inundation depth-area relationships in Pa-chang Creek Basin* by 2-D finite difference model and cell model (Time t = 24 hrs)

Inundation	Inundation Area (km^2)	
Depth, d (m)	Finite Difference Model**	Cell Model***
0.0 < d < 0.5	146	143
0.5 < d < 1.0	24	26
1.0 < d < 1.5	10	11
1.5 < d	9	8

* Total area = 187 km^2

** Total number of grids = 4,493

*** Total number of cells = 86

Table 3. Comparison of inundation depth-area relationships in Pa-chang Creek Basin by cell model at various times (Return period T = 10 years)

Inundation	Inundation Area (km^2)		
Depth, d (m)	t = 12 hrs.	t = 16 hrs.	t = 24 hrs.
0.0 < d < 0.5	158	143	148
0.5 < d < 1.0	20	26	16
1.0 < d < 1.5	9	11	12
1.5 < d	0	7	10

Table 4. Comparison of inundation depth-area relationships in Pa-chang Creek Basin by cell model at various return periods (Time t = 16 hrs)

Inundation	Inundation Area (km^2)		
Depth, d (m)	T = 10 yrs.	T = 50 yrs.	T = 100 yrs.
$0.0 < d < 0.5$	143	123	113
$0.5 < d < 1.0$	26	29	31
$1.0 < d < 1.5$	11	21	24
$1.5 < d$	7	13	16

Table 5. Comparison of simulation time needed by 2-D finite difference model and cell model

Simulated Area	Computation* Time (min)	
	Finite Difference Model	Cell Model
Fictitious square area	15.4	0.6
Gia-yi area	672.0	31.8
Pa-chang Creek Basin	367.8	14.3

* Computer used: SUN SPARC 10 MODEL 41

CONCLUSIONS

1. The cell model is developed in this study to simulate the inundation in Pa-chang Creek Basin efficiently and accurately. Previous studies indicated that the simulation results of the 2-D finite difference model is in good accuracy, however, it is not satisfactory for urgency owing to the long simulation time. For Pa-chang Creek Basin, the computational time needed by cell model is approximately 1/26 of that needed by finite difference model. Therefore, the cell model can be used as the inundation-forecasting model in the complete flood warning system.
2. All the parameters of cell model are calibrated according to the simulation results from the 2-D finite difference model under 10-year return period rainfall. The calibrated cell model is then applied to the simulations of inundation under 50- and 100-year return periods rainfall. The results for 50- and 100-year return periods rainfall are also in reasonably good precision.
3. Results from this study indicate that the range of inundation area expands from the lowlands along the Pa-chang Creek to higher areas through the drainage system. The range of inundation area is indeed increasing with the length of return period.
4. In order to ensure the quality of input data, the digital terrain model (DTM) and geographic information system (GIS) are employed to treat the ground elevation and the boundary of the simulated area.
5. The simulations of inundation in the mid and downstream Pa-chang Creek Basin under 10-, 50-, and 100-year return periods rainfall have been completed. The results are in good precision and can be used to find out the inundation area for the rainfall of various return periods. The inundation-forecasting model together with the rainfall-forecasting model constitute the most important part of the flood warning system.

ACKNOWLEDGMENT

The work presented herein was sponsored by the R.O.C. National Science Council under Grant NSC82-0414-P002-014B. The authors wish to express their sincere thanks to the Council for the financial support.

REFERENCES

Cunge, J.A., F.M. Holly and A. Verwey (1980). *Practical Aspects of Computational River Hydraulics.* Pitman Publishing Ltd., London.

Hromadka, T.V. II, T.R. Walker, C.C. Yen and J.J. DeVries (1989). Application of the U.S.G.S. diffusion hydrodynamic model for urban floodplain analysis. *Water Resour. Bull.*, 25, 1063-1071.

Hsu, M.H. (1992). Simulation of inundation with overflow on levee along Keelung River. In: Proceedings, The CCNAA-AIT Joint Seminar on Prediction and Damage Mitigation of Meteorologically Induced Natural Disasters. May 21-24, 1992. Natl. Taiwan Univ., Taipei, Taiwan.

Hsu, M.H., J.S. Lai and C.L. Yen (1990). Two-dimensional inundation model for Taipei city. In: Proceedings, The Fifth International Conference on Urban Storm Drainage, Osaka, Japan.

Hsu, M.H. and W.H. Teng (1993). Simulation of inundation in Pa-chang Creek Basin, III - Establishment of the relationship between precipitation and inundation area. R.O.C. NSC Damage Mitigation Res. Report No. 82-29.

Hsu, M.H., W.H. Teng and S.H. Lin (1994). A study on runoff and inundation models for Gia-yi area in Taiwan (II). Tech. Report No. 175, Hyd. Res. Lab., Natl. Taiwan Univ., Taipei, Taiwan.

Vongvisessomjai, S., T. Tingsanchali and C. Chaiwat (1985). Bangkok flood plain model. In: Proceedings, 21st IAHR Congress, Melbourne, Australia, 19-23 August, 1985.

Xanthopoulos, T. and C. Koutitas (1976). Numerical simulation of a two dimensional flood wave propagation due to dam failure. *J. Hydraul. Res.*, 14, 321-331.

Yen, C.L. and M.H. Hsu (1984). Flood routing models for Tanshui river system under present conditions. In: Proceedings, The CCNAA-AIT Joint Seminar on Res. for Natural Hazard Mitigation, Tainan, Taiwan.

Yen, C.L., M.H. Hsu and P.C. Hsieh (1990). Simulation of inundation in urban area with storm sewer systems. In: Proceedings, The International Symposium on Urban Planning and Stormwater Management, Malaysia.

FLOOD MITIGATION THROUGH JOINT FLOOD CONTROL OF PARALLEL RESERVOIRS

F. N. -F. Chou and D. -Y. Chen

Department of Hydraulics and Ocean Engineering, National Cheng-Kung University, Tainan 70101, Taiwan

ABSTRACT

This paper applied the optimal control theory to develop optimal coordinated flood control policies of reservoirs. The operation target was to reduce the flood peak discharge at a control point on downstream river channel. Reservoirs' operation, river flood routing and lateral inflows of downstream river channel were all considered in the analysis. The Tsengwen and Nanhwa reservoirs' system was selected as case study. The results demonstrated that the obtained flood control operation policies could effectively alleviate the flood peak at control point. Storage of the two reservoirs were also maintained within acceptable range during flood operation.

KEYWORDS

Flood mitigation; multireservoir; reservoir operation; flood routing; optimization; optimal control theory; minimum principle.

INTRODUCTION

Mitigating the flood disaster of downstream flood plain through joint operation of reservoirs had caught great attention in Taiwan in the last decade. Analysis of joint flood control operation of reservoirs has to deal with the complicate hydrologic and hydraulic system. The uncertainties of hydrologic process can not be completely documented in operation rules. A proper operation decision has to consider real-time on-site hydrologic condition. It is even more difficult to obtain a multi-reservoir coordinated operation rule than a single reservoir control under this situation. If the information of flood forecasting system can be utilized for decision, it would recommend a most appropriate operation policy with on-line optimization analysis.

The flood at the control point of floodplain is composed of releases from upstream reservoirs and the lateral inflows of downstream river channels. These peaks of different releases and lateral inflows might arrive the control point at the same time. It would increase the property damage of downstream floodplain. The only discharges that can be controlled are releases from reservoirs with gated spillway. This paper combined the optimization analysis and river flow routing to study the real-time joint flood control operation of reservoirs. The objective is to alleviate the flood peak of downstream river channel through appropriate adjustment of releases from different reservoirs.

LITERATURE REVIEW

To mitigate downstream flood by means of reservoir operation, one has to deal with the flood wave movement through river channels. The flood released from reservoir is attenuated and has a time lag when it arrives the downstream control point. This lag time between releasing water from reservoir and evaluating flood peak at control point come across a challenge to operation decision if several reservoirs are encountered in the system. This complexity of channel flow hydraulics limits the ordinary optimization algorithms, such as linear programming and dynamic programming, to be applied successfully. In most of the time, simulation is a common approach to evaluate different flood control operation policies. One example was that Unver and Mays (1987) developed a flood management system to determine the operation policy based on real-time rainfall forecasting and river stage.

Without considering river flow routing, most optimization algorithms adopted for reservoir flood control were linear programming (LP) and dynamic programming (DP). Windsor (1973) suggested a flood control system with recursive LP. He updated reservoir storages and other system variables in each iterative time step. However there was not any case study in his paper. Fergusson (1974) also applied separable programming to analyze the operation of reservoirs. DP can optimize the flood control policy of a drainage system, in where the influence of time lag is not significant. However, applying DP to multi-reservoir control would inevitably face the *curse of dimensionality*. It is therefore the computation burden prevented wide application of this algorithm. Jamieson and Wilkinson (1972) had used DP to optimize the release policy of a reservoir. Trotta *et al.* (1977) applied DP to analyze the control policy of urban storm of San Francisco.

Besides LP and DP, Chu and Yeh (1978) studied the daily and hourly real-time operation policy of reservoirs with non-linear programming. On the other hand, Papageorgiou (1983) adopted optimal control theory (OCT) and minimum principle to analyze the flood control of detention ponds for optimal operation of storm water treatment plants. The basic theories of the two approaches are similar, however, the solution approaches are different. The former is a type of direct method, the latter is classified as indirect method. As to the computation efficiency, the latter is usually superior than the former. To consider the stochastic inflow of reservoir, Wasimi and Kitanidis (1983) developed a linear quadratic Gaussian model for daily flood operation. It can also apply to real-time flood control operation of a multi-reservoir system.

If a system contains more than one reservoir, these algorithms would hardly obtain a solution within a reasonable time limit except the OCT for optimal flood mitigation analysis. OCT is very efficient for solving control problems. It usually takes less computation time than mathematical programming schemes. It is therefore more appropriate for analyzing multi-reservoir real-time flood control operation. This paper applied OCT to optimize the joint flood control policy of parallel reservoirs system. It combined reservoir operation and river flow routing. The lateral inflows of downstream river channels were also included. The objective of this joint flood control was to mitigate the flood peak of downstream river channel.

SYSTEM CONDITIONS OF RESERVOIR FLOOD CONTROL OPERATION

The continuity equation of a reservoir should be satisfied during flood control:

$$x_i(k+1) = x_i(k) - u_i(k) + q_i(k) \qquad i = 1, \ldots, N \tag{1}$$

where,

$x_i(k)$ = storage of i-th reservoir at time step k;

$u_i(k)$ = release of i-th reservoir at time step k;

$q_i(k)$ = inflow of i-th reservoir at time step k;

N = total time steps of flood control analysis.

The storages or states of all reservoirs should be maintained in acceptable ranges:

$$x_{i,\min}(k) \le x_i(k) \le x_{i,\max}(k) \tag{2}$$

where,

$x_{i,\min}(k)$ = minimum allowable storage of i-th reservoir at time step k;

$x_{i,\max}(k)$ = maximum allowable storage of i-th reservoir at time step k.

To guarantee the water supply after flood control operation, reservoir storage at the end of operation horizon is required to be close to a target level:

$$x_i(N+1) = FS_i \tag{3}$$

where,

FS_i = target storage of i-th reservoir at the end of operation.

The reservoir release, a variable that can be controlled, is a function of reservoir stage and opening of spillway gate. Releases can only be carried out within acceptable range during operation:

$$u_{i,\min}(x_i(k)) = u_{i,\min}(k) \le u_i(k) \le u_{i,\max}(k) = u_{i,\max}(x_i(k)) \tag{4}$$

where,

$u_{i,\min}(k)$ = minimum allowable release of *i*-th reservoir at time step *k*;

$u_{i,\max}(k)$ = maximum allowable release of *i*-th reservoir at time step *k*.

Discharges that can be released before and after inflow peak are subject to two extra legal constraints. Reservoir release can not exceed the maximum inflow discharge that has occurred before inflow peak and it can not exceed the peak discharge after inflow peak:

$$u_i(k) \le \max\{q_i(1), q_i(2), \cdots, q_i(k)\} \qquad k \le t_p \tag{5}$$

$$u_i(k) \le q_i(t_p) \qquad k > t_p \tag{6}$$

where,

t_p = peak time of inflow flood.

In addition, the increment of discharges between two adjacent releases before inflow peak to be less than the maximum increasing rate of inflow discharge is also required by law:

$$u_i(k) - u_i(k-1) \le \max\{[q_i(2)-q_i(1)], [q_i(3)-q_i(2)], \cdots, [q_i(k)-q_i(k-1)]\} \qquad k \le t_p \tag{7}$$

The objective of joint flood control operation of reservoirs is to reduce the peak discharge at downstream control point. Therefore the flow routing of river channel is considered to more precisely estimate the flood peak at downstream. One of the hydrologic approaches of river flow routing adopted in this paper is the Muskingum method:

$$O_i(k) = c_0 \cdot u_i(k) + c_1 \cdot u_i(k-1) + c_2 \cdot O_i(k-1) \tag{8}$$

where,

$O_i(k)$ = the flood discharge of release of i-th reservoir arrives at downstream end of a river reach at time step k;

c_0, c_1, c_2 = Muskingum coefficients.

OPTIMAL FLOOD CONTROL OF A TWO-PARALLEL RESERVOIR SYSTEM

The contents of joint flood control of a two-parallel reservoir's system include reservoir operational analysis and flood routing of downstream river channels. The objective of operation is to keep the discharge at downstream control point as minimum as possible or to be less than the design discharge of levee. To guarantee a satisfactory storage of water after operation and keep the release as smooth as possible, the objective function of joint control operation of two parallel reservoirs can be written as:

$$Min \quad \frac{S}{2}\sum_{k=1}^{N}\left[x_3(k)+x_4(k)+LF_1(k)+LF_2(k)\right]^2 + \frac{1}{2}\left\| X(N+1) - FS \right\|_E^2 + \frac{1}{2}\sum_{k=1}^{N}\left\| U(k) - U(k-1) \right\|_R^2 \tag{9}$$

where,

$x_3(k)$ = flood discharge, which is released from reservoir 1, at downstream control point after flood routing;

$x_4(k)$ = flood discharge, which is released from reservoir 2, at downstream control point after flood routing;

LF_1 = lateral inflow of the river reach between reservoir 1 and the control point;

LF_2 = lateral inflow of the river reach between reservoir 2 and the control point;

E, R = weighting factors, a symmetric matrix;

S = weighting factors, a constant.

and the system state and control variables are defined as:

$$X^T = [x_1, x_2, x_3, x_4] \tag{10}$$

$$U^T = [u_1, u_2] \tag{11}$$

The system states and control variables are subject to constraints such as system dynamics and feasible value of available ranges as described in the previous section. An unconstrained objective function, or called Lagrangian, can be obtained by combining the system state dynamics and penalty functions of state-space constraints into the original objective function:

$$J = \frac{S}{2}\sum_{k=1}^{N}\left[x_3(k)+x_4(k)+LF_1(k)+LF_2(k)\right]^2 + \frac{1}{2}\left\| X(N+1)-FS \right\|_E^2 + \frac{1}{2}\sum_{k=1}^{N}\left\| U(k)-U(k-1) \right\|_R^2 + $$
$$+\frac{1}{2}\sum_{k=1}^{N}\left\| \Psi[X(k)-X_{\min}] \right\|_Q^2 + \frac{1}{2}\sum_{k=1}^{N}\left\| \Psi[X_{\max}-X(k)] \right\|_P^2 +$$
$$+\sum_{k=1}^{N}\Lambda^T(k+1)\begin{bmatrix} x_1(k+1)-x_1(k)+u_1(k)-q_1(k) \\ x_2(k+1)-x_2(k)+u_2(k)-q_2(k) \\ x_3(k+1)-c_{01}u_1(k+1)-c_{11}u_1(k)-c_{21}x_3(k) \\ x_4(k+1)-c_{02}u_2(k+1)-c_{12}u_2(k)-c_{22}x_4(k) \end{bmatrix} \tag{12}$$

where,

P, Q = weighting factors, symmetric matrix;

$\Lambda^T = [\lambda_1, \lambda_2, \lambda_3, \lambda_4]$ = Lagrange multiplier, is also called co-state variable;

and the function Ψ is defined as:

$$\Psi(\eta) = 0 \quad if \quad \eta \ge 0;$$
$$\Psi(\eta) = \eta \quad if \quad \eta < 0 \tag{13}$$

The required computation time of optimal control theory (OCT) is less than the non-linear programming for most control problems. It is therefore applicable for real-time on-site operation analysis of a reservoir. By applying discrete OCT to this control problem (Kuo, 1992), the Hamiltonian function of this control system is defined as:

$$H(k) = \frac{S}{2}\left[x_3(k)+x_4(k)+LF_1(k)+LF_2(k)\right]^2 + \frac{1}{2}\left\| U(k)-U(k-1) \right\|_R^2 +$$
$$+\frac{1}{2}\left\| \Psi[X(k)-X_{\min}] \right\|_Q^2 + \frac{1}{2}\left\| \Psi[X_{\max}-X(k)] \right\|_P^2 +$$
$$+\Lambda^T(k+1)\begin{bmatrix} x_1(k)-u_1(k)+q_1(k) \\ x_2(k)-u_2(k)+q_2(k) \\ c_{01}u_1(k+1)+c_{11}u_1(k)+c_{21}x_3(k) \\ c_{02}u_2(k+1)+c_{12}u_2(k)+c_{22}x_4(k) \end{bmatrix} \tag{14}$$

and the terminal objective of this control system is:

$$G(N+1) = \frac{1}{2}\left\| X(N+1)-FS \right\|_E^2 \tag{15}$$

The necessary conditions of the minimum system objective are:

1. Feasibility conditions:

$$\frac{\partial H(k)}{\partial \lambda_i(k+1)} = x_i(k+1) \qquad i = 1, \cdots, 4 \tag{16}$$

These are exactly the system state equations:

$$x_i(k+1) = x_i(k) - u_i(k) + q_i(k) \qquad i = 1, 2 \tag{17}$$

$$x_j(k+1) = c_{0i}\, u_i(k+1) + c_{1i}\, u_i(k) + c_{2i}\, x_j(k) \qquad j = 3, 4 \tag{18}$$

where, $k = 1, \ldots, N$.

2. Co-state conditions:

$$\frac{\partial H(k)}{\partial x_i(k)} = \lambda_i(k) \qquad i = 1, \cdots, 4 \tag{19}$$

These are the recursive equations of co-state variables:

$$\lambda_1(k) = \lambda_1(k+1) + Q_{11}\left[\Psi(x_1(k) - x_{1,\min})\right] - P_{11}\left[\Psi(x_{1,\max} - x_1(k))\right] \tag{20}$$

$$\lambda_2(k) = \lambda_2(k+1) + Q_{22}\left[\Psi(x_2(k) - x_{2,\min})\right] - P_{22}\left[\Psi(x_{2,\max} - x_2(k))\right] \tag{21}$$

$$\lambda_3(k) = c_{21}\,\lambda_3(k+1) + S\left[x_3(k) + x_4(k) + LF_1(k) + LF_2(k)\right] \tag{22}$$

$$\lambda_4(k) = c_{22}\,\lambda_4(k+1) + S\left[x_3(k) + x_4(k) + LF_1(k) + LF_2(k)\right] \tag{23}$$

3. Optimal control conditions, or the minimum principle:

$$\frac{\partial H(k)}{\partial u_i(k)} = 0 \qquad i = 1, 2 \tag{24}$$

One can obtain the optimal control equations of this system as:

$$0 = R_{11}u_1(k) - \frac{R_{11}}{2}\left[u_1(k+1) + u_1(k-1)\right] - \lambda_1(k+1) + c_{11}\,\lambda_3(k+1) \tag{25}$$

$$0 = R_{22}u_2(k) - \frac{R_{22}}{2}\left[u_2(k+1) + u_2(k-1)\right] - \lambda_2(k+1) + c_{12}\,\lambda_4(k+1) \tag{26}$$

4. The terminal conditions of co-state variables are:

$$\frac{\partial G(N+1)}{\partial x_i(N+1)} = \lambda_i(N+1) \qquad i = 1, \cdots, 4 \tag{27}$$

The co-state variables at terminal stage are:

$$\lambda_i(N+1) = E_{ii}\left[x_i(N+1) - FS_i\right] \qquad i = 1, 2 \tag{28}$$

$$\lambda_j(N+1) = 0 \qquad j = 3, 4 \tag{29}$$

The E_{ij}, Q_{ij} and P_{ij} in the above equations represent the element at *i*-th row and *j*-th column of E, Q and P matrix.

NUMERICAL COMPUTATION ALGORITHM

The system states can be computed forwards from the first stage to the last with equations (17) and (18). All co-state variables can be computed from the last stage to the first with equations (20) to (23) after computed their values at the last stage with equations (28) and (29). Thus, the state variables are computed forward in time from initial known states of the first stage. The co-state variables are computed backwards in time from the terminal co-state variables that are computed with state variables. This forms a two-point boundary value problem.

With all state and co-state variables, the optimal values of system control variables can be computed from equations (25) and (26). However, the controls thus computed may not fall into feasible ranges. In this situation, the minimum principle should be directly applied to find an optimal control such that the Hamiltonian is at its minimum value. The Fletcher-Powell variable metric method is adopted to find the minimum value of Hamiltonian function in this paper (Wismer and Chattergy, 1978). The solution process starts from an identity matrix K and defines:

$$g^i = \nabla H(U(k)^i) \tag{30}$$

and

$$V(k)^i = -K^i \cdot g^i \tag{31}$$

where the superscript i stands for the index of iteration. A new optimal control is obtained from:

$$U(k)^{i+1} = U(k)^i + \delta^i \cdot V(k)^i \tag{32}$$

where, δ is determined such that $H(U(k)^i)$ is a minimum. Here the golden section method is used for one dimensional search. The next step is to update the search direction $V(k)$ by computing a new K matrix. Let:

$$\alpha(k)^i = \delta^i \cdot V(k)^i \tag{33}$$

$$\beta(k)^i = g^{i+1} - g^i \tag{34}$$

then K can be updated by:

$$K^{i+1} = K^i + A^i + B^i \tag{35}$$

where,

$$A^i = \left[\alpha(k)^i \cdot (\alpha(k)^i)^T\right] / \left[(\alpha(k)^i)^T \cdot \beta(k)^i\right] \tag{36}$$

$$B(k)^i = \left[K^i \cdot \beta(k)^i \cdot (K^i \cdot \beta(k)^i)^T\right] / \left[(\beta(k)^i)^T \cdot K^i \cdot \beta(k)^i\right] \tag{37}$$

Finally, the weighting factors P and Q are increased gradually to guarantee the state space constraints are satisfied during the iterative optimization process.

CASE STUDY: TSENGWEN AND NANHWA PARALLEL RESERVOIRS SYSTEM

The system of Tsengwen and Nanhwa reservoirs in Tsengwen river basin, see Fig. 1, is selected for case study. Yuching, a town located at the confluence of Tsengwen river and Hoku river, is selected as the control point for evaluating the optimal joint operation of the two upstream reservoirs. The objective of operation is to alleviate the peak discharge at this river confluence. There are four different components of total discharge at Yuching: the routed flood discharges of releases from Tsengwen and Nanhwa reservoirs, and the lateral discharge of river reaches between Yuching and Tsengwen and Nanhwa reservoirs. The limiting storages of flood control operation of Tsengwen and Nanhwa reservoirs are listed in table 1 (TTRCC, 1974; TWCB, 1988). The Muskingum floods routing coefficients of downstream Tsengwen river reach are -0.004, 0.951 and 0.053 with parameters k being 1.05 hour and x being 0.47. For Hoku river reach, the coefficients are -0.0616, 0.9342 and 0.1274 with parameters k being 1.22 hour and x being 0.47. The system objective function of joint flood control of these two parallel reservoirs can be written as:

$$J = \frac{S}{2}\sum_{k=1}^{N}\left[x_3(k)+x_4(k)+LF_1(k)+LF_2(k)\right]^2 + \frac{1}{2}\left\|\begin{bmatrix} x_1(N+1) \\ x_2(N+1) \\ x_3(N+1) \\ x_4(N+1)\end{bmatrix} - \begin{bmatrix} 616.9 \\ 158.05 \\ 0 \\ 0 \end{bmatrix}\right\|_E^2 + \frac{1}{2}\sum_{k=1}^{N}\left\|\begin{bmatrix} u_1(k) \\ u_2(k)\end{bmatrix} - \begin{bmatrix} u_1(k-1) \\ u_2(k-1)\end{bmatrix}\right\|_R^2 +$$

$$+\frac{1}{2}\sum_{k=1}^{N}\left\|\Psi\left[\begin{bmatrix} x_1(k) \\ x_2(k) \\ x_3(k) \\ x_4(k)\end{bmatrix} - \begin{bmatrix} 438.75 \\ 13.24 \\ 0 \\ 0\end{bmatrix}\right]\right\|_Q^2 + \frac{1}{2}\sum_{k=1}^{N}\left\|\Psi\left[\begin{bmatrix} 719.58 \\ 169.96 \\ 0 \\ 0\end{bmatrix} - \begin{bmatrix} x_1(k) \\ x_2(k) \\ x_3(k) \\ x_4(k)\end{bmatrix}\right]\right\|_P^2 +$$

$$+\sum_{k=1}^{N}\left[\lambda_1(k+1),\lambda_2(k+1),\lambda_3(k+1),\lambda_4(k+1)\right]\cdot\begin{bmatrix} x_1(k+1)-x_1(k)+u_1(k)-q_1(k) \\ x_2(k+1)-x_2(k)+u_2(k)-q_2(k) \\ x_3(k+1)+0.004u_1(k+1)-0.951u_1(k)-0.053x_3(k) \\ x_4(k+1)+0.0616u_2(k+1)-0.9342u_2(k)-0.1274x_4(k)\end{bmatrix} \tag{38}$$

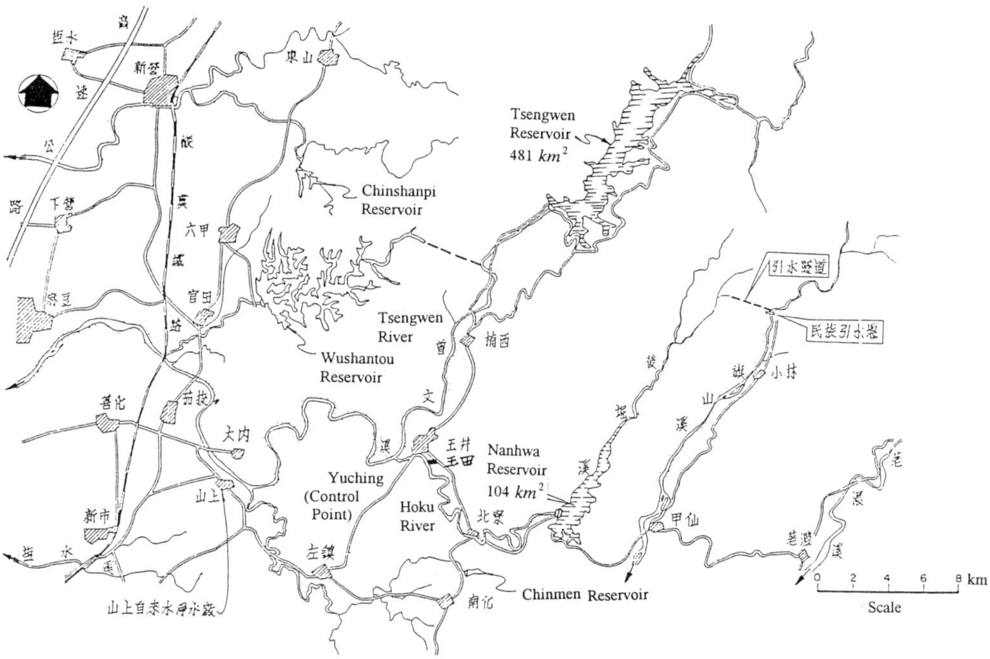

Fig. 1. Location map of Tsengwen and Nanhwa reservoirs

Table 1. The storage conditions of flood control operation

Reservoir	Tsengwen Reservoir		Nanhwa Reservoir	
	Storage (M.CM)	Stage m.s.l. (M)	Storage (M.CM)	Stage m.s.l. (M)
Maximum Surcharge Storage	719.58	232.5	169.96	185
Reservoir Design Capacity	671.63	230	158.05	180
Terminal Target Storage	616.90	227	158.05	180
Starting Operation Storage	581.23	225	158.05	180
Minimum Allowable Storage	438.75	216	13.24	134

A control by only one Tsengwen reservoir is also studied to compare the effectiveness of joint operation of two parallel reservoirs. In this case the Nanhwa reservoir is set to be free spill. The system objective function of this single Tsengwen reservoir control is:

$$J = \frac{S}{2}\sum_{k=1}^{N}\left[x_2(k)+O(k)+LF_1(k)+LF_2(k)\right]^2 + \frac{1}{2}\left\|\begin{bmatrix}x_1(N+1)\\x_2(N+1)\end{bmatrix}-\begin{bmatrix}616.9\\0\end{bmatrix}\right\|_E^2 + \frac{R}{2}\sum_{k=1}^{N}\left[u(k)-u(k-1)\right]^2 +$$

$$+\frac{1}{2}\sum_{k=1}^{N}\left\|\Psi\left[\begin{bmatrix}x_1(k)\\x_2(k)\end{bmatrix}-\begin{bmatrix}438.75\\0\end{bmatrix}\right]\right\|_Q^2 + \frac{1}{2}\sum_{k=1}^{N}\left\|\Psi\left[\begin{bmatrix}719.58\\0\end{bmatrix}-\begin{bmatrix}x_1(k)\\x_2(k)\end{bmatrix}\right]\right\|_P^2 +$$

$$+\sum_{k=1}^{N}\left[\lambda_1(k+1),\lambda_2(k+1)\right]\cdot\begin{bmatrix}x_1(k+1)-x_1(k)+u(k)-q_1(k)\\x_2(k+1)+0.004u(k+1)-0.951u(k)-0.053x_2(k)\end{bmatrix} \tag{39}$$

The flood of September 3, 1981 is selected to test the computation efficiency and evaluate the applicability of the model in real-time flood forecasting system. This is the largest flood that has occurred since the Tsengwen reservoir started to operate from 1974. The corresponded inflow hydrographs of Nanhwa reservoir and downstream river channels are synthetic 100-year floods. The optimal flood control analysis of a single Tsengwen reservoir and two parallel reservoirs are performed. The operation horizon is 50 hours in total with each time step of one hour. It takes 10 minutes to analyze the optimal policies of the two parallel reservoirs' control with a 80486-20 personal computer. It is acceptable in real-time operation compared to an operation time interval of one hour.

The operation process of Tsengwen reservoir of a single reservoir control is shown in Fig. 2. Hydrographs at Yuching is plotted in Fig. 3. For the joint control of two parallel reservoirs, Fig. 4 and Fig. 5 are the operation processes of Tsengwen and Nanhwa reservoirs. Figure 6 shows hydrographs at Yuching of this optimal control. The numeric results are listed in table 2. The peak times of released hydrographs of the two reservoirs are separated by the model. Thus the total peak discharge at downstream control point can be greatly reduced. As Nanhwa reservoir releases large quantity of water, Tsengwen almost does not release. However it does release water for adjusting purpose at the very beginning stage of operation.

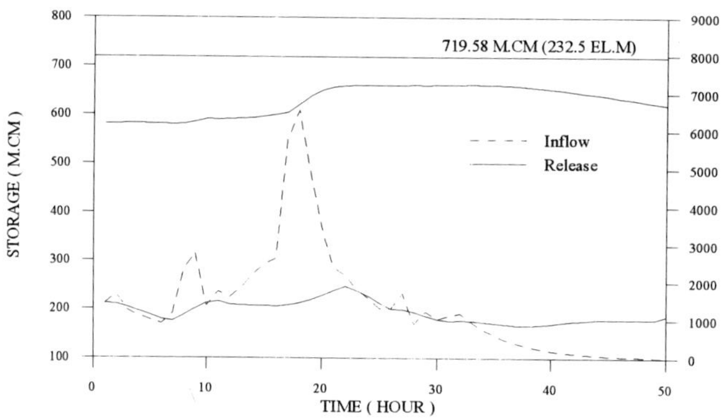

Fig. 2. Optimal operation process of Tsengwen reservoir of single reservoir control

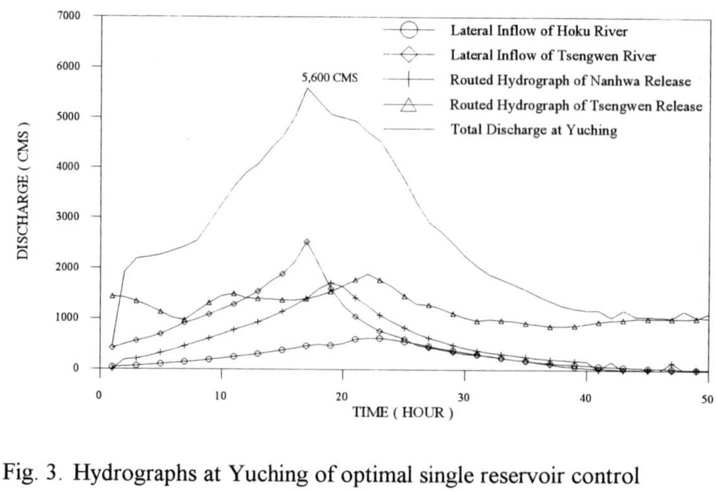

Fig. 3. Hydrographs at Yuching of optimal single reservoir control

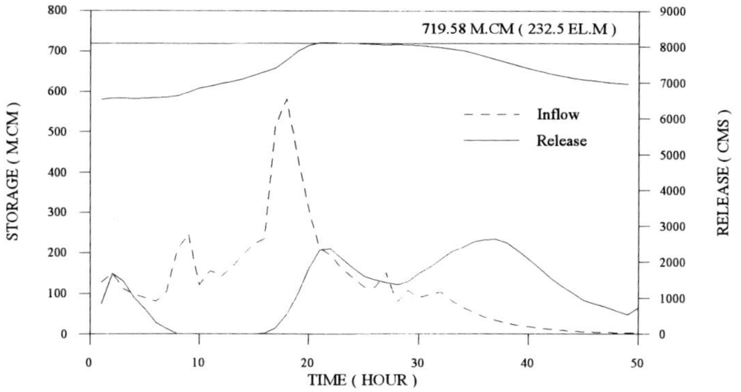

Fig. 4. Optimal operation process of Tsengwen reservoir of two parallel reservoirs' control

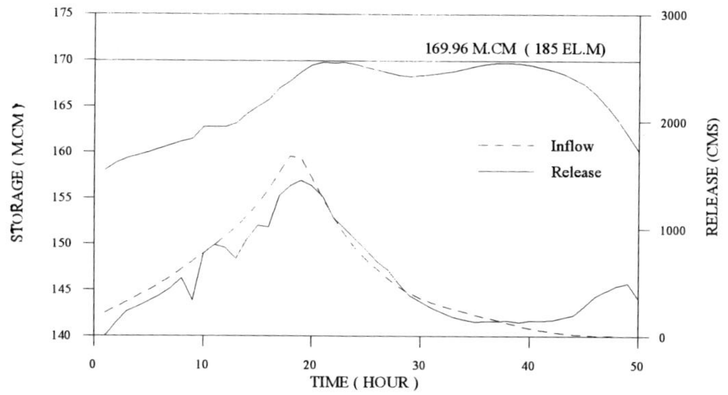

Fig. 5. Optimal operation process of Nanhwa reservoir of two parallel reservoirs' control

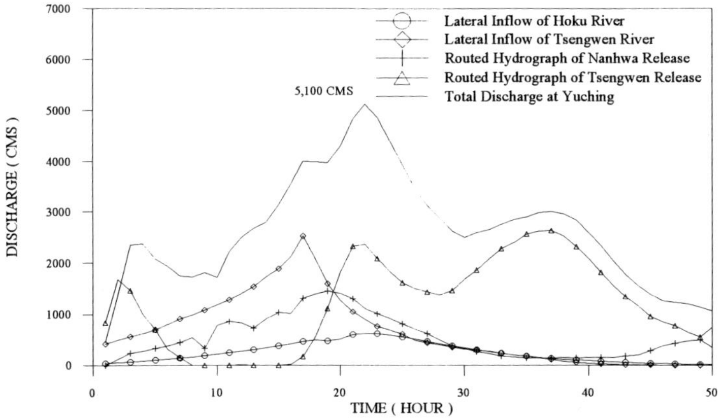

Fig. 6. Hydrographs at Yuching of optimal two parallel reservoirs' control

Table 2. Comparison of downstream flood peak at control point by a single and two parallel reservoirs control

Nanhwa Inflow	No Shifting		8 hours ahead		8 hours behind	
Control Type	One single	Two parallel	One single	Two parallel	One single	Two parallel
Peak Time (Hour)	17	22	9	24	25	25
Discharge (CMS)	5,590	5,130	4,865	4,475	5,729	4,769

Note: *One single* stands for only one Tsengwen reservoir control and *Two parallel* means controlled by both Tsengwen and Nanhwa reservoirs.

It assumes that the peaks of two reservoir inflows occur at the same time since one is synthetic. The Nanhwa inflow is shifted 8 hours ahead and behind for further analysis to investigate the influence of inflow peak occurred at different hours. Figure 7 and 8 are the hydrographs at Yuching of these cases. When the inflow hydrograph of Nanhwa reservoir is shifted 8 hours ahead, the peak of Nanhwa release will be ahead of Tsengwen's peak. The opposite condition is observed if the inflow peak of Nanhwa reservoir is 8 hours behind. The peak of Nanhwa release will be behind that of Tsengwen's. In other words, the peak hours of releases of the two reservoirs are switched each other in the two cases and the peak discharges are not overlapped.

From the above results, it is usually quite flexible to jointly control the releasing priorities of the two reservoirs of this system. The operation policies can effectively separate the peak hours of reservoir releases and downstream lateral inflows when they all reach the control point. The peak discharge at control point can thus be alleviated. The reservoir storages are also retained in allowable range to fully use the flood control capacity and guarantee the dam safety. The reservoir storage is always close to the terminal target storage at the end of operation.

There are five weighting factors in the objective function to weigh the relative importance of different items. Adopting different weights will effect the reservoir releases and stages. The weighting factor S can greatly change the peak discharge and time of control point by affecting the relative values between the objectives of total discharge and reservoir target storage. To fully utilize the reservoir capacity for operation can decrease the total discharge at control point to its maximum level. However, it will impose pressure on operator when storage is close to reservoir's maximum surcharge storage. This paper decreased the weighting factor

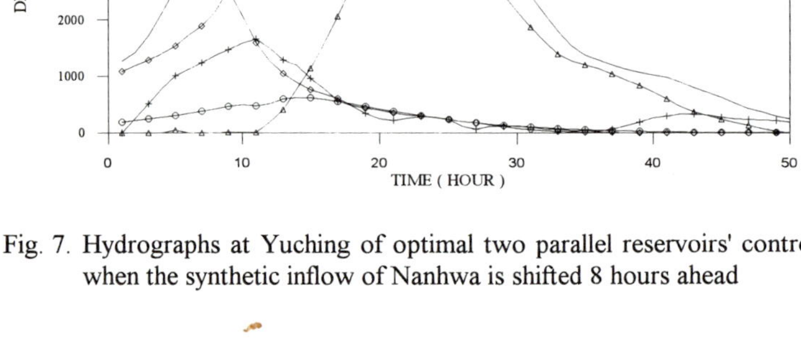

Fig. 7. Hydrographs at Yuching of optimal two parallel reservoirs' control when the synthetic inflow of Nanhwa is shifted 8 hours ahead

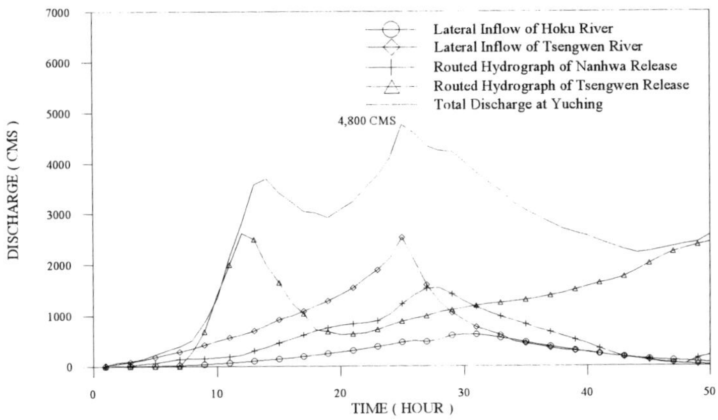

Fig. 8. Hydrographs at Yuching of optimal two parallel reservoirs' control when the synthetic inflow of Nanhwa is shifted 8 hours behind

Table 3. Comparing the influence of different weighting factors *S* to the flood peak mitigation at downstream control point by parallel reservoir control with 1981 flood as an example

Weighting Factor *S*	Peak Time (Hour)	Peak Discharge at Control Point (CMS)
1.0	22	5,130
0.8	25	5,440
0.3	21	6,562
0.1	21	6,848

S from 1.0 to 0.1 to check its influence. Table 3 shows the minimum peak discharge and peak time at control point of 1981 flood. Peaks of the two reservoirs' releases are not separated largely in time with a small value of *S*. The total discharge at control point is large due to overlapping of peaks.

Selecting proper values of weighting factors, it depends on the objective of flood control operation. The weighting factors will interfere one another. To decrease the maximum release is usually conflict to hold the reservoir storage within acceptable range. If reservoir storage is closes to its maximum surcharge storage, operational flexibility is usually not enough to most operators even the reservoir storage is safe. One needs to find a balance between maintaining safety storage and release when assigning values of different weighting factors. A better approach is to consider different weighting factors for different size of floods.

SUMMARY AND CONCLUSIONS

This paper studied the optimal joint flood control operation of a two parallel reservoirs system. It combined the analysis of reservoir operation and downstream river flood routing. The case study demonstrated that the proposed optimal control algorithm can schedule the timing of releasing flood of the two reservoirs to prevent peaks from overlapping at the control point. This can effectively alleviate the flood peak discharge at downstream. The reservoirs' storage can be maintained in acceptable range during operation. The terminal storages of all reservoirs are also close to their target levels. The results also demonstrated that not only the improvement of downstream flood peak is evident but the computation time is much less than applying dynamic programming analysis.

The developed control algorithm can be used for analyzing the optimal flood operation rule or be used on-line for joint flood control operation. To apply this optimal control approach in real-time, the forecasting of reservoir inflow should be included. The optimal coordinated policies can be progressively analyzed during operation. Proper values of weighting factors in objective function should be investigated for floods of different return periods or different size of historical floods before site application.

ACKNOWLEDGMENTS

The authors thank the personnel of Taiwan Water Conservancy Bureau and Tsengwen Reservoir Administration Bureau for providing the necessary data. This research was supported by the ROC National Science Council under grant number NSC80-0414-P006-02B.

REFERENCES

Chu, W. S. and W. W.-G. Yeh (1978). A nonlinear programming algorithm for real-time hourly reservoir operations. *Water Resour. Bull.*, *14*, 1048-1063.

Fergusson, J. R. (1974). *Methodologies for Operating Reservoirs for Short-term Flood Control. Technical Rep. No. 85*, Water Resources and Marine Sciences Center, Cornell University, Ithaca.

Jamieson, D. G. and J. C. Wilkinson (1972). River Dee research program 3. A short-term control strategy for multipurpose reservoir systems. *Water Resour. Res.*, *8*, 911-920.

Kuo, B. C. (1992). *Digital Control Systems*. 2nd Ed., Saunders College Publishing, Ft. Worth.

Papageorgiou, M. (1983). Automatic control strategies for combined sewer system. *J. of Envir. Engrg.*, *ASCE*, *109*, 1385-1402.

Taiwan Tsengwen Reservoir Construction Commission (1974). *Tsengwen Reservoir Construction Report, Vol.I*, Tainan (in Chinese).

Taiwan Water Conservancy Bureau (1988). *Nanhwa Reservoir Detailed Planning Report*. Taichung (in Chinese).

Trotta, P. D., J. W. Labadie and N. S. Grigg (1977). Automatic control strategies for urban stormwater. *J. of Hydr. Div.*, *ASCE*, *103*, 1443-1459.

Unver, O. and L. W. Mays (1987). Real-Time Flood Management Model for Highland Lake System. *J. of Water Resour. Plng. and Mgmt.*, *ASCE*, *113*, 620-638.

Wasimi, S. A. and P. K. Kitanidis (1983). Real-time forecasting and daily operation of a Multireservoir system during flood by linear quadratic Gaussian control. *Water Resour. Res.*, *19*, 1511-1522.

Windsor, J. S. (1973). Optimization model for the operation of flood control systems. *Water Resour. Res.*, *9*, 1219-1226.

Wismer, D. A. and R. Chattergy (1978). *Introduction to Nonlinear Optimization, A Problem Solving Approach*. Elsevier North-Holland, New York.

RAINFALL ESTIMATE FROM SATELLITE OBSERVATIONS

M. L. C. Wu

Laboratory for Atmospheres, NASA/Goddard Space Flight Center, Greenbelt, MD 20771, U.S.A.

1. INTRODUCTION

Goddard Laboratory for Atmospheres (GLA) has produced several years of global data sets (1979 to 1980 and 1986 to 1988) from the interactive-forecast-retrieval-analysis system (IFRA) and planned to produce more than 15-year data set in the near future. The GLA IFRA system consists of three components: a physical retrieval system for analyzing HIRS2/MSU satellite sounding data; a forecast system based on a state-of-the-art GCM (Kalnay et al., 1983), and an Optimum Interpolation analysis system (Baker, 1983). The GLA IFRA system produces temperature profiles, moisture profiles, and clouds (Susskind, et al., 1983) which can be used to compute clear sky and all sky outgoing longwave radiation (OLR).

Wu (1991) developed a rainfall retrieval algorithm based on the clear sky and all sky OLR (hereafter referred to as the longwave radiative forcing (LRF) algorithm) and demonstrated the validity of monthly rainfall estimates. In this paper, we present results of further application of the LRF algorithm to the IFRA data. Our objectives are: (1) to estimate global rainfall from satellite observations based on IFRA data, and (2) to investigate the seasonal, intra-seasonal and inter-annual variations of rainfall estimate.

In this paper, section 2 presents rainfall algorithm for the convenience of the reader. Section 3 presents monthly mean rainfall rates for January and July of 1988. Section 4 presents the water vapor residence time in the atmosphere. Section 5 presents potential field of vertically integrated water vapor flux and the potential function of rainfall estimate. In order to investigate whether the rainfall estimates can be used for the study of atmospheric variations, some characteristics of the rainfall estimate is presented in section 6, its principal modes of variation in section 7, and conclusions are presented in section 8.

2. RAINFALL ESTIMATE

The LRF algorithm is developed based on long-wave cloud radiative forcing and surface radiative forcing (for details see Wu, 1991). The long-wave cloud radiative forcing (ΔOLR) is defined as the difference between the clear sky OLR and all skay OLR (Ramanathan, 1987; Wu and Susskind, 1990). The clear sky and all sky OLR are computed using temperature, humidity and clouds retrieved from the IFRA system in conjunction with the Wu-Kaplan radiative transfer model (Krishnamurti, 1982). ΔOLR is zero over clear regions and is large over overcast deep convective cloud regions. Over ocean, the cloud radiative forcing is closely linked to rainfall. Over land, rainfall and ground wetness are also related, and ground wetness controls the surface temperature. Where the diurnal temperature range is large (small), the ground wetness or rainfall amount is small (large). The diurnal surface forcing due to long-wave radiation can be measured by the difference of the computed daytime and nighttime clear sky OLR (ΔCLR).

The LRF rainfall estimates are derived by correlating ΔOLR and ΔCLR with the daily rain-gauge measurements (Walker and Mintz, private communication). The correlation coefficient between the LRF rainfall estimates and GOES Precipitation Index (GPI) is 0.914 over the tropical oceans. Significant improvements of the LRF rainfall estimate are found over land regions in the extra-tropics because the LRF separate the background signal better than threshold methods. The correlation coefficient between the LRF rainfall estimate and rainfall estimate based on a visible channel of the HIRS2 is 0.887 in the tropics from 30°S to 30°N, and is 0.844 with rainfall estimate based on 37 GHz of Scanning Multi-channel Microwave Radiometer (SMMR) on Numbus-7.

3. MONTHLY MEAN RAINFALL ESTIMATE

Global distribution of rainfall estimates have been derived for 1979, 1980 and 1988 using the LRF algorithm. The global distribution of rainfall estimates for January and July 1988 are shown in Figures 1a and 1b, respectively.

In January, the major rainfall occurs over the eastern Indian Ocean, Maritime continent and northeast of Australia. The rain band associated with the Australian monsoon, the rainfall over central southern Africa and northern South American has a tendency to extend southeastward into the southern mid-latitudes. These features give rise to the characteristic shape of the South Pacific Convergence Zone (SPCZ) and the rain band over Brazil/Argentina and Southern Africa. The mid-latitude rain bands along the storm tracks are also properly revealed.

(a)

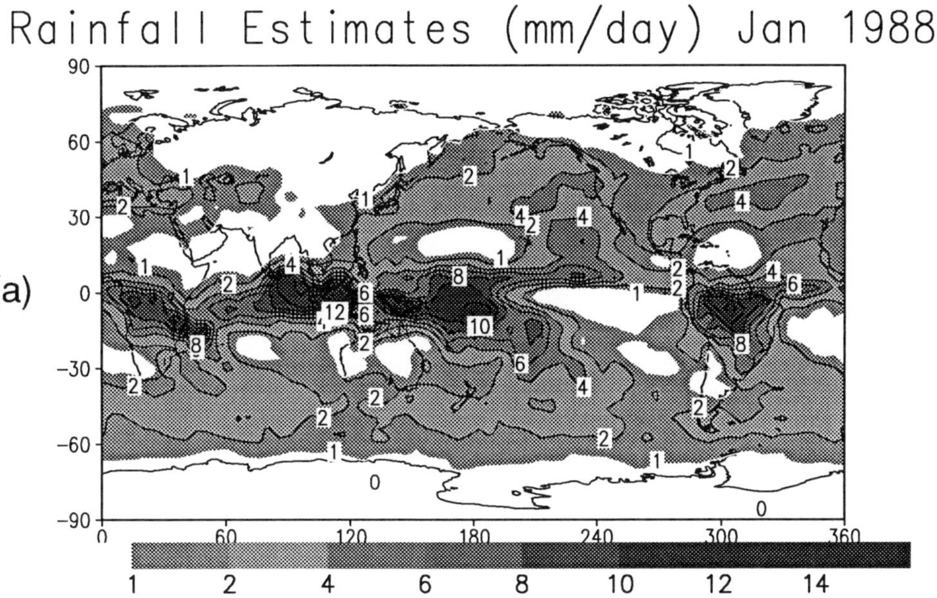

(b)

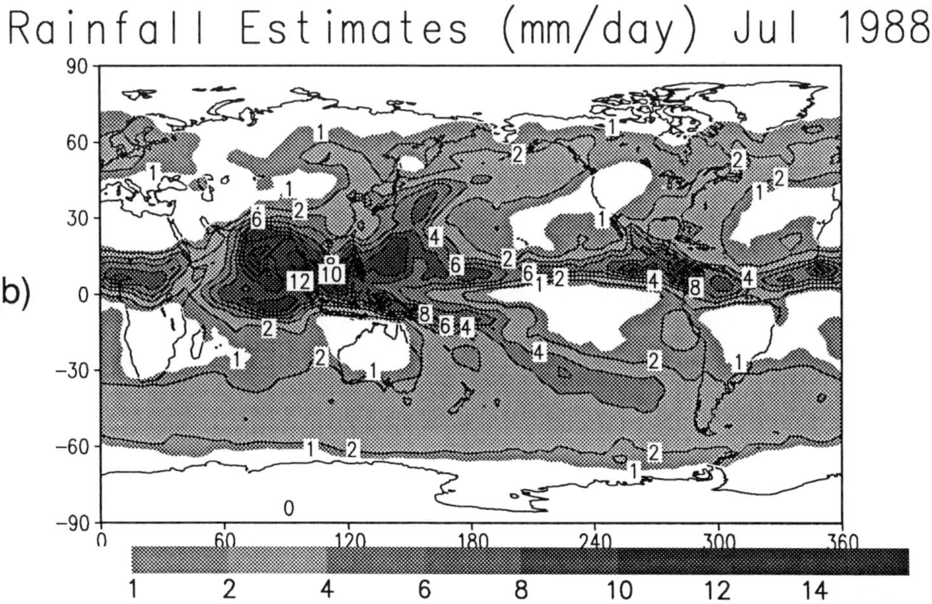

Figure 1. Mean rainfall estimate (mm day^{-1}) derived from LRF algorithm for 1980: (a) January; (b) July.

In July, the major ITCZ is found in the Northern Hemisphere between 5°-10°N. The equatorial rain belt associated with the ITCZ, the rain band associated with the East Asian and Indian monsoon are well depicted. Maximum monthly rainfall rate over the Indian monsoon regions reach up to 14 mm day^{-1}. The rain band associated with Baiu front over Japan in July are very well defined. The July dry season over eastern China is also revealed.

4. WATER VAPOR RESIDENCE TIME

A quantity that will provide useful information regarding the moisture cycling time scale is the moisture residence time defined as the ratio between the precipitation rate and the ambient atmospheric water vapor content. Equivalently the water vapor residence time is a measure of time a unit area atmospheric column takes to deplete its moisture content, at a rate equal to its climatological value. The actual residence time of water vapor will vary from this basic measure because of advection processes. Global distribution of water vapor residence time for winter and summer seasons of 1979 is computed.

Figure 2a shows that during the northern winter, residence times shorter than 10 days are found in the winter monsoon regions (maritime continents and northeast of Australia), the central part of southern Africa, and the northern part of South America, indicating the rapid recycling of atmospheric moisture due to convective processes in these regions. Figure 2b shows that during the northern summer, residence times shorter than 10 days are found over the summer monsoon areas, central America, and equatorial Africa.

Global and hemispheric mean water vapor residence time ranges from approximately 8 to 12 days. The summer hemisphere has longer residence time than the winter hemisphere, because of the more abundant moisture in the former. Furthermore, the water vapor residence time in the summertime northern hemisphere is over 25% longer than the southern hemisphere counterpart.

5. POTENTIAL FIELD OF VERTICALLY INTEGRATED WATER VAPOR FLUX AND THE POTENTIAL FUNCTION OF RAINFALL ESTIMATE

The divergent component of water vapor flux is driven by the large-scale divergent circulation (Chen, 1985). In this section, the potential (χ_Q) of vertically integrated water vapor flux ($Q = q \cdot \vec{v}$, where q is specific humidity and $\vec{v}$ is velocity vector) is computed by solving the Poison equation. The χ_Q is related to the divergence component of the water vapor flux.

(a)

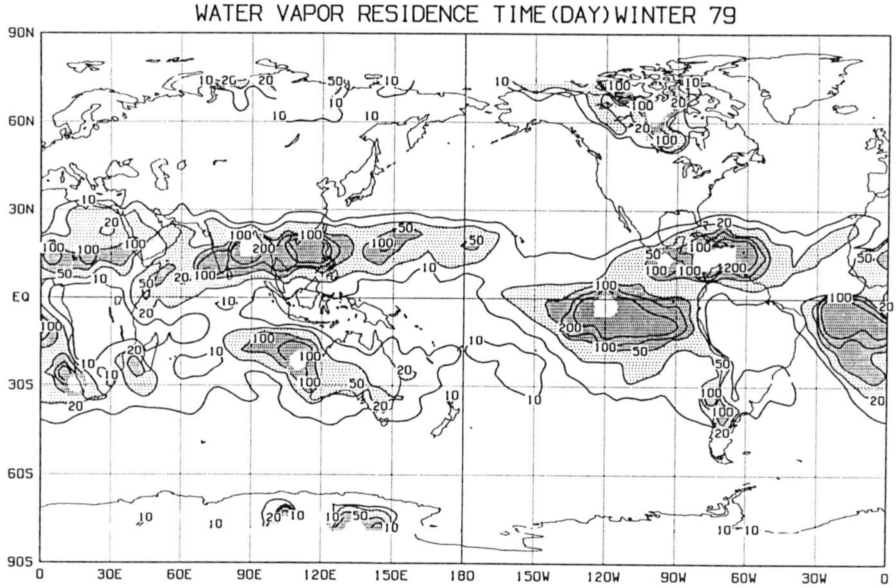

(b)

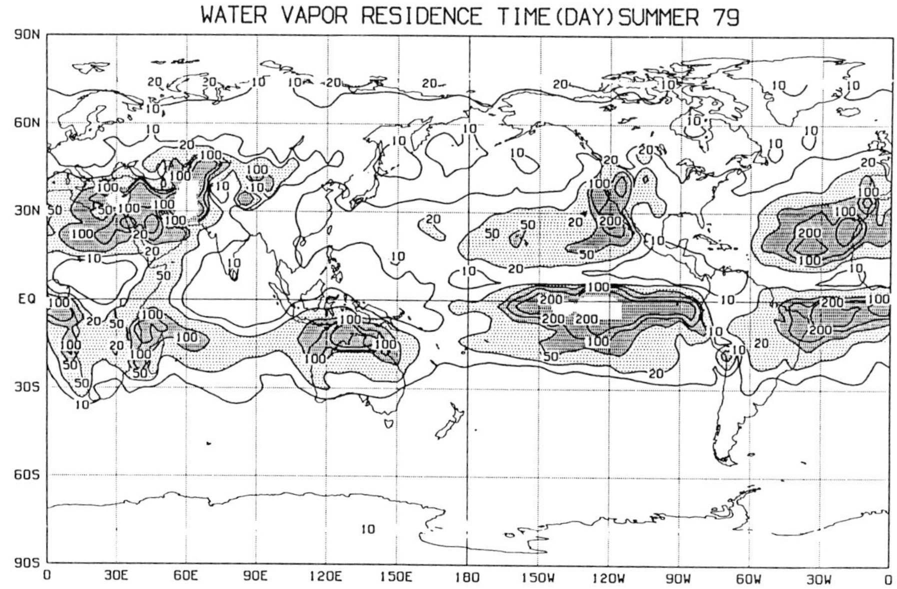

Figure 2. Seasonal mean water vapor residence time (day) for 1979: (a) winter season; (b) summer season.

In order to study the consistency between the vertically integrated water vapor flux with evaporation minus precipitation (E - P), global distribution of evaporation at the Earth surface is needed. But this field is difficult to derive using existing satellite data and current methodology. Instead, the consistency between the χ_Q and the potential function of rainfall estimate (χ_p) is presented in this section.

The potential of vertically integrated water vapor flux, χ_Q and gradient of χ_Q are presented in Figures 3a and 3b for the winter and summer of 1979, respectively, and the potential function of rainfall estimate, χ_p and gradient of χ_p are presented in Figures 3c and 3d, respectively. These Figures reveal that the potential fields of χ_Q and the gradient of χ_Q are consistent with the potential function of the rainfall estimate, χ_p, and gradient of χ_p. For example, the major water vapor sources for winter are over the equatorial western Pacific basin (Australian winter monsoon regions), the northern part of South America, and central part of the south Africa; and the major water vapor centers for summer are over Central America, equatorial Africa, and Indian summer monsoon areas. Those areas are also centers of heavy rainfall estimate. These features have been identified in Figure 12 of Chen (1985). His results are based on the Geophysical Fluid Dynamics Laboratory (GFDL) analysis.

The current results reveal that the water vapor converges toward three centers of high water water content in the tropics and indicate that the local Hadley circulation and the longitudinal Walker circulation perform the water vapor transport to maintain the high water vapor content over these three preferable tropical centers. The converged water vapor is then transported upward by the two types of circulation to increase the moisture content of the atmosphere and, in turn, to enhance the condensation, therefore, rainfall.

6. SOME CHARACTERISTIC OF RAINFALL ESTIMATE

The atmosphere is known to have variations over a wide range of spatial and temporal scales. Most of these variations are linked with latent heat release/rainfall. The following two sections present results of analyses of the ten-day mean rainfall estimate aimed at identify modes of seasonal and intra-seasonal variations in the rainfall estimate.

The LRF rainfall estimates have shown that intra-seasonal oscillations have strong regional characteristics, in particular over East Asia, India and the western Pacific (Wu, 1991). The following sections describe the intra-seasonal rainfall characteristics in more detail over these regions.

(a)

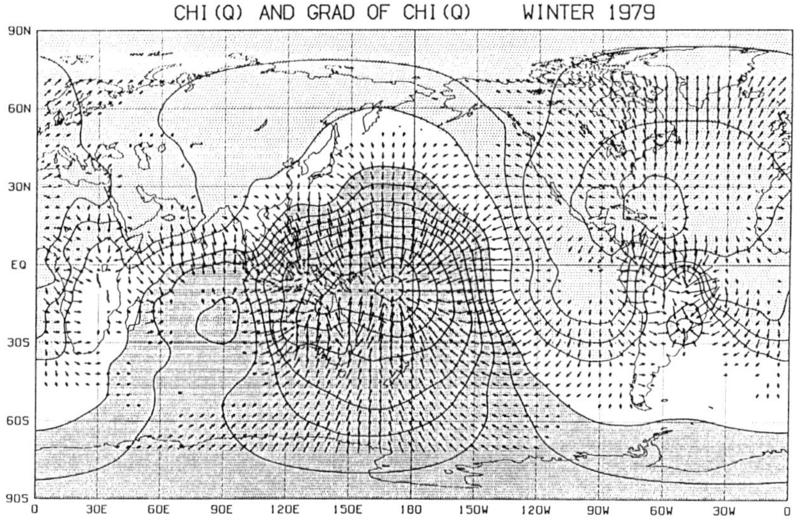

(b)

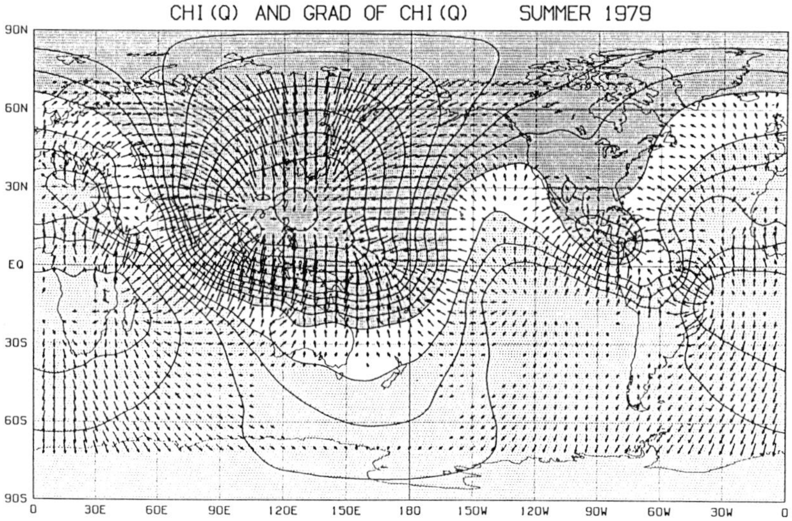

Figures 3a and 3b. Mean potential of vertically integrated water vapor transport χ_Q and its gradient for 1979: (a) winter season; (b) summer season.

(c)

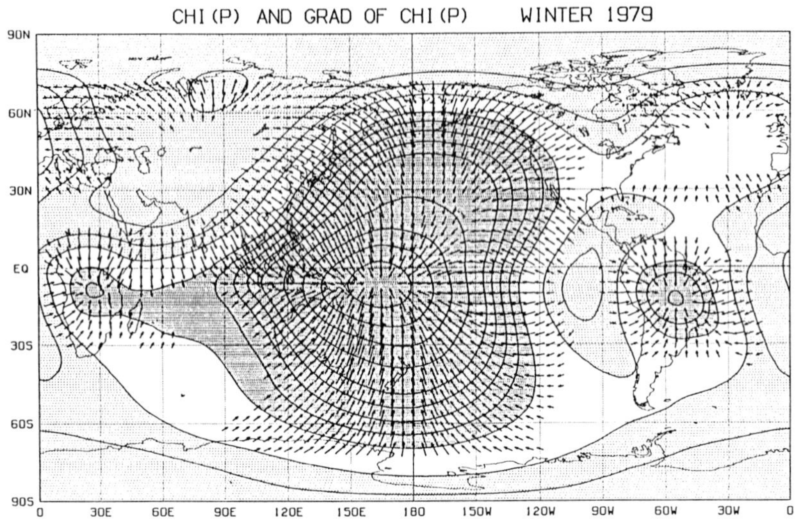

(d)

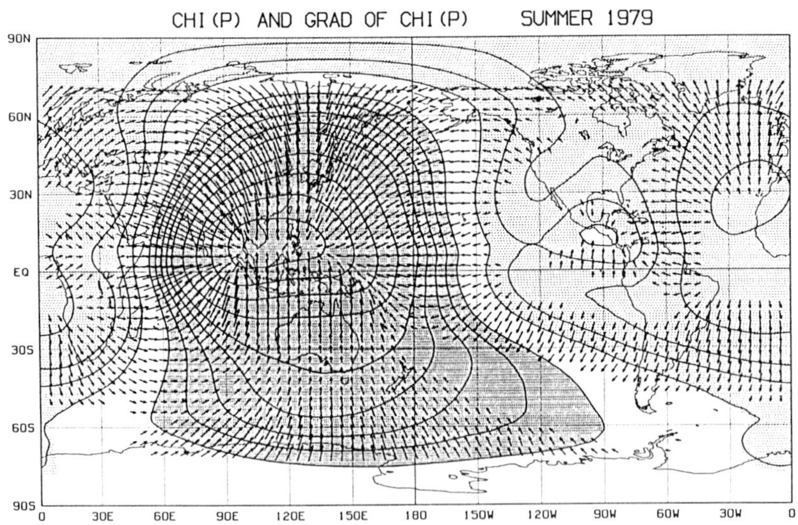

Figures 3c and 3d. Mean potential function of rainfall estimate χ_p and its gradient for 1979: (c) winter season; (d) summer season.

Latitude-time cross section

(i) East-Asian monsoon and adjacent regions (10°S - 50°N; 110°E - 120°E)

The latitude-time cross section of precipitation rate (mm per 10 days) for the 1980 over the East-Asia monsoon and adjacent regions is shown in Figure 4a. The rainfall variation between 10°S to 10°N at 30-60 day intervals is quite obvious.

Figure 4a indicates the regional fluctuation in rainfall over East Asia does not occur in isolation, but is clearly connected with the rainfall fluctuations in the equatorial regions to 10°S. In the large domain from 10°S to 50°N, the regional rainfall variation over East Asia is part of a large-scale rainfall pattern in which intra-seasonal oscillations are very pronounced. Many of the rainfall features found over East Asia appear to have their origin in near-equatorial regions.

(ii) Indian monsoon and adjacent regions (10°S - 38°N; 60°E - 90°E)

A similar latitude-time cross section has been plotted in Figure 4b for the Indian monsoon and adjacent region. The 30 - 60 day period oscillation around the equatorial region is clearly shown in the figure. The fluctuations of monsoon rainfall over the Indian subcontinent appear as northward excursions of the near-equatorial rain-belt. Similar to the East Asian region, the Indian monsoon fluctuations also emerge as part of the large-scale precipitation pattern related to the planetary scale 30-60 day oscillations in the equatorial regions.

7. PRINCIPAL MODE OF RAINFALL ESTIMATE

Empirical orthogonal function analysis has been applied to the rainfall estimates for 1979 and 1980 aimed at identify principal modes of atmospheric variation at seasonal in intra-seasonal time scales.

a. Seasonal variations

The seasonal variation of the global precipitation pattern is illustrated in Figure 5a which shows the 1st empirical orthogonal function (EOF) and its principal component of the monthly mean rainfall estimates for an area ranging from 50°S to 70°N. The 1st EOF which shows the annual cycle during the two years (1979 and 1980) explains 49.2% of the total variance. In the tropics, there are three centers of major rainfall variation zones; these centers represent a north-south standing oscillation of the principal

(a)
(b)

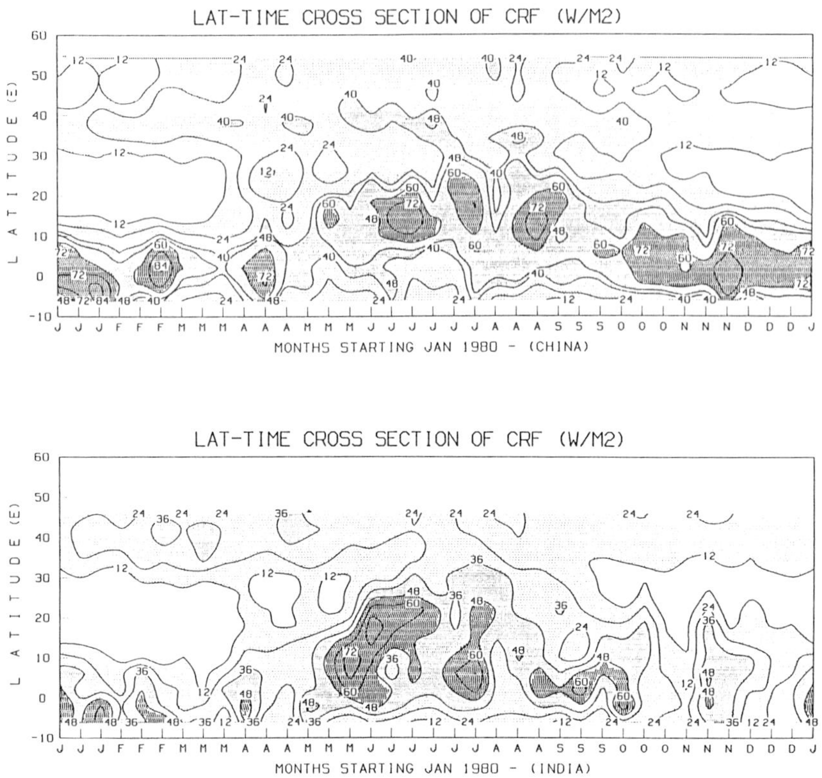

Figure 4. Latitude-time cross-section of rainfall estimate (mm per 10 days): (a) over the East-Asia monsoon and adjacent regions from 10°S to 50°N and from 100°E to 115°E; (b) over the Indian monsoon and adjacent regions from 10°S to 40°N and from 60°E to 90°E.

rainfall zones (Figure 5a) between the maritime continent/North Australia and Southeast Asia/India; Central Africa and the southern edge of the Sahel; central South America and Central America. The equatorial region, which has large rainfall through out the year (zeroth harmonic), has minimum amplitude of seasonal variation in this depiction. The dominance of the annual cycle (first harmonic) in this rainfall pattern is obvious from the principal component. This annual cycle is closely related to annual cycle of solar heating over land.

Figure 5b shows the 2nd EOF and its principal component. It shows an annual cycle largely over the ocean, it peaks in Spring and Winter instead of Summer and Winter as shown in the 1st EOF. This mode of oscillation is closely related to surface temperature over the trade wind areas.

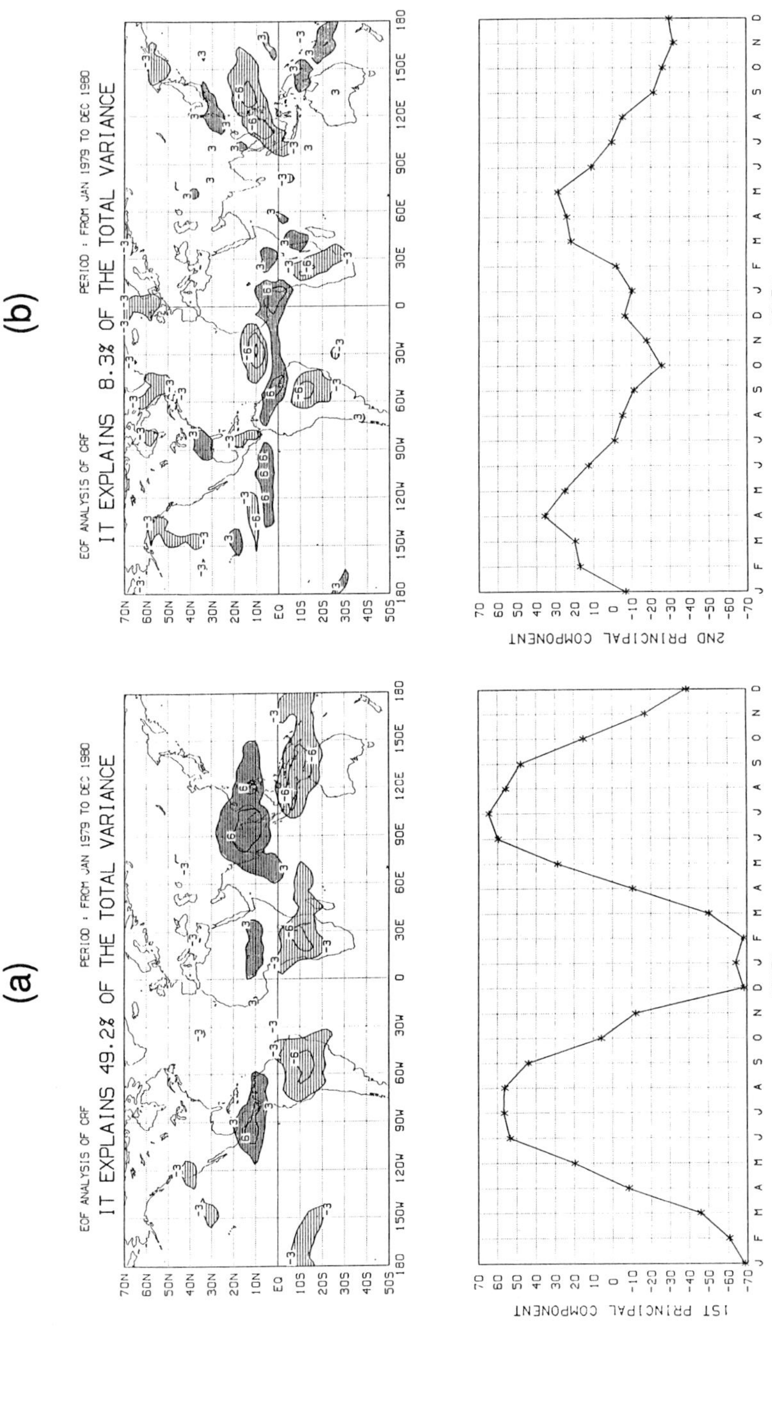

Figure 5. EOF and its principal component from 50°S to 70°N: (a) EOF 1; (b) EOF 2.

b. Intra-seasonal variation

The following is a description of the tropical rainfall pattern associated with the dominant modes of intra-seasonal variation in the tropics based on 10-day mean rainfall estimates.

Wu (1991) has shown that the 30-60 day oscillation prevails from 30°S to 50°N in LRF rainfall estimates for the FGGE year. This mode is present all year round in the equatorial region. EOF has been applied to the ten-day mean rainfall estimates for 1979 and 1980 aimed at identify principal modes of intra-seasonal variations.

Figures 6a and 6b show the 1st and 2nd truncated EOFs and their principal components in which variations with wave number less than 5 have been truncated. These two Figures show an eastward propagation mode of an east-west dipole around the equatorial Indian Ocean and Pacific with a 30 to 60 day period oscillations (Madden, R. A. and P. R. Julian, 1971). This mode of oscillation is considered as a tropical mode.

Figures 6c and 6d show the 3rd and 4th truncated EOFs and their principal components. The monsoon trough, the Mei-Yu/Baiu trough and the SPCZ are well revealed in the spatial pattern of the EOFs. Most of the EOF patterns are oriented northeastward in the Northern Hemisphere and southeastward in the Southern Hemisphere. This mode of oscillations with a period of 30 to 60 days is considered as an extra-tropical mode.

8. CONCLUSIONS

Global distribution of rainfall estimates, water vapor residence time and moisture transport have been estimated from IFRA data.

The current result reveals that the water vapor converges toward three centers of high water vapor content in the tropics and indicates that the local Hadley circulation and the longitudinal Walker circulation perform the water vapor transport to maintain the high water vapor content over these three preferable tropical centers. The converged water vapor is then transported upward by the two types of circulation to increase the moisture content of the atmosphere and, in turn, to enhance the condensation, therefore, rainfall. The seasonal cycle of global rainfall estimates consists of a north-south migration of major precipitation zones between northern Australia and India/East Asia and two secondary zones between central Africa and the southern edge of the Sahel, and Central South America and Central America.

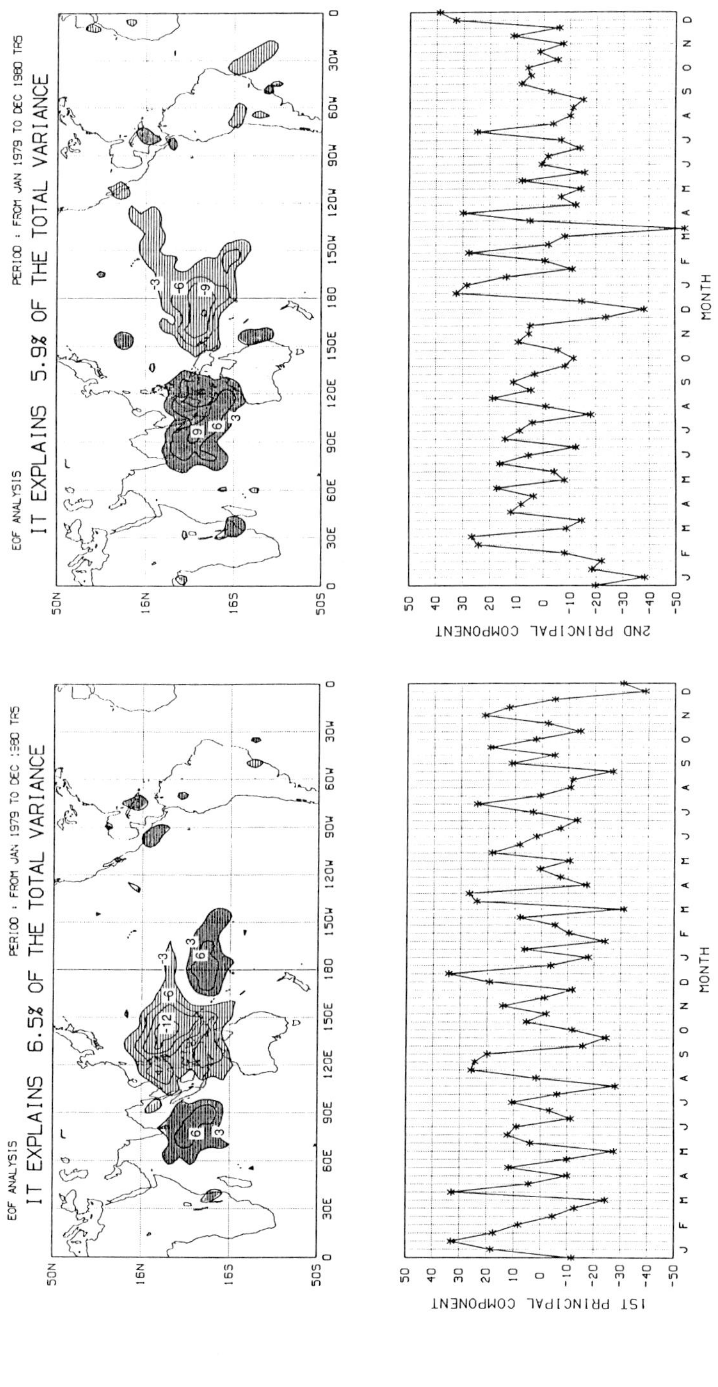

Figures 6a and 6b. Truncated EOF and its principal component from 50°S to 70°N: (a) EOF 1; (b) EOF 2.

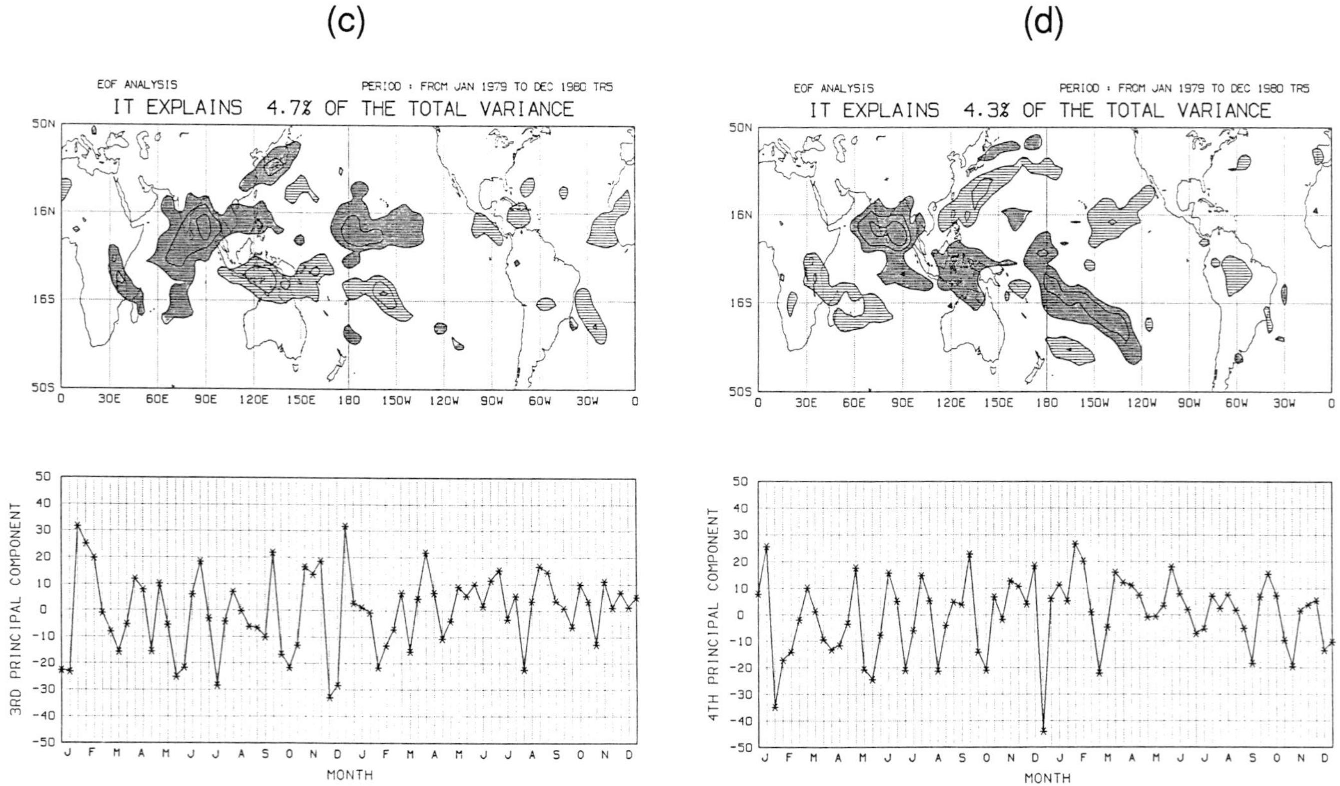

Figures 6c and 6d. Truncated EOF and its principal component from 50°S to 70°N: (c) EOF 3; (d) EOF 4.

In addition, the major precipitation pattern migrates eastward from northern summer to winter and retreat westward in spring in the equatorial regions is also clearly identified.

The ten day mean rainfall estimate reveals an ubiquitous 30-60 day period oscillation superposed on the seasonal cycle. The oscillation is particularly strong over the monsoon region of southeast Asia and India. Most important regional fluctuation in rainfall over East Asia is clearly connected with the rainfall fluctuations in the equatorial regions to 10°S. In the large domain from 10°S to 50°N, the regional rainfall variation over East Asia is part of a large-scale rainfall pattern in which intra-seasonal oscillations are very pronounced. Many of the rainfall features found over East Asia appear to have their origin in near-equatorial regions. Similar to the East Asian region, the Indian monsoon fluctuations also emerge as part of the large-scale precipitation pattern related to the 30-60 day oscillations in the equatorial regions.

Several principal modes of oscillation have been identify through the EOF analysis of the ten-day mean rainfall estimates: a seasonal cycle over land associated with solar heating over land, a seasonal cycle largely over the ocean associated sea surface temperature over the trade wind areas, a tropical mode of 30 to 60 day period oscillations, and an extra-tropical mode of 30 to 60 day period oscillations.

All these rainfall variations are closely linked to other hydrologic parameters such as total precipitable water and vertically integrated water vapor flux.

The results of this paper demonstrate that the ten-day mean LRF rainfall estimate possess valuable information in the seasonal and intra-seasonal time scales, and regional and global scales. Because the LRF is unique in the sense that it is a by-product of a comprehensive retrieval system, it is consistent with a host of other global radiational and hydrologic parameters derived from the same retrieval system. Therefore the LRF rainfall should not only be examined for its own sake, but should be analyzed and cross-correlated with other retrieved global fields. At present, work to produce a multi-year set of satellite retrieved and model assimilated hydrologic fields are underway at NASA/GLA. The multi-year LRF rainfall product will be a useful data-set of future climate research and model validation.

ACKNOWLEDGMENT

The author would like to thank Dr. J. Susskind for providing retrieval products from the analysis of the HIRS2/MSU sounding data.

REFERENCES

Baker, W. E., 1983: Objective analysis and assimilation of observational data from FGGE. *Mon, Wea. Rev., 111,* 328-342.

Chen, T. - C., 1985: Global water vapor flux and maintenance during FGGE, *Mon. Wea. Rev., vol. 113*, 1801-1819.

Kalnay, E., R. Balgovind, W. Chao, D. Edelmann, J. Pfaendtner, L. Takacs, and K. Takano, 1983: Documentation of the GLAS Fourth Order General Circulation Model, NASA Technical Memorandum 86064 (NTIS N8424028)

Krishnamurti, V., 1982: The documentation of the Wu-Kaplan radiation parameterization. NASA Tech. Memo 83926, NASA/Goddard Space Flight Center, Greenbelt, MD. 93pp.

Lau, K.-M. , and P. H. Chan, 1985: Aspects of the 40-50 day oscillation during the northern winter as inferred from outgoing long-wave radiation *Mon. Wea. Rev., 113,* 1889-1909.

Madden, R. A. and P. R. Julian, 1971: Detection of a 40-50 day oscillation in the zonal wind in the tropical pacific. *J Atmos. Sci., 28*, 702-708.

Ramanathan, V., 1987: The role of earth radiation budget studies in climate and general circulation research, *J. Geophys. Res., 92*, 4075-4095.

Susskind, J., J. Rosenfield, D. Reuter, and M. T. Chahine, 1984: Remote sensing of weather and climate parameters from HIRS2/MSU on TIROS-N. *J. Geophys. Res., 89D,* 4677-4797.

Wu, M. L. C., 1991: Precipitation estimates from Satellite: Using difference Fields of Outgoing Long-wave Radiation. (to appear in March issue of *Atmosphere-Ocean*).

_____, and J. Susskind, 1990: Outgoing long-wave radiation computed from HIRS2/MSU soundings. *J. Geophys. Res., 95, D6*, 7579-7602.

Socio-economic Impact

SOCIAL SCIENCE APPROACHES IN DISASTER RESEARCH: SELECTED RESEARCH ISSUES AND FINDINGS ON MITIGATING NATURAL HAZARDS IN THE URBAN ENVIRONMENT

J. M. Nigg

Disaster Research Center, University of Delaware, Newark, DE 19716, U.S.A.

ABSTRACT

This paper has two purposes: (1) to review the major issues that social scientists address with respect to reducing the risk from natural hazards in urban areas and (2) to provide an overview of the research findings associated with these issues. The paper focuses on the following issues: public threat perception and action; community-level disaster preparedness planning and response; community hazard mitigation actions. The paper concludes with a discussion of the importance of cross-national research to improve our understanding of social action and social organization with respect to disaster and hazard management.

KEYWORDS

Disaster research; comparative research; hazards; threat perception; disaster phases; preparedness; response; recovery; mitigation; planning.

The social consequences of disasters have been rigorously studied by social scientists, primarily sociologists and geographers, since the late 1940's (cf., Fritz and Marks, 1954; Barton, 1969; Kreps, 1981; Quarantelli, 1987, 1992; Nigg, 1994). More recently, additional social science disciplines--economics, political science, psychology, and policy analysis--have focused research attention on the social aspects of disasters.

Even in the earliest studies in this field, social science researchers rejected the popularly-accepted notion that societies affected by a disaster experience some disintegration, and that individuals in such situations engage in irrational and anti-social behavior. Instead of discovering support for this "disorganization" model of societal response to disaster events, sociological researchers argued that disaster-stricken communities resolve disaster problems through relatively normal collective or institutional approaches, and that these approaches allow them to rebound with some degree of resilience (Nigg et al., 1989).

Because early disaster research efforts focused principally on the impact and immediate post-impact periods of a disaster, a great deal was learned about reaction and response to a disaster, but little of the research was concerned with preventing the disaster or reducing its consequences (Nigg et al., 1994). Within the past 15 years, however, mitigation and hazard reduction have become very important topics of investigation for social scientists. No longer can we be satisfied only to provide relief and recovery assistance to the victims--citizens and governments alike--of a disaster. Disaster costs globally are escalating as the numbers of our populations continue to rise, as the density of populations in our cities increases dramatically, and as our infrastructural systems become more fragile and overburdened. Our attention, indeed, must be turned to discovering how to lessen the vulnerability of our human and material resources, to make our cities more "disaster-resistant" in socially and economically acceptable ways.

The purpose of my presentation today will be to identify some of the major research issues that social scientists address with respect to mitigating and reducing natural hazards, especially in urban areas. Before, however, I begin talking specifically about this body of research, I would like to make two points about social science approaches to disaster research; the first concerns the definition of a disaster; and the second categorizes the various strands of social science research on natural hazards and disasters within a typological framework.

A DEFINITION OF DISASTER

For social scientists, natural disaster events--even those that could be considered "great" earthquakes of magnitude 8.0 or more--are not particularly significant events that require investigation unless that seismic activity substantially disrupts the built or social environments. Conversely, a small, non-damaging earthquake that seismologists and engineers might disregard could have great research interest for social scientists if, for example, the general public believed that event indicated that a larger damaging quake were inevitable and began to prepare themselves.

In other words, from a social science perspective, it is not always true that disaster consequences are related to the magnitude or intensity of the disaster agent itself. Even when disaster events have similar physical characteristics they may not have similar social consequences. Take, for example two recent earthquakes--the 1988 earthquake in Armenia and the 1989 Loma Prieta earthquake in California. The earthquake in Armenia, which was 6.9 magnitude, killed approximately 25,000 people, injured more than 31,000, and left 514,000 homeless. By contrast, the slightly larger (7.1) magnitude Loma Prieta earthquake in the San Francisco Bay area killed only 62 people, injured 3,800, and left approximately 12,000 homeless. The relationship between the physical event and "social damage" it produces is quite tenuous (Dynes, 1992).

For this reason, social scientists define hazards and disasters not on their physical characteristics but on the basis of social disruption. A **hazard** can be defined as: "some aspect of the physical environment that threatens the well-being of individuals and their society." These threats to well-being include harmful or disruptive effects to social, economic, and political systems, or to the built environment (buildings, dams, power and water systems, bridges and roads) that interferes with normal, daily life of a community.

A hazard, then, only exists when it is capable of creating a **disaster**. A disaster can be said to occur only "when the built and social environments are so disrupted that the resources of the social system are overwhelmed and the system is unable to meet the demands placed on it for goods and services that are routinely expected by its citizens."[1]

---

[1] This definition is adapted from the classic definition developed by Charles Fritz (1961: 655).

A CATEGORIZATION OF RESEARCH EFFORTS

Figure 1 presents a method for categorizing different research efforts on two different dimensions: (1) by disaster phases; and by (2) the unit of analysis of the study.[2]

UNITS OF ANALYSIS	DISASTER PHASES			
	Preparedness	Mitigation	Response	Recovery
Individuals				
Groups/ Organizations				
Organizational Networks/ Communities				
Regions/ Societies				

Fig. 1. Typological Scheme For Categorizing Research Topics on Earthquake Hazards

Disaster Phases

Four phases of disaster-related activities have been used by emergency management specialists to develop hazard reduction and disaster response plans. Although there is some overlap between these phases (that is, they are not mutually exclusive), each generally occurs at a different time in relation to the impact of a disaster event and consists of different types of activities taken in order to protect the social system from that impact or its consequences.

Social science research topics can also be categorized under one of these four phases. **Preparedness** refers to those actions taken prior to the impact of a disaster that will improve the ability of the system to respond to the event. Preparedness activities would include: the development of emergency response plans for organizations; training of emergency response personnel; interorganizational coordination of communication, responsibility, and authority; identification and provisioning of temporary mass shelters for victims; the development of warning systems; and stockpiling of medical resources for immediate treatment of victims.

The **response** phase involves putting into effect the plans developed during the preparedness phase. It may actually begin immediately before the disaster strikes--if, for example, there are indications that a potential disaster could occur--and continue for days or weeks, depending on the severity of the disaster impact. Response activities would include: issuing warnings to emergency response organizations and

---

[2] This typology was adapted from Thomas Drabek (1986).

to the general public; establishing emergency operations centers to direct the post-impact activities; directing evacuation of threatened areas; undertaking search and rescue efforts; firefighting or controlling hazardous materials incidents; restoring needed lifeline systems (electricity, water, sewerage, communications, transportation); and providing shelter, food, and medical care for disaster victims during the immediate post-impact period.

The response phase, which usually ends within hours or days after impact, leads directly into the **recovery** phase. Recovery activities may last days, weeks, or even years depending on the severity of the disaster and the size of the geographic region affected. Recovery of a community or region refers to returning the social systems and the built environment to pre-impact levels of functioning. Recovery efforts would include: reconstruction of buildings and infrastructural systems; provision of long-term replacement housing for victims whose homes were destroyed; economic assistance to businesses to resume operation and production; relief assistance to families to recover from loss of household goods and lost wages.

Mitigation refers to actions taken to lessen the actual impact of the disaster agent--and earthquake, a flood, a hurricane or typhoon, a tsunami--on the population or the built environment. Mitigation efforts may begin taking place even during the early response periods and continue until the next disaster event occurs. Mitigation actions would include: the development of building codes with seismic, high water, or wind design elements for both new and existing structures; the establishment of construction practices and regulatory systems to ensure that these provisions are correctly incorporated during the construction process; the development of land use plans that would restrict development in areas of high geologic or hydrologic risk; the identification of practices to protect non-structural systems in and the contents of buildings (such as manufacturing equipment and computers); and the identification of construction practices to protect lifeline systems (electrical, gas, water, and sewerage systems).

Units of Analysis

The second dimension of this typology refers to the systemic level of analysis that is of interest to the researcher; that is, what level of **behavior** is the researcher interested in explaining. We can identify four general types of research foci along a micro-macro continuum--from the behavior of individuals to that of societies, large geographic regions, or nation states.

Depending on the unit of analysis selected by the researcher, different theoretical frameworks are used to formulate hypotheses and to develop the research design. For example, studies of individuals are most likely to use theories from social psychology. Studies of groups (including both existing groups and groups that emerge during a crisis) or formal organizations often use one of several organizational theories or theoretical approaches from the field of collective behavior and social movements. When interorganizational or community studies are conducted, theoretical frameworks often come from network analysis, urban sociology, urban planning, intergroup relations, and management sciences. For macro-level studies of societies or nation states, theoretical perspectives from political science, political economy, and policy analysis are often applied.

EXAMPLES OF RESEARCH ON MITIGATION TOPICS

Mitigation refers to those actions undertaken to reduce a community's vulnerability from possible future impacts of a hazard agent which would result in a disaster. Although individuals can also undertake mitigation measures, most of the emphasis in mitigation efforts as well as in mitigation research is on professional and governmental attempts to reduce vulnerability. Mitigation activities are generally thought of as collective solutions to a threat which could have widely felt effects in a community, region, state, or country.

How these governmental units decide which mitigation techniques to use is central to mitigation research efforts. Why do some communities adopt certain mitigation measures when other communities that have a similar objective risk do not, opting either for a different mix of techniques or none at all? What characteristics of the community and its history of experience with a particular hazard agent (or agents) are related to the acceptability of some hazard reduction measures over others? How do the community's political and economic climates affect the adoption of different mitigation strategies? These questions highlight the importance of the need to understand the context in which the decision-making process takes place as key actors (i.e., community leaders or elected officials) consider different mitigation actions.

Early mitigation research was conducted by social geographers who were concerned about the success of flood mitigation efforts (White, 1964, 1975). More recently, questions have been raised by planners and political scientists about the success of the United States' National Flood Insurance Program--the federal government's primary policy tool to reduce flood risks (Burby et al., 1985).

Similar questions have also been raised about local, state, and federal attempts to reduce community vulnerability from earthquakes. Before communities can make decisions about what to do, they must have some clear understanding of the seismic hazard and risk in their area. Sociologists began investigating how the public assessed earthquake threat in the mid 1970's when geoscientists in California felt they were on the verge of being able to predict earthquakes and governmental officials were concerned about how the public would react to such announcements (Mileti et al., 1981; Turner et al., 1986). Contrary to the belief that people would become overly fearful and be likely to engage in mass evacuation if such announcements were issued, these research efforts concluded that individuals wanted to be kept informed, even of very low probability events, and that the public was positively disposed toward continued efforts to educated them about what measures could be undertaken to reduce potential harms if such announcements were made. Subsequent assessments of how people did in fact respond to such announcements confirmed the sociological contention (Nigg, 1982) that people would not panic, but would instead continue to function normally while seeking out additional information about the prediction and hazard reduction measures (Goltz, 1984).

Although considerable research attention has been focused on California because of its high degree of seismic activity, social science researchers are now turning their attention toward understanding the adoption of earthquake mitigation efforts in other areas of objectively high seismic risk, high seismic vulnerability (because the built environment has not been constructed with seismic concerns in mind), but a low probability of the occurrence of a damaging earthquake event--areas such as Puget Sound in Washington State; the Central United States; South Carolina, and the intermountain West, Alaska, and Hawaii. Mushkatel and Nigg (1987a, 1987b) discovered that almost no mitigation actions had been undertaken in either high or moderate seismic risk areas in six states in the New Madrid Fault zone, although key actors' awareness of the earthquake threat was moderately high.

Research has confirmed that the adoption of mitigation measures is strongly associated with the high frequency or occurrence of a specific disaster agent. Possibilities for initiating new mitigation policies and risk-reduction programs are highest during the post-impact period; but this "window of opportunity" closes rather quickly. Desire to get the community back to "normal" and to get businesses "back on their feet" as quickly as possible puts pressure on local government officials to rebuild at the same sites and in the same manner as before the disaster, even if they may be merely reproducing the conditions that caused this disaster. But even when mitigation programs are implemented following a major disaster, it has been found that some communities and states have rescinded previously adopted mitigation policies both due to internal economic constraints and to external pressures for development.

Constraints during "normal" times work against adoption or enthusiastic implementation of measures, even in areas of high seismic activity, areas routinely inundated by riverine flooding or storm surges, or areas imperilled by hurricanes or firestorms. For example, it is extremely difficult to introduce hazard-relevant changes in land use patterns in already-developed areas. Economic development interests are often in conflict with improvements in structural safety, whether for new construction or the rehabilitation of existing structures, or for new development in hazardous areas. The most frequently heard reasons for this lack of interest in or commitment to hazard reduction from key actors is either that they are not convinced that the hazard in their area is sufficient to warrant such measures, or they are concerned that businesses will move elsewhere where it is cheaper to buy and build if such policies are adopted in their community.

And when disaster events are very infrequent--such as a damaging earthquake in an area that has not experienced a seismic event in 100 years or more--but their occurrence could result in significant life loss and economic and social disruption to a community or region, what conditions are likely to lead to the adoption of mitigation adjustments? Mileti (1980) identified two factors that he argues are related to the adoption of risk mitigation adjustments across natural hazard agents--the capacity to implement various policies; and the perceived costs of the implemented policy.

Capacity refers to the resources a governmental or administrative entity needs in order to undertake new adjustment-related tasks in relation to their on-going responsibilities. The capacity to implement any new mitigation effort is high if it does not require much change in resource allocation and utilization from the status quo. Higher levels of adoption are also related to the availability of economic resources and expertise, both of which are necessary to implement new programs.

Even if decision makers believe that their community is exposed to a significant threat, they may believe that the cost involved in attempting to mitigate that hazard is unacceptably high. Cost are not limited to economic concerns solely but also include calculations of social and political costs. Elected officials, especially those at the local level of government, must weigh the social costs of not taking action (possible life loss and economic disruption) against political costs of supporting mitigation actions (unfavorable reactions from interest groups who would be required to bear economic costs for changing their current practices or who would lose support for non-hazard related programs, such as social welfare programs). Also, mitigation research has demonstrated that, unless carefully thought out, adoption of some measures may result in negative changes in the character of neighborhoods or the community itself, affecting the overall well-being of the community. For example, the City of Los Angeles passed its first retrofit ordinance for pre-1933 unreinforced masonry buildings in 1981. Many activist citizen groups raised concerns that the 8,000 buildings of this type--which were predominately in the older, inner-city areas of Los Angeles--would create hardships for the people who lived and worked in these structures. Many of these structures provided low-cost housing for the poor and elderly; they also housed several minority-owned businesses. Using a variety of incentive and loan programs, the city tried to limit the negative consequences on both the building owners and tenants as the long-term process of retrofitting took place.

With respect to natural hazard mitigation efforts, a great deal of research is currently underway in the United States on a number of different issues associated with different hazard agents. The Great Midwest Floods of 1993 provide an extraordinary opportunity for social scientists to analyze and evaluate a new federal philosophy underlying flood hazard reduction--a shift in federal policies from structural solutions to flooding problems (i.e., building dams and levees) to non-structural solutions (i.e., moving parts of almost 200 communities out of the floodway and allowing the land to return to marshlands. Little research has been done on the relocation of communities due to natural hazards; this situation provides a wonderful "natural laboratory" to determine how communities of different types are affected by this new policy. Similarly, work is continuing on different types of earthquake hazard reduction measures to determine what incentives, regulatory strategies, professional actions, or insurance coverage will enhance the likelihood of their adoption and implementation by local governments. More recently we have seen evaluations of existing policies and programs being undertaken in order to determine whether they are having the intended outcomes--the reduction of loss, damage, and risk in different parts of the country. The National Earthquake Hazard Reduction Program and the National Flood Insurance Program are two of these assessments now in progress.

But the findings of social science research in the United States may not be applicable in other parts of the world. Our governmental style, the extreme influence of the private sector on governmental policy, our historically uncritical acceptance of technological solutions to problems, and our extremely diverse population and class system, may affect the ways we have developed and implemented our natural hazard mitigation strategies and solutions. Because so little cross-cultural, comparative research has been undertaken--not just with respect to natural hazards, but also with respect to technological, and environmental hazards--we do not know how generalizable our research findings or our theoretical explanations are to other cultures and societies. I would hope that one of the outcomes of this workshop will be to develop a set of socially important, as well as theoretically relevant, issues that could be addressed by social scientists in our respective countries in order to broaden our understanding of human response to situations of threat and disaster in urban areas.

REFERENCES

Barton, A. (1969). Communities in Disaster. Doubleday, New York.

Burby, R.J., S.P. French, B.A. Cigler, E.Kaiser, D.H. Moreau and B.Stiftel (1983). Flood Plain Land Use Management: A National Assessment. Westview Press, Boulder, CO.

Drabek, T. (1986). Human System Response to Disaster: An Inventory of Sociological Findings. Springer-Verlag, New York.

Dynes, R.R. (1992). "Disaster Reduction: The Importance of Adequate Assumptions About Social Organization." Proceedings of the Seminar "Training for Disaster Reduction". pp.14-25. Disaster and Emergency Reference Center, Delft University of Technology, Delft, The Netherlands.

Fritz, C.E. and E.Marks (1954). "The NORC Studies of Human Behavior in Disaster." Journal of Social Issues. 10, 3, 26-41.

Goltz, J. (1984). "Are the News Media Responsible for the Disaster Myths? A Content Analysis of Emergency Response Imagery." International Journal of Mass Emergencies and Disasters. 2, 345-368.

Kreps, G.A.(1981). In: "The Worth of the NAS-NRC (1952-1963) and DRC (1963-present) Studies of Individual and Social Responses to Disasters." (J. D. Wright and P. H. Rossi, eds.), Social Science and Natural Hazards. ABT Books, Cambridge, MA.

Mileti, D.S. (1980). "Human Adjustment to the Risk of Environmental Extremes." Sociology and Social Research. 64, 3, 327-347.

Mileti, D.S., J.Hutton and J. Sorenson (1981). Earthquake Prediction Response and Options for Public Policy. Institute of Behavioral Science, University of Colorado, Boulder, CO.

Mushkatel, A.H. and J.M. Nigg (1987a). "Opinion Congruence and the Formulation of Seismic Safety Policies." Policy Studies Review. 6, 645-656.

Mushkatel, A.H. and J.M. Nigg (1987b). "The Effect of Objective Risk on Key Actor Support for Seismic Mitigation Policy." Environmental Management. 11, 77-86.

Nigg, J.M. (1982). "Communication Under Conditions of Uncertainty: Understanding Earthquake Forewarnings." Journal of Communication 32, 27-36.

Nigg, J.M. (1994). "Influences of Symbolic Interaction of Disaster Research." Self, Collective Action and Society: Essays Honoring the Contributions of Ralph H. Turner (G. Platt and C. Gordon, eds.), JAI Press, Greenwich, CT. (In Press)

Nigg,J.M. and D.Mileti (1994). "Natural Hazards and Disasters." In: Handbook of Environmental Sociology (R.E. Dunlap and W. Michaelson, eds.), Greenwood Press, Westport, CT. (forthcoming)

Nigg, J.M. and Ronald W. Perry (1989). "Influential First Sources: Brief Statements with Long-Term Effects." International Journal of Mass Emergencies and Disasters. 6, 3, 311-343.

Quarantelli, E. L. (1987). "Disaster Studies: An Analysis of the Social Historical Factors Affecting the Development of Research in the Area." International Journal of Mass Emergencies and Disasters. 5, 285-310.

Quarantelli, E. L. (1992). "Disaster Studies: An Analysis of the Consequences of the Historical Use of a Sociological Approach to the Development of Research in the Area." Unpublished manuscript.

Turner, Ralph H., J.M. Nigg and D.H. Paz (1986). Waiting for Disaster. University of California Press, Berkeley.

White, G.F. (1964). Choice of Adjustment to Floods. Department of Geography, University of Chicago, Chicago.

White, G.F. (1975). Flood Hazard in the United States. Institute of Behavioral Science, University of Colorado, Boulder.

ECONOMIC ASPECTS OF URBAN VULNERABILITY AND DISASTER MITIGATION

B. G. Jones* and S. E. Chang**

** Department of City and Regional Planning,*
Cornell University, Ithaca, NY 14850, U.S.A.
*** EQE International, Inc., 18101 Von Karman Avenue,*
Suite 400, Irvine, CA 92715, U.S.A.

ABSTRACT

The economic impacts of natural disasters are becoming greater for events of the same severity because of increasing urbanization and levels of investment in the built physical environment. Interest in the topic has grown in recent years, and contributions to our understanding are widely scattered over a variety of sources. A conceptual framework is developed for research on this subject, and the literature is briefly reviewed. The purpose of this survey article is to provide an overview of the current state of the field implicitly indicating areas that have received greater and lesser attention. While the material covered is extensive it does not claim to be complete, and the emphasis is on the use of methods of regional economic analysis.

KEYWORDS

Economic impact; natural disasters; urban vulnerability; regional economics.

INTRODUCTION

The major role of the social sciences in disaster research is that of identifying, quantifying, and analyzing the impacts natural disasters have on the populations and socioeconomic systems of impacted regions. They focus on the interaction in disaster situations between two systems: the social and economic system and the built physical environment system that accommodates it. How do failures in one system affect the other and how can understanding of these relationships advance the goal of vulnerability reduction?

We have become increasingly aware that economic impacts of disasters are getting greater at the same time that casualties have been decreasing. For example, based on available data for 1950 to 1979 for the U.S. from the *Catalogue of Significant Earthquakes, 2000 B.C. - 1979*, by Robert Ganse and John Nelson, N.O.A.A., 1981, Figure 1 shows that direct economic losses from moderate to large earthquakes have been increasing over time, even when normalized for earthquake magnitude, population growth, and Gross National Product. This increase reflects in part the expansion in the amount of built physical environment at risk over time. Unfortunately, because of changes in data collection methodologies, the analysis cannot be continued with data beyond 1979.

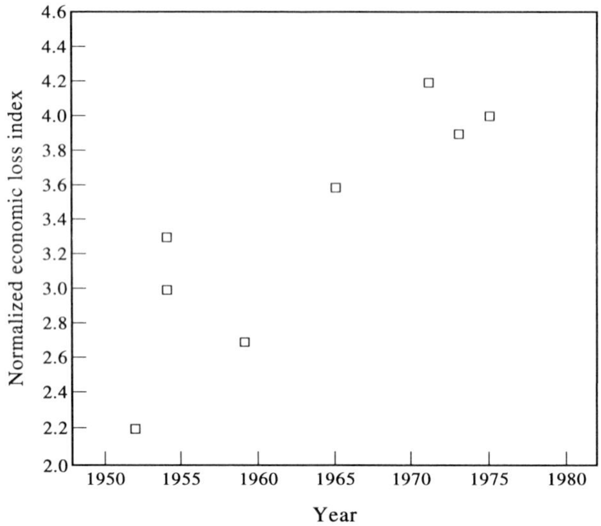

Figure 1. Normalized Economic Loss: U.S. Earthquakes, Magnitudes 6.0 to 7.9, 1950-1979.

At the urban and regional level, the economic aspects of natural disasters can be classified into two subject areas for research: impact on stocks or fixed capital investments, and impact on flows or production and income. In addition, natural disaster mitigation is influenced by behavioral decisions at the microeconomic level. This paper discusses recent research in each of these three areas, focusing on the application of empirical methodologies for assessing urban vulnerability and disaster mitigation.

THE BUILT PHYSICAL ENVIRONMENT

The immediate economic impact of disasters is destruction of or damage to the enormous accumulated investment in the built physical environment, or social overhead capital (Jones, 1981). Very little systematic information exists about these elements and the processes that create them, yet information on the exposure to natural hazards is essential for the assessment of an urban area's vulnerability to natural disasters. Furthermore, following a disaster, knowledge of the building stock inventory is critical for calculating damage rates and establishing the severity of the disaster; for example, the destruction of 10 school buildings is much more catastrophic for a region that had only 100 schools before the disaster than for one that had 1,000.

While substantial attention has been given to the accumulation of capital stocks, much of this has been concerned primarily with equipment used directly in production processes. There is a relatively small literature on the economics of capital investment in the physical built environment of buildings and infrastructure. We have made some strides in expanding our understanding of the determinants of investment in different types of structures and the extent to which these investments can be predicted based upon changes in the economy, particularly sectoral composition, by developing econometric estimates of investment functions for different structural types (Chang, 1994b). The evidence shows, in particular, that because different industries have different requirements for physical infrastructure, a region's capital accumulation pattern in different types of physical infrastructure is directly related to the industry mix of its economy.

The accumulated stock at any point in time consists of the survivors of investments over the years minus those that have been discarded or demolished. Little attempt has been made to investigate systematically demolition processes of built physical elements. Recently the small existing literature has been reviewed, data on demolitions gathered, and methods developed for estimating demolition rates, resulting in considerably longer survival rates than usually supposed. Studies that have been made for urban areas where reliable data have been available for reasonably long periods indicate annual rates of demolition in the range of .09% to .4%. Estimated probable service life of buildings constructed this year range from 150 years to more than 500 years (Steinmeyer, 1994).

We have made strides in recent years in developing methods for estimating and inventorying the surviving stock of elements at risk. Since Del Tosto and Eguchi (1980) multiplied population weighted by income by a "building wealth-income" coefficient to estimate the value of the building stock at risk in a region. We have learned more about factors influencing such investments and the ways in which they accumulate over time so that with modest rates of addition and extremely low rates of attrition their replacement would require some multiple of Gross National Product. Replacement cost estimates that have been developed indicate that for the total built environment they are a little less than 4 times GNP. The replacement cost of lifeline infrastructure alone seems about equal to GNP. Gross sectoral product in construction, in comparison, averages about 10% of GNP, and only about half is for new construction (Jones *et al.*, 1991).

The analysis of empirical regularities in the relationship between urban population, population density, urban population distribution and building patterns has enabled development of indirect methods for estimating urban building stock (Jones *et al.*, 1976; Jones *et al.*, 1987). A conceptual framework for compiling a baseline inventory, determining relationships, analyzing patterns of distribution in space, and developing indirect methods from a detailed study of one medium-sized metropolitan area, Wichita-Sedgwick County, Kansas, has been summarized and findings reported by Jones (1994).

Field tests of indirect methods for estimating existing stocks of buildings and infrastructure indicate they are general and applicable in different places and cultures. They have been shown to provide reasonable estimates for many types of building and infrastructure for Memphis, Tennessee (Jones and Chang, 1994). With adjustments for urban size, population density, and regional housing preferences, the methods have also proven applicable to as large and heterogeneous a metropolitan area as Los Angeles, for example, providing estimates to within 5 percent of actual totals for the number of residential buildings at the county level (Chang, 1994a). Applied to Kyoto, Japan, the indirect methodology appears to be highly robust for aggregate quantities of infrastructure and buildings, although at a more disaggregated level expected cultural differences (such as in the distribution of roadways between highways, streets, and alleys, and population densities) between the U.S. and Japan are evident (Kanda, 1994).

DISRUPTION OF THE ECONOMY

The secondary economic impact of disasters derives from the disruption of flows resulting from damage to and destruction of the physical elements that facilitate them. Knowledge of how destruction and damage affect local economies and reduce income and product is very useful to assess the magnitude of an event, to plan assistance, and to assign priorities to mitigation measures and in recovery processes.

Some recent research has focused on the economic vulnerability and impact of disasters on selected businesses and industries. For example, one survey of business impact from disasters has shown that in the 1993 Midwest floods, damage to lifelines such as electricity, water and telephone service were a more important cause of business disruption than was the actual physical flooding itself (Tierney, 1994). Potential losses to lifeline industries themselves in natural disasters has been investigated in one study which uses information on market and pricing structure to assess the potential revenue losses to the natural gas industry of earthquake-induced pipeline breaks (Seligson and Eguchi, 1994).

The usefulness and applicability of many methods of urban and regional economic analysis have been investigated. Over the last two decades of research, we have seen an increase in the number and sophistication of analytical methods applied to estimating the regional economic effects of natural disasters. Many of these methods and their applicability to assessing the impact of a catastrophic earthquake are discussed in various issue papers included in the proceedings of a conference held by the National Research Council (NRC, 1992).

For example, a comprehensive economic loss estimation methodology based upon regional income and product accounts has been developed by the United Nations Economic Commission for Latin America and the Caribbean (ECLAC) and applied to a broad range of natural disasters. ECLAC has successfully utilized this methodology to estimate the impact on key aspects of the macroeconomy, including the impact on gross domestic product (GDP), industrial production, the balance of trade and the balance of payments, the level of indebtedness and of monetary reserves, the state of public finances, and the amount of gross capital investment (ECLAC, 1991; Jovel and Caballeros, 1993). These estimates are produced rapidly after a disaster to inform public policy in assessing relative magnitude, mobilizing resources, and seeking external aid. Joseph G. Hirschberg, Peter Gordon, and William J. Petak related direct economic loss from damage and secondary loss of income to per capita income in the impacted area (Petak *et al.*, 1978).

Regional input-output models present another approach that has been widely used to assess economic impact, particularly at the urban and regional level. Input-output models provide a means for estimating how direct physical losses create indirect economic losses (in terms of reductions in regional production, income, and employment) as a result of linkages between different sectors of the economy. Since Harold Cochrane (1974) first applied regional input-output models to forecasting the impact of a hypothetical natural disaster twenty years ago, this method has been developed as a basic tool and has been applied to a variety of actual as well as scenario events. Pate (1980a, 1980b) used input-output analysis to estimate loss of production from damaged facilities and losses in the national economy resulting from the lack of availability of inputs. Rose (1982) applied a regional input-output model to estimate forgone production of goods and services resulting from interruption of economic activities because of loss of energy inputs from damaged utility generation or transmission systems. Scawthorn, Khater, and Rojahn (1993) combined national data for several lifeline networks inventoried on a regioanl basis with the national input-output model to estimate direct lifeline damage and indirect economic losses for eigth scenario earthquakes. Eguchi and colleagues (1994) at EQE used these models to assess the regional economic impact of earthquake-related disruption of crude oil delivery in the New Madrid Seismic Zone. Input-output models have also been used to assess the indirect and induced business losses from direct losses to lifelines using damage estimates for eight hypothetical earthquakes applying the national input-output model (Boisvert, 1992). The models have also been used to determine the impact of lifeline disruption resulting from Hurricane Andrew (Wiggins and Eguchi, forthcoming). In one interesting application, researchers at the Public Works Research Institute in Japan applied regional input-output analysis to evaluate the economic effects of the 1983 Nihonkai-chubu earthquake on the product and income of Akita Prefecture (Kawashima and Kanoh, 1990). Another innovative approach has been to combine input-output and spatial allocation modeling to measure business disruption resulting from lifeline losses, so that regional economic impact can be distributed among constitutent cities as well as economic sectors (Gordon and Richardson, 1992). While the uses of input-output models in disaster vulnerability and impact assessment are clear, their appropriateness in the economic disequilibrium following a disaster has been questioned. This issue is addressed in recent work which attempts to solve for a new post-disaster equilibrium within an input-output framework (Cochrane, 1994).

Regional econometric models have also been applied to assessing disaster impact. Examples include the work by Richard Ellson, Jerome Milliman and Blaine Roberts (1984) comparing the economic impact on income losses and recovery period for hypothetical anticipated and unanticipated natural disasters in the Charleston, South Carolina region. Research by Tapan Munroe and Kenneth Ballard (1983) at Pacific Gas and Electric Company used a multi-county econometric model. Recently, Matthew Drennan (1993) used an econometric model of the New York City economy to assess the economic effects of a man-made disaster, the 1993 bombing of the World Trade Center.

Social accounting matrices provide another promising method for assessing regional economic effects of disasters. Social accounting matrices enable analysis of how a disaster affects different socio-economic groups in the impacted region, thereby providing insight into the distributional aspect of the economic impact. Cole (1993a, 1993b, 1994) has conducted substantial research in this area. Computable general equilibrium models based on data from input-output tables and social accounting matrix institutional accounts which have the advantage of an explicit role for prices have been proposed as a useful technique for analyzing impacts of natural hazards (Brookshire and McKee, 1992). Yezer and Rubin (1987) presented a general equilibrium model of a city subject to natural hazard risks and found the chief effect of a natural disaster was on land values rather than product or income and that long run effects were attributable to changes in expectations of future productivity. West and Lenze (1994) call attention to the fact that different models produce different results for complex events such as natural disasters and develop a general framework and regional impact model which they apply to Hurricane Andrew.

The results of the applications of these methods indicate the ways social and economic systems are vulnerable to destruction resulting from natural disasters. However, they seldom provide specific guidance for assigning priorities in undertaking mitigation measures and in planning for reconstruction and recovery. The application of other techniques often in combination with more general models of economic systems assist in filling this gap. In this regard, we have been developing and adapting methods for analyzing the effects of damage to and destruction of specific built physical elements including the infrastructure of lifeline systems, buildings and other structures, nonstructural elements, disruption of services provided by lifelines, and interruptions of the linkages between suppliers and users, sellers and customers from damage to structures and lifelines. For example, the earthquake that struck the Baguio region in the Philippines destroyed roads and railroads cutting all transportation linkages with the rest of the country. The undamaged Hewlett Packard manufacturing plant had to arrange for supplies to be delivered and product transported by helicopter. The impact of outside government assistance following a disaster has been explored through the use of econometric models of the farm sector after the 1993 Midwest floods (Loveridge, 1994).

Some attempts have been made to apply mathematical programming to modify input-output analysis so that it can serve as a planning tool for post-disaster reconstruction and recovery. In particular, Rose and Benavides (1993) have combined input-output analysis and linear programming to develop a method for assessing the impact of utility lifeline disruption in a disaster and determining optimal strategies for recovery. The application of pricing and rationing as tools for optimal resource allocation following a disaster have been explored with respect to electricity availability following a disaster (Rose and Benavides, 1994; Benavides and Rose, 1994).

Another method that has been used is cost/benefit analysis. Gordon, Banerjee and Wingo (1986) used Delphi techniques with a panel of experts to estimate damage to buildings and non-building structures in sub-areas of the City of Los Angeles to evaluate mitigation policies's costs and benefits and their incidence on various income groups. Schulze, Brookshire, Hageman, and Tschirhart (1987) developed a methodology for estimating the costs of earthquake mitigation measures in buildings and the benefits from the increased resistance and also applied it to Los Angeles. Another example of an application is a study sponsored by the National Committee on Property Insurance which compared the anticipated losses from two scenario earthquakes with the anticipated benefits of different types of mitigation strategies including changes in seismic building codes and retrofit of hazardous buildings (Litan *et al.*, 1992). Another application of cost/benefit analysis consists of research being performed by Scawthorn and Zadeh on mitigation strategies for damage to nonstructural components in buildings, specifically fire protection systems (Scawthorn *et al.*, 1993; Zadeh *et al.*, 1994).

MITIGATION AND ECONOMIC BEHAVIOR

Another important area of economic research that is conceptually related to cost/benefit analysis and linear programming deals with the study of micro-behavior. It attempts to determine how individuals perceive risk, how they set thresholds of risk as a basis of decisions, how they respond to it by avoidance, mitigation, discounting value, and insurance.

Engineering practice has emphasized designing for life safety, but building owners and occupants have become increasingly concerned with post-event functionality and operability of facilities since the economic costs of down-time are likely to be substantially greater in many instances than direct losses from damage. Little is known about thresholds of acceptable risk, and this is becoming a subject for study.

The US Congress is considering a comprehensive earthquake insurance program, but nothing is known about how the availability of insurance will affect the decisions of property owners and local planning and zoning departments regarding microzone management. This question arose with coastal management early in the flood insurance program and was not resolved (Kunreuther, 1974). The question is the impact of information about risk of natural hazards on behavior, and the theory of expected utility has been developed analytically and applied to studies in the field with varying results. Brookshire, Thayer, Tschirhart, and Schulze (1985) find a hedonic price index for safety in the Los Angeles and San Francisco areas as an indication of self insurance. Earthquake and volcano hazard notices were found to alter risk perceptions and investment decisions in a study of Mammoth Lakes, California (Bernknopf, Brookshire, and Thayer, 1990). However, Cochrane (1993) found evidence from real estate markets both within and outside hazard zones indicated that households respond minimally to published information about earthquake risk. Hedonic methods have been used to isolate causal factors and determine whether population growth reflects the availability of hazard insurance or simply the provision of public infrastructure. On the other hand, microeconomic utility models have been developed that demonstrate the contribution of natural hazard information on individual decision-making related to disaster mitigation and risk reduction for landslide hazards (Bernknopf, 1994).

SUMMARY

Since disasters are space specific and have impacts on limited geographical areas, we must adapt analytical methods developed for other purposes to the study of the social and economic consequences of disasters on regional systems. We have been developing data and methodological models to advance our understanding of the subtleties and complexities of social and economic losses resulting from natural disasters.

While we are better able now than a few years ago to identify and measure relationships between physical damage and destruction and impacts on regional social and economic systems, much more progress is necessary. While we have gained much knowledge regarding the principal source of urban disaster vulnerability, the built physical environment system at risk from natural hazards, further research is needed into how this environment changes over time in conjunction with socioeconomic development and how these changes increase or reduce disaster vulnerability. In addition, while many tools of regional economic analysis have been applied to the study of urban disaster impact, their reliability and usefulness depend critically on the quality of the inputs that drive the models. Research is particularly needed into how different elements of the built environment are used by the various sectors of the regional economy and how the structure of economic production and consumption change in response to a disaster event. For example, we have little systematic knowledge about the sources of crucial bottlenecks in the flows of economic services, about the possibilities for substitution in production processes given disaster-related constraints, about changes in consumption demand following disasters, or about the disaster-induced changes in regional imports and exports through which economic impacts are transmitted outside the physically impacted region. Research into such areas can enable more accurate modeling of regional economic impacts; in addition, it can provide a crucial link to public policy analysis by identifying specific sources of economic vulnerability to disasters. Finally, while progress has been made in the microeconomic study of risk and mitigation behavior, a very important gap remains in knowledge of risk perception and acceptable risk at the societal level.

REFERENCES

Benavides, J. and A.Z. Rose (1994). Lifeline Capacity Expansion Planning Under Uncertainty. Presented at *41st North American Meetings of the Regional Science Association International*. Niagara Falls, Ontario, November 16-20, 1994.

Bernknopf, R.L. (1994). Earthquake Hazard Mitigation: Using Science for Safetylogical Survey. Open File Report 94-172.

Bernknopf R.L., D.S. Brookshire, and M.A. Thayer (1990). Earthquake and Volcano Alerts: an Economic Evaluation of Risk-perception Changes. *Journal of Environmental Economics and Management*, 18(1), 35-49.

Boisvert R.N. (1992). Indirect Losses from a CAtastrophic Earthquake and the Local, Regional, and National Interests. *Indirect Economic Consequences of a Catastrophic Earthquake*. Washington, DC: National Earthquake Hazards Reduction Program, CEMA.

Brookshire D.S. and M. McKee (1992). Other Indirect Costs and Losses from Earthquakes: Issues and Estimation. *Indirect Economic Consequences of a Catastrophic Earthquake*. Washington, DC: National Earthquake Hazards Reduction Program, CEMA.

Brookshire D.S., M.A. Thayer, J. Tschirhart, and W.D. Schulze (1985). A Test of the Expected-unity model: evidence from earthquake risks. *Journal of Political Economy*, 93(2), 369-389.

Chang, S.E. (1994a). Building Inventory Estimation with Regional Characteristics: Application to Earthquake Vulnurability Assessment in Los Angeles. Presented at *41st North American Meetings of the Regional Science Association International*, Niagara Falls, Ontario, November 16-20, 1994a.

Chang, S.E. (1994b). Reional Infrastructure Investment: An Analysis Of Capital Accumulation in the Built Physical Environment Disaggregated by Industry. Ithaca, New York: *Program in Urban and Regional Studies, Cornell University.*

Cochrane, H.C. (1994). An Expert Model of Regional and Secondary Losses from Disasters. Presented at *41st North American Meetings of the Regional Science Association International*, Niagara Falls, Ontario, November 16-20, 1994.

Cochrane, H.C. (1993). Earthquake Insurance and Mitigation Measures, presented at *40th North American Meetings of the Regional Science Association International*, Houston, Texas, November 11-14, 1993.

Cochrane, H.C. (1974). Predicting the Economic Impact of Earthquakes, *Working Paper No. 15*, Institute of Behavioral Science, University of Colorado, Boulder, Colorado.

Cole, H.S. (1993). Awa di Bida Social Accounting for Lifelines and Livelihood. Presented at *40th North American Meetings of the Regional Science Association International*, Houston, Texas, November 11-14, 1993.

Cole, H.S. (1994). The Impact of the Retsov-Akzo Mine Collapse on Livingston County and its Neighbors. Presented at *41st North American Meetings of the Regional Science Association International*, Niagara Falls, Ontario, November 16-20, 1994.

Cole, H.S. (1993b). Social Accounting for Dissaster Preparedness and recovery Planning. Buffalo, New York: National Center for Earthquake Engineering Research.

Del Tosto, R. and R.T. Eguchi (1980). Loss Estimation from Earthquake Shaking in California. *Proceedings of the Seventh World Conference on Earthquake Engineering, Vol. 9, Socio-Economic Aspects*. Istanbul: Seventh World Conference on Earthquake Engineering, 353-359.

Drennan, M. (1993). The Cost of Ill Intentions: Measuring the Economic Impact of the World Trade Center Explosion. Presented at *40th North American Meetings of the Regional Science Association International*, Houston, Texas, November 11-14, 1993.

ECLAC (1991). Manual para la Estimacion de los Efectos Socio-Economicos de los Desastres Naturales. Santiago: *ECLAC*.

Eguchi, R.T., H.A. Seligson, and J.H. Wiggins (1993). Estimation of Secondary Economic Losses Associated with Lifeline Disruption. Presented at *40th North American Meetings of the Regional Science Association International*, Houston, Texas, November 11-14, 1993.

Ellson, R.W., J.W. Milliman, and R.B. Roberts (1984). Measuring the Regional Economic Effects of Earthquakes and Earthquake Predictions. *Journal of Regional Science*, 24, No. 4, 559-579.

Gordon, P., T. Banerjee, and L. Wingo (1986). Evaluation of Mitigation Strategies for Disaster Events. Los Angeles, California: *School of Urban and Regional Planning, University of Southern California*.

Gordon, P. and H.W. Richardson (1992), Business Interruption Effects of a Major Earthquake in the Newport/Inglewood Fault Zone (NIFZ). A report submitted to the National Committee on Property Insurance, Boston, Massachusetts, by The Planning Institute, School of Urban and Regional Planning, University of Southern California, Los Angeles.

Jones, B.G. (1981). Planning for the Reconstruction of Earthquake Stricken Communities. In: *Proceedings of the P.R.C.-U.S.A. Joint Workshop on Earthquake Disaster Mitigation Through Architecture, Urban Planning and Engineering*, (Ye Yaoxian and Henry H. Lagorio, eds.), Beijing.

Jones, B.G. (1994). Development of a Methodology for Making Indirect Estimates of the Built Physical Environment. *Research Accomplishments: 1986-1994*. Buffalo, New York: National Center for Earthquake Engineering Research, 107-120.

Jones, B.G. and S.E. Chang (1994). A Comparison of Indirect and Direct Estimates of the Built Physical Environment in the Memphis Region. *Proceedings of the Fifth U.S. National Conference on Earthquake Engineering, Volume III*. Oakland, CA: Earthquake Engineering Research Institute.

Jones, B.G., B.D. Lewis, and D.G. Westendorff (1991). The Economic Impact on the Built Physical Environment of Damage by Natural Disasters: A Model and Empirical Evidence. *International Symposium on Building Technology and Earthquake Hazard Mitigation*. CIB Publication 142. Rotterdam, The Netherlands: International Council for Building Research Studies and Documentation.

Jones, B.G., D.M. Manson, J.E. Mulford, M.A. Chain (1976). The Estimation of Building Stocks and Their Characteristics in Urban Areas: An Investigation of Empirical Regularities. Ithaca, New York: *Program in Urban and Regional Studies, Cornell University*.

Jones, B.G., D.M. Manson, C.M. Hotchkiss, M.J. Savonis, and K.A. Johnson (1987). Estimating Building Stocks and Their Characteristics. Ithaca, NY: *Program in Urban and Regional Studies, Cornell University*.

Jovel, J.R. and R. Caballeros (1993). Estimation of Macroeconomic Effects of Natural Disasters in Latin America and the Caribbean. Presented at *40th North American Meetings of the Regional Science Association International*, Houston, Texas, November 11-14, 1993.

Kanda, H. (1994). Indirect Methods for Estimating the Built Physical Environment: An Application to Kyoto, Japan. Ithaca, NY: *Program in Urban and Regional Studies, Cornell University*.

Kawashima, K. and T.Kanoh (1990). Evaluation of Indirect Economic Effects Caused by the 1983 Nihonkai-chubu, Japan Earthquake. *Earthquake Spectra*, 6, No. 4, 739-55.

Kunreuther H. (1974). Economic Analysis of Natural Hazards: An Ordered Choice Approach. In: *Natural Hazards: Local, National, and Global* (G. White, ed.). Oxford University Press, New York.

Litan, R., F. Krimgold, K. Clark and J. Khadilkar (1992). Physical Damage and Human Loss: The Economic Impact of Earthquake Mitigation Measures. Boston, Massachusetts: National Committee on Property Insurance.

Loveridge, S. (1994). Early Warning System for Weather Related Agricultural Disasters. Presented at *41st North American Meetings of the Regional Science Association International*, Niagara Falls, Ontario, November 16-20, 1994.

Munroe, T., and K.P. Ballard (1983). Modeling the Economic Disruption of a Major Earthquake in the San Francisco Bay Area: Impact on California. *The Annals of Regional Science*, 17, No. 3, 23-40.

Pate, M.E. (1980a). Evaluation of the Economic Consequences of Natural Catastrophes. *Proceedings of the International Conference on Engineering for Protection from Natural Disasters*. Bangkok: Asian Institute of Technology.

Pate, M.E. (1980b). Assessment and Mitigation of Earthquake Effects on Economic Production. *Proceedings of the Seventh World Conference on Earthquake Engineering, Vol. 9, Socio-Economic Aspects*. Istanbul: Seventh World Conference on Earthquake Engineering, 301-306.

Petak, W.J., A.A. Atkission, and P.H. Gleye (1978). Natural Hazards: A Public Policy Assessment. Redondo Beach, California: *J.H. Wiggins Company*.

Rose, A.Z. (1982). Utility Lifelines and Economic Activity in the Context of Earthquakes. *Social and Economic Impact of Earthquakes on Utility Lifelines* (Jeremy Isenberg, ed.), New York: American Society of Civil Engineers, 107-120.

Rose, A.Z., and J.Benavides (1994). Optimal Recovery of Utility Lifelines after Major Earthquakes: Market Mechanisms vs Rationing. Presented at *41st North American Meetings of the Regional Science Association International*, Niagara Falls, Ontario, November 16-20, 1994.

Rose, A.Z., and J.Benavides (1994). Interindustry Models for Analyzing the Economic Impacts of Earthquakes and Recovery Policies. Presented at *40th North American Meetings of the Regional Science Association International*, Houston, Texas, November 11-14, 1993.

Scawthorn C., S. Eder, M. Zadeh, and G. Johnson (1993). Economic Impacts of Earthquake Damage to Nonstructural Components. Presented at *40th North American Meetings of the Regional Science Association International*, Houston, Texas, November 11-14, 1993.

Scawthorn, C., M. Khater, and C. Rojahn (1993). The Economic Impact of Earthquake Disruption to Lifelines in the Coterminous United States. Presented at *40th North American Meetings of the Regional Science Association International*, Houston, Texas, November 11-14, 1993.

Schulze W.D., D.S. Brookshire, R.K. Hageman, and J. Tschirhart, (1987). Benefits and Costs of Earthquake Resistant Buildings. *Southern Economic Journal*, 53, 934-951.

Seligson, H.A., and R.T. Eguchi (1994). Estimation of Indirect Losses Caused by Disruption of Lifeline Service: A Pilot Study of Memphis Light, Gas, and Water System. Presented at *41st North American Meetings of the Regional Science Association International*, Niagara Falls, Ontario, November 16-20, 1994.

Steinmeyer, J.R. (1994). The Survival of Urban Building Stock: An Analytical Technique Applied to Ithaca, New York, and Wichita, Kansas. Ithaca, NY: *Program in Urban and Regional Studies, Cornell University*.

Tierney, K. (1994). Business Vulnerability and Disruption: Data from the 1993 Midwest Floods.

Presented at *41st North American Meetings of the Regional Science Association International*, Niagara Falls, Ontario, November 16-20, 1994.

West, C. and D.G. Lenze (1994). Modeling the Regional Impact of Natural Disaster and Recovery: A General Framework and a Specific Application to the Case of Hurricane Andrew, presented at *41st North American Meetings of the Regional Science Association International*, Niagara Falls, Ontario, November 16-20, 1994.

Wiggins, J.H. (1994). Estimating Economic Losses Due to an Interruption in Crude Oil Deliveries Following an Earthquake in the New Madrid Seismic Zone. *Proceedings of the Fifth U.S. National Conference on Earthquake Engineering*, III, 1077-1086.

Wiggins, J.H. and R.T. Eguchi (forthcoming). Economic Impacts as a Result of Lifeline Disruption from Hurricane Andrew. Buffalo, New York: National Center for Earthquake Engineering Research.

Yezer, A.M.J. and C.B. Rubin (1987). The Local Economic Effects of Natural Disasters. *Paper #61*. University of Colorado, Institute of Behavioral Sciences.

Zadeh, M., D. Mirfendereski, and C. Scawthorn (1994). Economic Impacts of Earthquake Damage to Fire Protection Equipment and Systems in High-rise Buildings. Presented at *41st North American Meetings of the Regional Science Association International*, Niagara Falls, Ontario, November 16-20, 1994.

INDEX OF CONTRIBUTORS

KEYWORD INDEX